First Ecology: ecological principles and environmental issues

It is true: Man is the microcosm: I am my world.
Ludwig Wittgenstein

First Ecology

ecological principles and environmental issues

THIRD EDITION

Alan Beeby

Anne-Maria Brennan

London South Bank University

OXFORD

UNIVERSITY PRESS

OXFORD

UNIVERSITY PRESS

Great Clarendon Street, Oxford OX2 6DP

Oxford University Press is a department of the University of Oxford.
It furthers the University's objective of excellence in research, scholarship,
and education by publishing worldwide in

Oxford New York

Auckland Cape Town Dar es Salaam Hong Kong Karachi
Kuala Lumpur Madrid Melbourne Mexico City Nairobi
New Delhi Shanghai Taipei Toronto

With offices in

Argentina Austria Brazil Chile Czech Republic France Greece
Guatemala Hungary Italy Japan Poland Portugal Singapore
South Korea Switzerland Thailand Turkey Ukraine Vietnam

Oxford is a registered trade mark of Oxford University Press
in the UK and in certain other countries

Published in the United States
by Oxford University Press Inc., New York

© Alan Beeby and Anne-Maria Brennan 2008

The moral rights of the authors have been asserted
Database right Oxford University Press (maker)

First edition published 1997
Second edition published 2004
Third edition published 2008

British Library Cataloguing in Publication Data

Data available

Library of Congress Cataloging in Publication Data

Data available

Typeset by Graphicraft Limited, Hong Kong
Printed in Italy
on acid-free paper by
Legoprint S.p.A

ISBN 978-0-19-929808-2

1 3 5 7 9 10 8 6 4 2

BRIEF CONTENTS

DETAILED CONTENTS

SECOND THOUGHTS

We used Wittgenstein's assertion about the nature of reality to set the theme for the first edition of this book, and continue to misinterpret its meaning in this new edition. Confident that he would have forgiven us then, we quickly dismissed any thought of replacing it here. We still wish to keep humankind centre stage and try again to introduce ecology from a human perspective, using our species as both example and as principal actor.

Human beings are, each of us, a microcosm illustrating some fundamental ecological principles. With the right perspective, we can learn much ecology simply by comparing ourselves with other species. Like all living organisms we are the product of natural selection, shaped by environmental pressures. Even the most technologically advanced peoples are subject to the ecological processes of the world as they find it. Our species is beginning to realize that an understanding of such ecological checks is critical to its future.

The human context is the starting point for every chapter. In each case we seek to introduce one or more key ecological principles using common experience, drawing on examples with which most people will be familiar, or touch upon some aspect of human evolution or cultural development. Each chapter goes on to consider an application or an environmental issue in relation to which humankind has been a central player.

The prologue (*First words*) introduces the reader to the general organization of the book and its approach. It uses a current pollution problem from which our species may itself be under threat, but with a very different ecological scenario. The reader is encouraged to adopt a questioning attitude, with a warning that few ecological questions have simple or clear-cut answers. The prologue ends with a brief set of directions, describing some of the features of the book and its associated website.

Alongside human ecology, mediterranean-type ecosystems are used as a recurring example throughout. These are introduced in Chapter 5, to allow the reader to study one community in some detail. The mediterranean biome is closest to most peoples of the world and one of the most diverse. Its discrete and repeated ecosystems allow us to assess whether communities are self-organizing. Additionally they are one of the most threatened of habitats, and in their different locations, suffer different forms of human impact.

A number of other themes are revisited through the book, not least the polar bear, in this case to permit a detailed review of the autoecology of one species. Our web page ('Routes') gives explicit listings of these recurring topics and the associated activities students might undertake to make the connections for themselves.

Our primary aim has been to produce a book that is accessible and easy to read. Rather than a textbook that is highly sectionalized, we seek to provide a narrative that encourages the reader to complete a chapter. To maintain the flow, we have chosen not to cite supporting references in the text but name key authors and provide sectionalized references at the end of the book. In this edition we have added a series of case studies to illustrate the proper citation of references. The case studies also introduce current research, provide detail on one or more ecological techniques and serve to extend one or more topics in the associated chapter.

Additionally, boxes are used throughout to provide illustrative examples or extension material. New exercises have been included at the end of each chapter. The answers to these and additional exercises are available on the website. This also offers a virtual field course for students who would like to see how ecologists collect and analyse data. This can be followed either as a self-taught package or as part of an instructor-led programme, with exercises for students to practise basic skills in data analysis and presentation. These have been devised to illustrate some key ecological principles and provide images and short films that support the text.

First Ecology online resource centre

First Ecology online resource centre

First Ecology is supported by a range of electronic resources, available through the book's online resource centre at: www.oxfordtextbooks.co.uk/orc/beebybrennan3e/. The website is divided into two basic sections:

(i) Student support materials

This includes:

- **Answers to exercises** posed in the book, to allow students to self-check their answers as they attempt the exercises
- A section entitled '**Routes**' that provides a detailed map of how some key themes are developed through the book. This aims to demonstrate the connectivity of ecological principles, an awareness that allows students to become confident of their understanding of ecology. 'Routes' includes several **study titles**, related to each theme, that could direct student-based research using *First Ecology*.
- A **web link library** including all the URLs listed at the end of the chapters in the book. Additional web links to other sites for specific topics are also listed on a sectionalized basis. Please feel free to suggest links that should be added to this site, preferably offering a commentary on the content (see below for contact details). We hope to keep the web links up to date, though of course neither we, nor the publishers, can accept responsibility for any of the links posted here. We will endeavour to review these links periodically and offer our own commentary.
- A **hyperlinked bibliography**—direct links to on-line articles cited in the book

(ii) Tutor support materials

Access to the tutor support materials is free, but is password-protected to enable tutor-only access.

To obtain a password simply register with us by visiting the *First Ecology* online resource centre, clicking on the 'lecturer resource' button, and following the 'Not yet registered?' instructions.

The tutor-only part of the site includes:

- **Additional exercises** to supplement those in the book
- **Answers** to the text-based and web-based exercises
- **Downloadable figures**—those for which we hold copyright in the book
- **Powerpoint slide presentations.** There are nine presentations, corresponding to the chapters. These include a small number of animations and are complete with lecturer directions and suggestions.
- A **virtual field course.** Each exercise is accompanied with tutor notes and suggestions for further analysis, and in some cases, references enabling the work to be extended. Some of the exercises are suited to group work and presentations, others practise basic skills appropriate to an introduction to ecology.
- Lecturers might also consider some of the suggested topics for further study using the '**Routes**' function that picks out themes and connections across the different chapters.
- A **feedback** site where corrections or suggestions can be posted. There are better questions, and undoubtedly better answers to the questions, than those we have posed and we would welcome suggestions on both counts. Similarly, any comments on the text *per se* would also be very useful, as would any suggestions or datasets that could be used to extend the range of exercises in the virtual field course.

Contact us

To contact us with comments and suggestions, simply click on the 'feedback' link at: www.oxfordtextbooks.co.uk/orc/beebybrennan3e/

Grateful thanks

We remain indebted to Ambio, Martin Angel, Blackwell Scientific Publications, the Broads Authority, Bob Carling, Cambridge University Press, Luca Cavalli-Sforza, John Currey, John Dodd, John Feltwell, the French Ministry of Culture, Ken Giller, Elliot Gingold, Lee Hannah and Conservation International, Mike Harding, Gary Haynes, Alan Hopkins, John Hopkins, T. Jaffre, Steve Jansen, the University of Leuven (Belgium), B. Jedrzejewska, Susan Jenks, Joint Nature Conservation Committee, Tim Johns, Hefin Jones, Nancy Laurenson, Rachel Leech, George Lees, the Linnean Society of London, Joe Lopez-Real, Lynn Margulis, Lord May of Oxford, the late Chris Mead, Patrick Morgan, NASA, Natural History Museum, Kate Neale, Mike Newman, Mike Nicholls, Simon Parfitt, Panos Institute, Arthur Penhally, Val Porter, George Potts, Dominic Recaldin, Roger Reeves, John Rodwell, the Royal Entomological Society of London, Peter Saville, Candy D'Sa, Helen Sharples, Liz Sestito, Roger Tidman, Martin Tribe, Will Wadell, and Robert Wayne.

Specifically for subsequent editions we are grateful to Institutio Veneto Scienze Lettre ed Arti (VSLA), Cathy Bach, Eastern Michigan University (USA), J. Franklin, Bill Baker, Royal Botanic Gardens, Kew (UK), Guy Beaufoy (WWF), Tony Bradshaw (UK), Andy Bright (UK), the late Paddy Coker (UK), Eleanor Cohn, Wolverhampton University (UK), Marty Condon, Cornell University (USA), Mike Dobson, Manchester Metropolitan University (UK), Terence D. Fitzgerald, Cortland College (USA), Kevin Fort, University of California Davis (USA), Alastair Grant, University of East Anglia (UK), Russ Greenberg, The Smithsonian Migratory Bird Center (USA), Philip Grime, University of Sheffield (UK), Ian Hutton, John Lambshead, Natural History Museum (UK), Rex Lowe, Bowling Green State University (USA), Stuart McRae (UK), Alex Meinesz (University of Nice), Jonathan Mitchley, Imperial College (UK), Margaret Ramsey, Royal Botanic Gardens, Kew (UK), Vincent Savolainen, Royal Botanic Gardens, Kew (UK), Mark Seaward (University of Bradford), Dolph Schluter, University of British Columbia (Canada), Walker Smith, Virginia Institute of Marine Science (USA), Walter J. Tabachnick, University of Florida (USA), Catherine Toft, University of California (USA), and Barry Yates (UK). We particularly wish to thank Walker Smith for providing Box 8.6 (iron fertilization of marine phytoplankton) and Erik Scully, Towson University, for his work on the PowerPoint slides of the second edition. Our thanks also to Richard Dawkins and especially to Jonathan Crowe and Laura Hodgson of OUP who have been immensely supportive throughout the work.

The third edition has materialized in large part because of the enthusiasm and work of Ross Bowmaker of OUP, to whom we are immensely grateful. He and Jonathan Crowe have been a dedicated and supportive editorial team. We are also indebted to the team of reviewers they assembled to help shape the new edition: Reidar Andersen, Norwegian University of Science and Technology, Tony Andrew, University of Ulster (UK), Simon Cragg, University of Portsmouth, (UK), Andrew R. Dyer, University of South Carolina (USA), Mark Grover, Southern Utah University (USA), Leanne Hepburn, University of Essex (UK), Robert James, University of East Anglia, (UK), Chris Joyce, University of Brighton (UK), Andy Le Brocque, University of Southern Queensland (Australia), Zhi-Qing Lin, Southern Illinois University, Edwardsville (USA), Duncan McCollin, University of Northampton (UK), Thomas Meagher, University of St Andrews (UK), Topa Petit, University of South Australia, Andrew Powling, University of Portsmouth (UK), Ted Schuur, University of Florida (USA), Alejandro Serrano, Universidad Nacional Autónoma de México, Brita Svensson, Uppsala University (Sweden), Bill Tonn, University of Alberta (Canada), Joe von Fischer, Colorado State University (USA), Bethan Wood, University of Glasgow (UK), and John Zhou, University of Sussex (UK). We thank them and the OUP staff for their considerable help.

Our thanks also to our families, friends, and colleagues for their support, especially Larry Richmond, Jackie, Ralph, and Kate Beeby.

ACKNOWLEDGMENTS

There are instances where we have been unable to trace or contact the copyright holder. If notified, the publisher will be pleased to rectify any errors or omissions at the earliest opportunity.

Cover

Courtesy of NASA Visible Earth/Earth Observatory (Corbis)

First words

Frontispiece—Dr Mette Mauritzen, Norwegian Polar Institute

Figures

1—Photograph courtesy of Visible Earth, NASA, Jacques Descloitre and the MODIS Land Rapid Response Team. From the satellite Terra, taken June 7, 2001.
2—Swingley, a ringed seal—courtesy of Jason Wettstein, Alaska SeaLife Center.
3—Drawn from data in Carlsen, E. *et al.* 1992. Evidence for decreasing quality of semen during past 50 years. *British Medical Journal 305*, 609–613.
5—Mike Brinkley, Oakland Nature Preserve, Florida
6—Data redrawn from Rauschenberger, R. H. *et al.* 2004. Achieving environmentally relevant organochlorine pesticide concentrations in eggs through maternal exposure in *Alligator mississippiensis*. *Marine Environmental Research 58*, 851–856.

Chapter 1

Frontispiece–Kind permission of *Bone Clones* (http://www.boneclones.com/BH-KRO-1.htm).

Figures

1.1—Patches, a Great White Shark. James Moskito and *Shark Diving International* (http://www.greatwhiteadventures.com).

1.5—Modified, from Cavalli-Sforza, L. L. 2001. *Genes, Peoples and Languages*. Penguin.
1.6—Alan Beeby
1.9—Alan Beeby
1.11—Alan Beeby
1.13—*Homo floresiensis* endocasts redrawn from photographs with the supporting online material associated with Falk, D. *et al.* 2005. The Brain of LB1, *Homo floresiensis*. *Science 308*, 242–245.

Chapter 2

Frontispiece—Anne-Maria Brennan

Figures

2.1—The Linnean Society of London
2.2—After Woese, C., Kandler, O., and Wheelis, M. 1990. Towards a natural system of organisms: proposal for the domains Archaea, Bacteria and Eucarya. *Proceedings of the National Academy of Sciences USA 87*, 4576–4579.
2.3—Andy Bright (European Blackbird), Roger Tidman (American Blackbird), Anne-Maria Brennan (European Bluebell), John Feltwell (Californian Bluebell)
2.4—(1) © 2005 Zasavica; (2) © Karlheinz Knoch 2005; (3) © Bob Gibbons (www.ardea.com); (4) © Stephen Dalton (www.nhpa.co.uk); (5) © David Dixon (www.ardea.com).
2.5—Natural History Museum, London
2.6—Anne-Maria Brennan
2.7—Anne-Maria Brennan
2.8—Anne-Maria Brennan
2.14—Eric Cooper, supplied by Dolph Schluter
2.15—Based on Schluter, D. 1994. Experimental evidence that competition promotes divergence in adaptive radiation. *Science 266*, 798–801.
2.16—Alan Beeby
2.17—Anne-Maria Brennan
2.18—Anne-Maria Brennan

2.19—After Antonovics, J., Bradshaw, A. D., and Turner, R. G. 1971. Heavy metal nutrient tolerance in plants. *Advances in Ecological Research 7*, 1–85.

2.20—Redrawn from Ammerman, A. and Cavalli-Sforza, L. L. 1971. Measuring the rate of spread of early farming in Europe. *Man 6*, 674–688.

2.21—Bill Baker, Royal Botanic Gardens, Kew—supplied by Vincent Savolainen.

2.22, 2.23, and 2.24—Redrawn from Savolainen, V. *et al.* 2006. Sympatric speciation in palms on an oceanic island. *Nature 44*, 210–213.

Chapter 3

Frontispiece—Photograph courtesy of Greenpeace/Kate Davison. A team from the Greenpeace ship MV Esperanza documents a catch being landed on board a Spanish flagged bottom-trawler, the Ivan Nores, in the Hatton Bank area of the North Atlantic, 410 miles northwest of Ireland. Bottom-trawling boats, the majority from EU countries, drag fishing gear weighing several tonnes across the sea bed, destroying marine wildlife and devastating life on underwater mountains—or 'seamounts'. © Greenpeace/Kate Davison.

Figures

3.4—Redrawn from data from DG Fisheries (European Union) 2003.

3.7—(a, b) Alan Beeby

3.8—Data redrawn from Jedrzejewska, B. *et al.* 1994—Effects of exploitation and protection on forest structure, ungulate density and wolf predation in Bialowieza Primeval Forest, Poland. *Journal of Applied Ecology 31*, 664–676.

3.9—Kathrin P. Lampert

3.11—Mette Mauritzen, Norwegian Polar Institute

3.12—Redrawn from Mauritzen, M. *et al.* 2002. Using satellite telemetry to define spatial population structure in polar bears in the Norwegian and western Russian Arctic. *Journal of Applied Ecology 39*, 79–90.

3.13—(a) Corel; (b) Nancy Laurenson

3.14—Royal Botanic Gardens, Kew

3.16—Martin Withers/FLPA

3.17—Nancy Laurenson

3.18—Diagram compiled from data from Ashley, M. V. *et al.* 1990. Conservation genetics of the Black Rhinocerus (*Diceros bicornis*). 1. Evidence from the mitochondrial DNA of three populations. *Conservation Biology 4*, 71–77 and Worldwide Fund for Nature.

3.19 and 3.20—Compiled and redrawn from Scott, B. E. *et al.* 2006. Effects of population size/age structure, condition and temporal dynamics of spawning on reproductive output in Atlantic cod (*Gadus morhua*). *Ecological Modelling 191*, 383–415.

Table 3.1—Table compiled from data supplied by the Worldwide Fund for Nature (http://www.worldwildlife.org) and the International Union for Nature Conservation (http://www.iucn.org).

Chapter 4

Frontispiece—Nancy Laurenson

Figures

4.3—Dr Joyce Gross, University of California, Berkeley

4.4—Anne-Maria Brennan

4.5—Redrawn from Bond, W. and Slingsby, P. 1984. Collapse of an ant–plant mutualism: the Argentinian ant (*Iridomyrmex humulis*) and myrmecochorous proteaceae. *Ecology 65*, 1031–1037.

4.6—Christoph Sheidegger

4.7—Alan Beeby

4.8—Roger Tidman

4.9—Anne-Maria Brennan

4.10—Redrawn from Crombie, A. C. 1946. Further experiments on insect competition. *Proceedings of the Royal Society of London, Series B 133*, 76–109.

4.11—J. D. Wilson, University of Georgia

4.12—After Southwood, T. R. E. 1977. Habitat, the templet for ecological strategies. *Journal of Animal Ecology 46*, 337–365.

4.13—Philip Grime

4.14—Dr Patrick Roper, supplied by Barry Yates

4.17—Redrawn from Potts, G. R. and Aebischer, N. J. 1989. Control of population size in birds: the grey partridge as a case study. In P. J. Whittaker and J. B. Grubb (eds), *Towards a More Exact Ecology*. Blackwell, Oxford.

4.19—Anne-Maria Brennan

4.20—Anne-Maria Brennan

4.21—Alex Meinesz

4.22—After Meinesz (personal communication) and Meinesz, A. *et al.* 2001. The introduced green alga *Caulerpa taxifolia* continues to spread in the Mediterranean. *Biological Invasions 3*, 201–210.

4.22—Bob Perry

4.23—Forestry Authority

4.24—After Tompkins, D. M. *et al.* 2002. Parapoxvirus causes a deleterious disease in red squirrels associated with UK population declines. *Proceedings of the Royal Society of London Series B, 269,* 529–533.

Chapter 5

Frontispiece—Alan Beeby and Ralph Beeby

Figures

5.2—Alan Beeby

5.3—Alan Beeby

5.4—Alan Beeby

5.5—(a) Kate Neale; (b) Jan Malan, Platinum Planet

5.6—(a) Sebastian Teillier; (b) Karen J. Carter 2003

5.8 and 5.10—Modified after Trabaud L. 1981. Man and fire: impacts on Mediterranean Vegetation. In F. Di Castri *et al.* (eds), *Mediterranean-type Shrublands*. Elsevier, Amsterdam.

5.11—Alan Beeby

5.13—Drawn using data from Cody, M. L. and Mooney, H. A. 1978. Convergence versus non-convergence in Mediterranean-climate ecosystems. *Annual Review of Ecology and Systematics 9*, 265–351.

5.14—Redrawn from Raunkiær, C. 1934. *The Life Forms of Plants and Statistical Plant Geography*. Oxford University Press, Oxford.

5.17—John Feltwell

5.18—Anne-Maria Brennan

5.20—Alan Beeby and Anne-Maria Brennan

5.21—Redrawn from Moore, P. D. 1986. Site history. In P. D. Moore and S. B. Chapman (eds), *Methods in Plant Ecology*. Blackwell, Oxford.

5.22—Larry Richmond

5.24—After Rubio-Casal, A. E. *et al.* 2001. Nucleation and facilitation in salt pans in Mediterranean salt marshes. *Journal of Vegetation Science 12*, 761–770.

5.25—Anne-Maria Brennan

5.26—Alan Beeby

5.27—Anne-Maria Brennan

5.28—John Rodwell/Cambridge University Press—Rodwell, J. S. (ed.). 1992. *British Plant Communities*, Vol. 3, *Grasslands & Montane Communities*. Cambridge University Press, Cambridge.

5.29 and 5.30—After Botes, A. *et al.* 2006. Ants, altitude and change in the northern Cape Floristic Region. *Journal of Biogeography 33*, 71–90.

Chapter 6

Frontispiece—Anne-Maria Brennan

Figures

6.1—Rachel Leech & Alan Beeby

6.3—NOAA

6.4—Alan Beeby

6.5—Redrawn from Zscheile, F. P. and Comar, C. L. 1941. Influence of preparative procedure on the purity of chlorophyll components as shown by absorption spectra. *Botanical Gazette 102*, 463–481.

6.7—OUP/data derived from Schultz, E. D. 1970. Der CO_2–Gaswechsel der Buche (*Fagus sylvatica* L.) in Abhängigkeit von den Klimafaktoren in Feiland. *Flora Jena 159*, 177–232; Schultz, E. D., Fuchs, M., and Fuchs, M. I. 1977a. Spatial distribution of photosynthetic capacity and performance in a mountain spruce forest in Northern Germany. I. Biomass distribution and daily CO_2 uptake in different crown layers. *Oecologia 29*, 43–61; Schultz, E. D., Fuchs, M., and Fuchs, M. I. 1977b. Spatial distribution of photosynthetic capacity and performance in a mountain spruce forest in northern Germany. III. The significance of the evergreen habit. *Oecologia 30*, 39–248.

6.8—After Orshan, G. 1963. Seasonal dimorphism of desert and Mediterranean chamaephytes and its significance as a factor in their water economy. In A. J. Rutter and F. H. Whitehead (eds), *The Water Relations of Plants*. Wiley, New York.

6.10—Ms Huerta

6.14—After Swift, M. J., Heal, O. W., and Anderson, J. M. 1979. *Decomposition in Terrestrial Ecosystems*. Blackwell, Oxford.

6.15—Anne-Maria Brennan

6.17 and 6.18—Drawn from data from Golley, F. B. 1960. Energy dynamic of an old-field community. *Ecological Monographs 30*, 187–200.

6.19—Data from Whittaker, R. H. 1975. *Communities and Ecosystems*, 2nd edn. Macmillan, New York.

6.21—Data from Varley, G. C. 1970. The concept of energy flow applied to a woodland community. In A. Watson (ed.), *Animal Populations in Relation to their Food Resource*. Blackwell, Oxford.

6.22—Drawn using data from Polischuk, S. C., Nortstrom, R. J., and Ramsay, M. A. 2002. Body burdens and tissue concentrations of organochlorines in polar bears (*Ursus maritimus*) vary during seasonal fasts. *Environmental Pollution 118*, 29–39.

6.23—Redrawn with modifications from Duckham, A. N. 1976. Environmental constraints. In A. N. Duckham, J. G. W. Jones, and E. H. Roberts (eds), *Food Production and Nutrient Cycles*. North Holland Publishing Company, Amsterdam.

6.24—After Tivy, J. 1990. *Agricultural Ecology*. Longman, Harlow.

6.25—Anne-Maria Brennan

6.26—After Balch, C. C. and Reid, J. T. 1976. The efficiency of conversion of animal feed and protein into animal products. In A. N. Duckham, J. G. W. Jones, and E. H. Roberts (eds), *Food Production and Nutrient Cycles*. North Holland Publishing Company, Amsterdam.

6.28—After Tivy, J. 1990. *Agricultural Ecology*. Longman, Harlow.

6.29—Based on data from Slesser, M. 1975. Energy requirements of agriculture. In J. Lenihan and W. W. Fletcher (eds), *Food Agriculture and the Environment*. Blackie, Glasgow; and Tivy, J. 1990. *Agricultural Ecology*. Longman, Harlow.

6.30—After Tivy, J. 1990. *Agricultural Ecology*. Longman, Harlow.

6.31—Guy Beaufoy

6.32—After Caraveli, H. 2000. A comparative analysis on intensification and extensification in Mediterranean agriculture: dilemmas for LFAs policy. *Journal of Rural Studies 16*, 231–242.

6.33—John Feltwell

6.34—(a) Dirk Morusi, Wikipedia; (b) Russ Greenberg

6.35 and 6.36—After Cruz-Angon, A. and Greenberg, R. 2005. Are epiphytes important for birds in coffee plantations? An experimental assessment. *Journal of Applied Ecology, 42*, 150–159.

Tables

6.1—Data from Whittaker, R. H. and Likens, G. E. 1973. The primary productivity of the biosphere. *Human Ecology 1*, 299–369.

6.2—Data from Cooper, J. P. 1972. *Photosynthesis and Productivity in Different Environments*. Cambridge University Press, Cambridge; and Hay, R. K. M. and Walker, A. J. 1989. *An Introduction to the Physiology of Crop Yield*, Longman, Harlow.

Chapter 7

Frontispiece—Morgan/A. Pengally

Figures

7.5—Courtesy of Google Inc.

7.6—After Reinoso, J. C. M. 2001. Vegetation changes and groundwater abstraction in SW Donana. Spain. *Journal of Hydrology 242*, 197–209.

7.7—After Ghassemi, F., Jakeman, A. J., and Nix, H. A. 1995. *Salinisation of Land and Water Resources*. New South Wales Press, Sydney.

7.8—Modified from Machita, L. 1973, 'Prediction of CO_2 in the atmosphere'. In G. M. Woodwell and E. V. Pecan (eds), *Carbon in the Biosphere*. United States, NTIS, Washington DC.

7.10—John Feltwell

7.11—Redrawn from Odum, E. P. 1989. *Ecology and our Endangered Life-support Systems*. Sinauer Associates, Sunderland, MA.

7.12—John Dodd

7.14—(a) Ken Giller; (b) Joe Lopez-Real

7.15—Anne-Maria Brennan

7.17—The Broads Authority

7.18—Venice Institute of Science, Letters and Arts

7.19—After Bettinetti, A., Pypaet, P., and Sweerts, J.-P. 1996. Application of an integrated management approach to the restoration project of the lagoon of Venice. *Journal of Environmental Management 46*, 207–227.

7.20—NASA

7.21—George Lees

7.22—After European Investment Bank 1990. *The Environmental Program for the Mediterranean: Preserving a Shared Heritage and Common Resource.* Report number 8504. International Bank for Reconstruction and Development/World Bank and European Investment Bank, Luxembourg.

7.23—Alan Beeby

7.24—Mike Griggs, supplied by Jonathan Mitchley

7.25—Redrawn from Bradshaw, A. D. 1984. Ecological Principles and land reclamation practice. *Landscape Planning 11*, 35–48.

7.26—T. Jaffre, supplied by Roger Reeves.

7.27—Anne-Maria Brennan

7.29—From Pan, J.-X. 1591. *A Review of River Flooding Control* (in Chinese).

7.30, 7.31, and 7.32—After Kechavarzi, C. *et al.*, 2007. Root establishment of perennial ryegrass (*L. perenne*) in diesel contaminated subsurface soil layers. *Environmental Pollution 145*, 68–74.

Table 7.3—After Bradshaw, A. D. 1987. Restoration: an acid test for ecology. In W. R. Jordan, M. E. Gilpin, and J. D. Aber (eds), *Restoration Ecology: A Synthetic Approach to Ecological Research.* Cambridge University Press, Cambridge.

Chapter 8

Frontispiece—Mike Hollingshead, http://www.extremeinstability.com

Figures

8.2—John Feltwell

8.3—Alan Beeby

8.5—Alan Beeby

8.7—(a) Panos Picture; (b) NASA image courtesy of Jeff Schmaltz, MODIS Rapid Response Team, NASA-Goddard Space Flight Center.

8.9—Photograph taken by the crew of the space shuttle (STS037–152–091), taken April 1991, courtesy of NASA/Johnson Space Center.

8.11—After Whittaker, R. H. 1975. *Communities and Ecosystems*, 2nd edn. Macmillan, New York.

8.12—Alan Beeby

8.17—John Feltwell

8.18—Nancy Laurenson

8.20—Anne Maria Brennan

8.21—Candy D'Sa

8.23—John Feltwell

8.25—National Oceanic and Atmospheric Administration

8.26—National Oceanic and Atmospheric Administration

8.27—Alan Beeby

8.28 and 8.30—Redrawn using data from Houghton, J. T. *et al.* (eds). 1990. *Climate Change: The IPCC Scientific Assessment.* Cambridge University Press, Cambridge.

8.29—Redrawn using data principally from the UK Natural Research Council 1989. *Our Future World: Global Environmental Research.*

8.31—Earth Observatory, NASA GSFC, Jacques Descloitre and the MODIS Land Rapid Response Team. From the satellite Acqua, taken April 20, 2003.

8.32—Data from Whittaker, R. H. and Likens, G. E. 1973, cited by Paul Colinvaux in *Ecology 2*, John Wiley, New York.

8.33—Redrawn from Stott, P. A. *et al.* 2000. External control of 20th century temperature by natural and anthropogenic forcings. *Science 290*, 2133–2137.

We are grateful to Dr Tim Johns in helping to supply the figures from the Hadley Research Centre of the UK Meteorological Office.

8.34—Redrawn from data published by the World Glacier Monitoring Service (http://www.geo.unizh.ch/wgms/), 2006.

8.35—Alan Beeby

8.36—Data from Osterkamp, T. E. 2005. The recent warming of the permafrost in Alaska. *Global Planetary Change 49*, 187–202.

8.37—Drawn from data in Behling, H. 2002. Carbon storage increases by major forest ecosystems in tropical South America since the Last Glacial Maximum and the early Holocene. *Global Planetary Change 33*, 107–116.

Tables

8.1—Based on Forman, R. T. T. 1995. *Land Mosaics. The Ecology of Landscapes and Regions.* Cambridge University Press, Cambridge.

8.2—Data from Houghton, J. T. *et al.* (eds). 1990. *Climate Change: The IPCC Scientific Assessment.* Cambridge University Press, Cambridge, and the *Latest Assessment* recent data updates (http://www.ipcc.ch/).

Chapter 9

Frontispiece—Mike Harding

Figures

9.1—Courtesy of NASA. Image ISS01E6765 from the Earth Sciences and Image Analysis Laboratory. Photograph taken by the crew of expedition 1 of the International Space Station.

9.3—After MacArthur, R. H. and Wilson, E. O. 1967. *The Theory of Island Biogeography.* Princeton University Press, Princeton, NJ.

9.5—Drawn from data compiled by the World Resources Institute for 2000–2001 using various sources including the UN and the International Union for the Conservation of Nature (IUCN).

9.6—Ove Hoegh-Guldberg

9.8—Image courtesy of Serge Andrefouet, University of South Florida, based on data from the USGS EROS Data Center, from the NASA Earth Observatory.

9.11—Photodisc

9.12—Redrawn from Tilman, D. and Downing, J. A. 1994. Biodiversity and stability in grasslands. *Nature* 367, 3633–3635.

9.13—M. V. Angel of the Southampton Oceanographic Centre

9.14—D. J. Currie and the Editor of the *American Naturalist*, University of Chicago Press.

9.16—Used with kind permission of the Editor of the *American Naturalist*, University of Chicago Press.

9.18—Redrawn from Tuckwell, H. C. and Koziol, J. A. 1992. World population. *Nature 359*, 200.

9.19—Digital Vision

9.20—Redrawn from Graham, N. A. J. *et al.* 2006. Dynamic fragility of oceanic coral reef ecosystems. *Proceedings of the National Academy of Sciences USA 103*, 8425–8429.

First words

Four- or five-year-old female polar bear on drift ice in the northwest Barents Sea, August 1999.

What the polar bear tells us

Polar bears are the largest land carnivores alive today yet they live on a part of the globe where there seems to be little for any animal to eat. It is, of course, the sea beneath their feet that provides their food. Curiously, these are terrestrial animals living in a marine habitat.

In fact, the bears are only forced onto land when the ice begins to melt and life on the floes becomes too precarious. In the summer they can no longer hunt seals from the ice and have to live off the reserves accumulated during the winter. Half the body weight of a well-fed polar bear may be fat but by the end of the summer this will fall to just 10 per cent.

Despite their isolation in the far north, well away from most humans, it seems that the bears may be affected by our industrial activity. A survey around Svalbard, a group of islands 800 km off the northern coast of Norway, found a small percentage of female bears with malformed genitalia. Their external reproductive organs have partially developed male characteristics. And it seems this may be our fault: researchers from the Norwegian Polar Institute suggest that industrial pollutants in the bear's tissues could be the cause of these changes—compared to those in Alaska, the bears captured in Svalbard have 20 times the concentration of some industrial chemicals.

If nothing else, this research shows that the pristine whiteness of the northern ice sheet belies the contaminants in the waters below. However, it is far from proven that the bears are affected by their burden. Since the initial report other researchers have described a single female from Greenland with low pollutant levels, but with comparable deformities, almost certainly a result of inflammation. Perhaps the Svalbard bears have been misdiagnosed. Yet other recent studies find pollutant levels are correlated with levels of a key female hormone in the blood plasma of a large number of nursing females from Svalbard, perhaps an indication of some disruption of their reproductive physiology.

We do not know if the fertility of the bears is affected by their abnormalities. The Institute reports catching fewer older bears in recent years, but this only indicates a shortening of their lifespan and says nothing about the number of cubs born. However, other aquatic mammals and a variety of reptiles and fish with elevated contaminant levels also show abnormal sexual characteristics. Many of these examples come from waters closer to major sources of pollution and may be more indicative of our own pollutant burden. Indeed, some scientists suggest that we too are showing the effects: compared to 50 years ago, human sperm counts have been halved and, on average, men today produce less semen.

We could go on, but this example may have already raised some questions in your mind. Such as—could this low level of deformity be 'normal' in polar bears? Perhaps it is only a short-term physiological response, with no reproductive significance. Perhaps it is a genetic defect, a variation that has arisen spontaneously in the Svalbard population, quite independently of its pollutant exposure. Can we demonstrate that the pollutants in its tissues are indeed responsible?

And why are the pollution levels so much higher in the Svalbard population anyway? How did industrial pollutants find their way to the Arctic? If it is something in the water, are all mammals affected in the same way? Are human beings at greater risk living further south, closer to the pollution sources?

We could go on (and do . . . see Box 1). We could ask questions about the ecology of the polar bear and whether this makes them more susceptible to the pollutants. Does the malformation have any effect—can the females still give birth and if so, is there any reduction in the fertility of the Svalbard bears? Is the population likely to decline when these changes affect only a small number of females? And if the polar bear has a poor reproductive future, do we also?

There are few answers and the questions, as you may have noticed, continue to multiply. Some of the answers come from understanding the adaptations of polar bears to their environment and their relationship to other species. The Svalbard bears have high concentrations because of the location of their islands (Figure 1) but all polar bears will accumulate organic pollutants in their fat. Part of the bear's

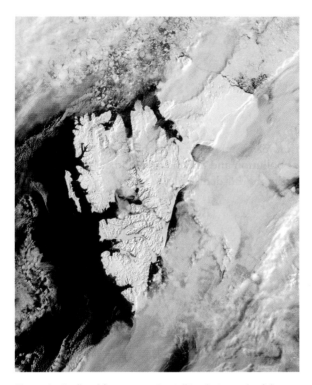

Figure 1 Svalbard from space. A satellite photograph of the islands with the wind direction picked out by the cloud formations passing over the archipelago. As a measure of the scale, the long thin island off the west coast, Forlandet, is 85 km long.

adaptation to survive the Arctic cold is to maintain a layer of fat beneath the skin, to act as insulation. They feed on animals which themselves concentrate pollutants in their insulating fat or blubber and it is such features of a polar bear's biology and ecology that put them at particular risk.

Ecology is the science that seeks to describe and explain the relationship between living organisms and their environment: in this case, everything that determines the distribution and numbers of the polar bear, from what they eat to their position in the global traffic of wind and water. Their anatomy and physiology is adapted to the conditions where they live, the waters in which they swim, and the long periods they have to survive without food. Continually subject to natural selection, the bears are fitted to their habitat and the prey they hunt.

To understand how the Svalbard bears receive the pollutants, we need to describe the movement of pollutants through oceans and food chains, in the currents and winds that move across the planet, and the plants and animals that move with them. Why some pollutants move along food chains—and others do not—is partly explained by the pollutants' chemistry but also by the habits, physiology, and biochemistry of the organisms at each step. To explain the exposure of the Svalbard bears requires an appreciation of everything from the global climate to cell metabolism.

Starting with the cell. At the moment we do not know whether the bears are showing a genetic abnormality, or simply a variation some individuals acquire with age. However, the presence of similar malformations in a range of aquatic animals suggests a common factor. It is unlikely that seals, bears, alligators, and whelks all share the same genetic factor, but it is probable that some basic feature of cellular structure or function—fat metabolism, perhaps—is impaired by these contaminants. Some groups are clearly more susceptible than others, possibly because of their genetic constitution, but also because of their physiology or feeding habits.

The pollutants most often blamed are the polychorinated biphenyls (PCBs), which are thought to mimic the mammalian sex hormones, the androgens and oestrogens (Box 2). Like natural hormones, PCBs are fat-soluble and accumulate in the fat of animals. At key stages in the growth of a young mammal the levels of the sex hormones govern the development of the gonads (ovaries and testes) and the genitalia. In experiments, common seals fed fish contaminated with PCBs have lower reproductive rates than control animals. Seals from the waters around Svalbard have high concentrations of PCBs in their fat and these are readily transferred to predators consuming them. Indeed, polar bears often consume the blubber of ringed seals preferentially, selecting this tissue for its high energy and high vitamin content (Figure 2).

At the global level, the affected Svalbard bears are on a principal route of pollution into the Arctic—the major air and water movements around the islands bring pollutants from northern Europe. Contaminated

Figure 2   The ringed seal (*Phoca hispida*) is the main prey species for polar bear (*Ursus maritimus*).

air arrives from the industrialized regions of Europe and North America (Figure 3) but most importantly, these pollutants arrive in the tissues of fish and seals migrating from the south. The isolation of the bears on these islands close to the North Pole is not as great as it seems, and the levels in their tissues testifies to the ease with which some contaminants move around the globe. Thus, the ecology of the Svalbard population, where it lives and what it eats, exposes it to higher concentrations of PCBs.

The story of the Svalbard bears suggests that our wastes could be affecting ourselves. We are also mammals, with essentially the same sex hormones and reproductive physiology, so the polar bears may be an important warning signal for us. There are, of course, major differences between the cold-adapted

polar bears and humans: we are, after all, a tropical or subtropical species, and only recent colonists of the higher latitudes. But the biology we share with the bears, and the possible effects of contaminants in their tissues, naturally prompts a series of questions (Box 1).

Even so, the case against the pollutants is not proven and we may be on thin ice. We have chosen this example to begin our survey of ecology because there are so many uncertainties—the data are not clear-cut and the connections have not been confirmed.

Scientists need to question and evaluate the information available, as part of a process of continual revision. For this reason, we want to encourage you to adopt a critical and evaluative approach to the ideas described in this book. Effective science proceeds by ruling out possibilities and generating alternative explanations. This is how we design experiments and collect data. Identifying and evaluating alternative explanations is crucial to the scientific process.

This book is a primer in the science of ecology and the major environmental issues. The bears of Svalbard have already introduced some of the key concepts. Ecology seeks to describe and quantify the connections between an organism and its environment. The biology and behaviour of the polar bear can only be fully understood by seeing how it is adapted to its living and non-living environment—how it survives the cold, finds a mate, catches its prey. Ecology describes how its population changes with time and its effects on the species around it. The bear's success in catching seals means these prey are being selected to evade capture, and the bears, in turn, must adapt to remain effective hunters. Ultimately, ecologists seek to explain all the features of polar bear ecology as a product of evolution through natural selection.

Through such interactions, the fortunes of the bears can be linked to those of the fish on which the seals feed, and the invertebrates on which the fish feed, and so on. In this way, collections of species are bound together to form an ecological community. Many of these connections may be tenuous or circuitous and it is difficult to decide which are critical to the nature of an ecological community. How would the community change if the polar bears were lost? Is the

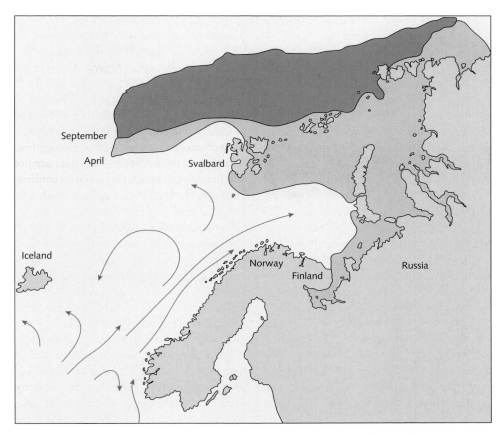

Figure 3 The major movements of wind and water currents into the seas around Svalbard in the Arctic circle. Migratory fish and mammals follow these currents and are a major source of some organic pollutants. The figure also shows the extent of the pack ice in April and September, from which polar bears hunt their main prey, seals.

collection of species within a community predictable for a given set of environmental conditions—and if so, why are polar bears only found in the Arctic?

Populations and communities grow and shrink in space and time, communities change across landscapes and between seasons, whilst life on the planet has a 4-billion-year history. Understanding the mechanisms by which species evolve, coexist and disappear will give us some insight into our own future and our impact on the Earth.

How to use this book

First Ecology provides a sequential development of ecology, running from the subcellular—the chromosomal basis of inheritance and evolution—through populations, communities, and ecosystems, to landscape and planetary ecology. The value of this hierarchy in describing some key ecological patterns becomes apparent when we start to explore high-level processes in Chapter 8. Some of the basic terms of ecology are given in Box 2 and a glossary is also provided at the end of the book.

Our starting point throughout the book is to introduce ecology from a human perspective. Whilst we assume little background biology, we hope to draw upon your everyday experience and indeed, your own biology. Each chapter begins with familiar ecological processes and ends with some example of how ecology can be used to manage or safeguard our environment.

Placing human beings in the foreground should help us to see humanity as part of the ecology of the planet. Over the last 100 years we have learnt much about our origins and the changes in the environment that prompted our evolution (Chapter 1). We can also trace the environmental changes that have followed in our wake. Mediterranean-type communities are as old as we are and, for the most part, each region with this type of climate has been greatly influenced by our activity. These ecosystems occur in five widely separated parts of the plant, and because of this, provide an insight into how ecological communities are put together (Chapter 5). They are also a community type most likely to change with significant global warming over the next 100 years. Throughout the book we use them as an exemplar ecological community.

Ecological processes that take place over large areas tend to happen very slowly but, as we saw with the polar bear, ecology encompasses processes that operate from the cellular to the planetary level. Part of our task is to explain how these interact (Chapters 8 and 9). Between bacteria and biome are highly complex communities of species that divert and delay the transfer of energy and nutrients (Chapters 6 and 7). Our release of fossil carbon over the last 300 years, and the reduction in the capacity of the biosphere to capture this, is causing a rise in atmospheric temperatures, and ecologists are now asked to predict the consequences (Chapter 8).

Ecology is a relatively young science and formal experiments on communities and ecosystems are confined to the last 100 years. This is a small sample of the Earth's history—important to remember when ecologists are asked to predict global changes over the next 100 years. The fossil record helps us to appreciate the current rate of species extinctions (Chapter 9), comparable to the scale of losses which have followed major environmental upheavals in the planet's history. Today we have named and catalogued (Chapter 2) only a small fraction of the species that are alive on the Earth and now we are losing them faster than we can count them. In the process, we may be undermining the resources we need to support ourselves (Chapters 3 and 4).

In a book of this kind some topics are not covered neatly in one chapter. Themes recur and concepts are developed in several places. This book does not offer a convenient set of notes—it is not a reference book—but rather a narrative which revisits a topic and views it from several perspectives. Seeing the connections and achieving the overview is essential to fully grasping the ideas and adopting an evaluative approach. To aid your navigation, each chapter begins with a brief contents list and finishes with a summary. **Emboldened text** highlights key terms and points to a definition in the glossary.

Embedded in the text are directions to related sections where further detail on a topic can be found. The companion website also has a section termed 'Routes', which allows you to find where key themes are revisited. Many of the suggested tutorial and seminar topics can be researched using this facility. This section also offers topics for independent study, and references to the sections that will support your work. For example, using the 'Routes' feature will allow you to become familiar with particular aspects of the ecology of polar bears or human evolution and development. The website also provides additional web addresses for specific topics, additional exercises and the answers to the exercises at the end of each chapter.

Directions for further reading are given at the end of each chapter, with a brief commentary on their content or particular merits. The websites given here provide useful support material, from sources that we judge to be reliable. However, we caution against relying on web-based sources and encourage you to become familiar with the major ecological journals.

Each chapter has a number of boxes to extend or clarify material or to provide overviews of key concepts. Several are diversions which allow us to pick up some of the recurrent themes. In each chapter there is a case study providing one or two items of current research in greater detail, to give some insight into the methods used by ecologists. There is no realistic prospect of properly introducing ecological methods in a basic text, but we hope that our choice of these topics imparts some flavour of the ecological techniques relevant to that chapter.

The pressure from several of our reviewers to include fully referenced text has been almost unbearable. As we have said in previous editions, our aim has been to keep the text uncluttered and to allow the ideas to develop without distraction. Our policy is to name key authors, and thereby allow sources to be identified from the sectionalized references at the end of the book. However students need to cite and use references fully and we set a poor example. Our compromise is to use the case studies to demonstrate how citations and reference lists are used in the formal scientific literature. We encourage you to follow this example.

Finally, the website now features a virtual field course that allows you to visit and analyse data from some of the ecosystems we describe. We provide a series of exercises which can be completed independently, and which are designed to allow you to complete some basic ecological data analysis and presentation. They also support several key ideas, and invite you to observe the world through an ecologist's eyes. We would very much like to include *bona fide* material or exercises from other teachers, especially if they add and support the topics in *First Ecology*.

You can find out more about Svalbard at:

www.unis.no

and the Norwegian Polar Institute at:

http://npiweb.npolar.no

 BOX 1   **Nature or nurture . . . or neuter**

What is it to be a man? In mammals, maleness means having a Y-chromosome which carries a gene that codes for a protein that triggers the development of particular cells which regulate sperm production and produce a hormone that interacts with other genes to drive the development of secondary sexual characters, such as a penis and a low voice. This sequence of switches all need to be set to on, early in the life of the embryo to stop it becoming female.

All mammals share the X–Y chromosome mechanism for sex determination, but the switches can be different in other animals. A variety of chromosome combinations govern sex in insects, birds, and reptiles. For alligators, it is the prevailing temperature during incubation that determines the eventual sex of the embryo—at relatively low temperatures females are produced. In mammals, gender is determined by the embryo's *nature*—the characters it has inherited with the sperm—but in alligators it is *nurture*—the temperature at which it is raised.

As with much of the rest of the plant and animal kingdoms, being male or female says much about the individual's ecology. However, being a man at the beginning of the twenty-first century is proving increasingly difficult. Sperm counts have fallen by half amongst United States males since 1940 (Figure 4). Around the globe, males in

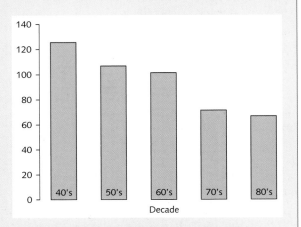

Figure 4   Average sperm counts (millions per millilitre) in US males by decade, 1940–1990.

(continued overleaf)

some populations show increasing genital variability, with more cases of undescended testicles and lower testicular weights. This has been described as 'feminization', a tendency observed in several other mammals and thought by some scientists to be associated with the pollution burden many people carry.

In the USA declining sperm counts are most pronounced in men exposed to certain agricultural pesticides and are less obvious in urban populations. However, the link with these organic chemicals and other so-called 'oestrogen-mimics' is far from proven. Part of the difficulty is establishing cause and effect—a number of surveys show that a high exposure is associated with low counts but we have difficulty allowing for other factors and demonstrating that the pollutants are indeed the major cause.

Neither is the response consistent: amongst the Svalbard bears the females are showing masculinization but feminization has not yet been described for the male bears. It may be that these pollutants act generally as 'endocrine disrupters', since the feminizing oestrogens and the masculine androgens share very similar chemical structures. Endocrine disrupters have been blamed for the increased incidence of reproductive disorders in humans—prostrate and testicular

cancers in the United Kingdom, the proportion of ectopic pregnancies in the United States, and a more general increase in breast cancer. Again, this is matching trends to rises in environmental levels of the pollutants, but not establishing a connection. Indeed, some surveys of mammals from contaminated habitats have failed to show any such effects.

We need to establish that the Svalbard bears are responding to the pollutant rather than suffering an inflammatory condition or a spontaneous (non-induced) genetic change originating in one female several generations ago. Describing the relationship between the dose of a poison and the response of the bears would require a series of controlled experiments. Fortunately, perhaps, the bears are neither sufficiently abundant nor cooperative enough to allow us to do this properly. We can, however, draw on experimental evidence from other mammals—grey and common seals show disorders of their immune system and skeletal and reproductive disorders when fed fish contaminated with PCBs.

Some of the best evidence for a link comes from alligators. PCBs are passed from mother to offspring in the alligator eggs. Lake Apopka is one of several lakes in Florida (Figure 5) with high levels of persistent organic pesticides and in one survey,

Figure 5 Lake Apopka in Florida.

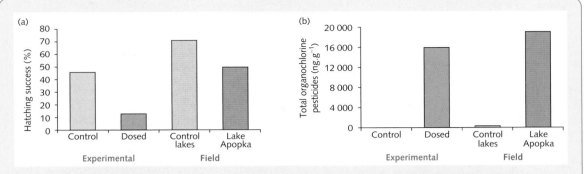

Figure 6 Evidence for the effect of pesticides on hatching rates and levels of organochlorine pesticides in the eggs of *Alligator mississippiensis*. This data, from the work of Heath Rauschenberger and colleagues, shows the results of a dosing experiment and a field survey of two uncontaminated and one of the three contaminated lakes (Apopka) they surveyed in Florida.

(a) Treated females produce eggs with a significantly lower hatching rate compared to those from the experimental controls. However, these controls do not differ in their hatching success from the contaminated eggs from Lake Apopka. This, even though the yolks of eggs from experimentally dosed mothers have levels of pesticides close to those of Apopka (b).

Across the survey of the three contaminated lakes there is no simple relationship between hatching success and total pesticides in the eggs. Thus, whilst the experimental dose does depress hatching rates, factors other than the concentration in the egg must be important in the field.

over 80 per cent of its alligator eggs failed to hatch (compared to an average of 20–30 per cent in nearby lakes). Apopka's hatchling survival rate was only one-tenth that of other lakes. Female juveniles had oestrogen levels twice those considered normal and produced twice as many eggs. In contrast, males had low levels of testosterone. Adult males showed signs of ovary development and their penises were around one-third their normal size. In experiments, painting alligator eggs with the two pesticides contaminating the lake (DDT and DDE) caused the hatchlings to develop abnormalities seen in the local population. Dosing female alligators with a cocktail of pesticides at ovulation and then feeding dosed food during their reproductive cycle reduced hatching rates of their eggs. The eggs also achieved pollutant levels over 300 times those from a control group of females (Figure 6).

There are enough examples of human beings changing their gender using hormones for us to know that what is written in the genes, even those defining sex, only represent possibilities. The bears and the alligators emphasize a very important general point—our genetic nature defines our potentialities and not our outcomes. The success or otherwise of an individual to reproduce depends on the interaction of these genes with the environment in which they find themselves, and for the alligators of Apopka, such nurture may lead to neutering.

Certainly, the more species we find with reproductive abnormalities combined with high pollutant levels, the more likely it is that the connection is real and that these compounds are responsible for our own falling sperm counts. This would be a sharp reminder that humans share physiologies with other mammals and may respond in the same way to these toxins. We cannot take ourselves out of the food chain or absent ourselves from the ecological processes which sustain the environment. We are poisoned for the same reasons that polar bears are poisoned and we rely on an ice sheet at the poles just as much as they do.

BOX 2   **A few definitions**

The formidable collection of names in biology can be a major barrier to understanding. A glossary is provided at the end of the book, but a few important definitions are needed here to support your study from the outset.

A *species* is any group of individuals that can actually or potentially breed with each other to produce viable and fertile offspring. Because there are degrees of genetic difference between all individuals, deciding where one species ends and another begins can be very difficult. *Hybrids* derived from two different (if closely related) species are possible, but these are rarely fertile and therefore represent a genetic dead-end. This definition is fine for sexually reproducing species, but becomes problematical in others. We look at the concept of the species in greater detail in Chapter 2.

Within a species there may be several *populations*—individuals of a single species occupying a particular location at a particular time. Notice that both place and time need to be defined carefully (Chapter 3).

Populations of different species are collected together into *communities* (Chapters 4–7). Within a community species interact, feeding on each other, competing for resources or cooperating through special relationships. Various types of community with highly predictable groupings of plant species, *biomes*, are found in different parts of the globe (Chapter 8).

A community together with its physical environment is an *ecosystem*. Very often, physical features, such as light, moisture and so on, define the sort of community that can be supported. Within the system, nutrients and energy move between the living (*biotic*) components and the non-living environment (its *abiotic* components) (Chapters 6 and 7).

We have just described a hierarchy moving from the species to the ecosystem. This is the sequence we use in this book to explore the science of ecology.

First Ecology online resource centre

First Ecology doesn't end with the last words of this book. There is also an online resource centre at:

www.oxfordtextbooks.co.uk/orc/ beebybrennan3e/

The site includes:

- Answers to exercises in the book so you can check your responses to the exercises as you work through them. We encourage you to attempt the exercises, to recap on the factual material, and to test your understanding of the concepts.

    There are also additional exercises (with answers) provided in a separate section of the site for your tutors to use.

- A '**Routes**' section, providing a map of how some key themes are developed through the book and including some case study titles, relating to each theme. This facility allows you to direct your own research into these key areas. You may find this useful if you wish to research or improve your understanding of these themes, and it should also demonstrate how the various principles in ecology need to be considered as a whole to see the complete picture.

- A **web link library**, including all the URLs listed at the end of chapters in the book. Additional web links to other sites for specific topics are also listed on a sectionalized basis. Please feel free to suggest links that should be added to this site, preferably offering a commentary on the content (see below for contact details). We hope this will serve to keep the web links up to date, but cannot accept responsibility for any of the links posted here. We will endeavour to review these links periodically and offer our own commentary. Additionally, **hyperlinks to all of the references** in the text, allow you (your institutional licence permitting) to view all of the cited papers electronically.

- A **virtual field course** comprising a series of basic exercises using real data collected by the authors from several sites during 2006. In each case, there is a brief introductory presentation supported by a video or slide sequence, setting out the aims of the exercise and links to the data files and instructions. Each exercise also has an accompanying set of tutor notes. These exercises could form the basis of an assessed element in a basic course in ecology. We hope that students and tutors will post their data analysis on the website, to suggest revisions and also to add additional datasets. Details of these facilities are available on the web page.

- A **hyperlinked bibliography**—direct links to online articles cited in the book.

    Finally, we would welcome your general comments, especially concerning those topics that need to be clarified in the text. To contact us with comments and suggestions, simply click on the 'feedback' link at:

www.oxfordtextbooks.co.uk/orc/beebybrennan3e/

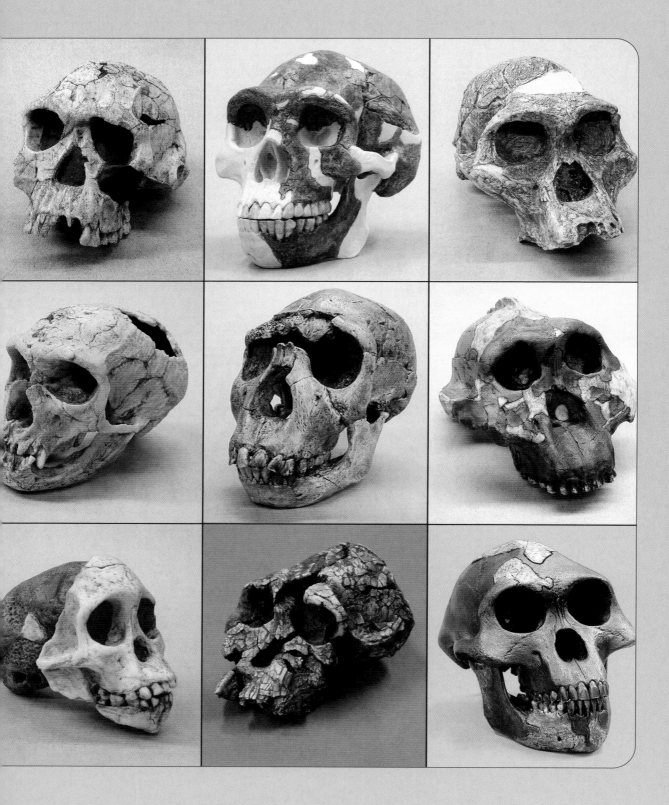

Origins

'Nature . . . or survival of the fittest, cares nothing for appearances, except in so far as they are useful to any being.'

Charles Darwin: *The Origin of Species*

CHAPTER OUTLINE

- Human beings as products of their environment and of natural selection.

- A brief history of human ancestry.

- The principles of evolution by natural selection.

- Why sexual reproduction is advantageous in a changing world.

- How genetic information is coded, read, and copied.

- Transmitting this information from generation to generation.

- The significance of variation for evolution.

← Meet the family. Reconstructed fossils skulls from our near relatives in the family Hominidae.

Figure 1.1 'Patches' the Great White shark (*Carcharodon carcharias*).

The beautifully streamlined profile of a shark imme-diately suggests its way of life. The finely shaped head, lifted on the wings of its pectoral fins and driven by the powerful motor of its long tail, reflects the lines of an open-water predator built for speed and manoeuvrability. As sleek as a jet plane, the shark contrasts with the heavy armour and forward-facing weaponry of the lobster. Lobsters are the tanks of the sea bed, armed and protected to stand and fight, rather than for speed.

Human beings are not so obviously shaped for one way of life. We have no distinct defensive struc-tures and our small teeth do not suit one particular diet. We are terrestrial animals, primarily a tropical species needing a warm habitat with fresh water. Yet our species has ranged far more widely. We have burrowed into the ocean floor, marked our territory on the moon, and sniffed around active volcanoes. Even without the armour plate of the beetles, we have visited more extreme environments than any other animal. And we can do this because of a single over-sized organ—the large programmable machine in our head. It is our brain that enables us to survive deep oceans or lunar landscapes. The brain is our most adaptable feature and its capacity to store and process information, to learn, to accept new instructions, or to develop new solutions, is the key to our success.

We are products of our environment in the same way as sharks or lobsters, but our adaptability comes from a capacity to modify environments to suit our needs. We have freed ourselves from the constraints of a single habitat and one way of life. Given the advantages our species enjoys, we inevitably ask why other animals have not also developed their mental capacity in the same way. Or to put the question the other way around, why should a thinking machine become so significant for one group of animals, at some particular location, at some point in the past? What was so different about our ancestors' ecology that made a powerful mental capacity so advanta-geous? Or indeed, possible?

We may be able to protect ourselves in submarines or spaceships but it was the pressures of an environ-ment selecting our ancestors that made us what we are today. Features we take to be characteristically human, features with which we are all familiar, have been shaped from our mammalian biology. In the process we have become a very significant species. Understanding our origins, and the processes in-volved, begins to explain how we have come to make such significant changes to the ecology of our planet. For this reason, we start our exploration of ecology by looking at our own evolutionary history, as products of an African grassland.

Later in this chapter we look at the principle of evolution and the mechanism of natural selection. We go on to consider how genetic information is coded and inherited, and see how this information can help reconstruct both our animal ancestry and our cultural history.

1.1 Origins of humanity

New species are produced when environments change. We know that early humans evolved at a time of considerable climate change in tropical Africa and these changes almost certainly created the genetic distance between gorillas, chimpanzees, and ourselves. Our ancestors were those apes that became adapted to life beyond the forest, where the trees gave way to grass.

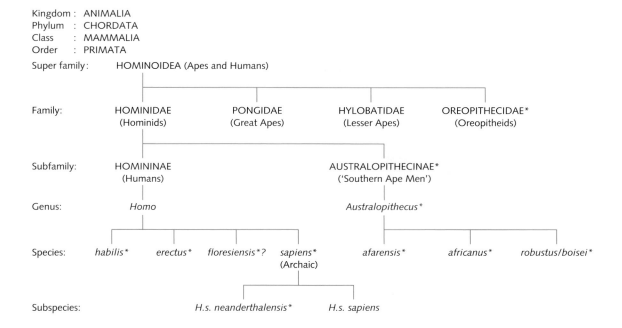

Kingdom : ANIMALIA
Phylum : CHORDATA
Class   : MAMMALIA
Order   : PRIMATA

Super family: HOMINOIDEA (Apes and Humans)

Family: HOMINIDAE   PONGIDAE   HYLOBATIDAE   OREOPITHECIDAE*
           (Hominids)   (Great Apes)   (Lesser Apes)   (Oreopitheids)

Subfamily: HOMININAE   AUSTRALOPITHECINAE*
              (Humans)   ('Southern Ape Men')

Genus: Homo   Australopithecus*

Species: habilis*   erectus*   floresiensis*?   sapiens*   afarensis*   africanus*   robustus/boisei*
                                                   (Archaic)

Subspecies: H.s. neanderthalensis*   H.s. sapiens

* Extinct group

Figure 1.2   A simplified classification of hominids, setting out the phylogenetic relationships of our species. Species which are closer together on a phylogenetic tree share more of their ancestry than those far apart— so our nearest relatives were the Neanderthals. The different levels in a hierarchy and their significance are described in Chapter 2.

The earliest human fossils are found in Eastern and Southern Africa in deposits dating from around 5 million years ago. Six million years ago we shared an ancestor with chimpanzees, but already the climate had started to become cooler and drier, and the forests in East Africa were retreating. In the West the forests persisted and a variety of apes could maintain a life in the trees. It was here, in the last million years or so, that gorillas and chimpanzees appeared, adapted to a variety of forest habitats. Other apes, those most able to survive the open grassland, had already begun to dominate in the drier East.

Bipedalism, standing and walking on two legs, is the feature of the apes we now designate as human. All hominoid apes (gibbons and orang-utans as well as the great apes of Africa—Figure 1.2) will stand and walk upright on the ground, but it was *Australopithecus*—southern ape-men—which first walked persistently on two legs. Interestingly, *Australopithecus* had not entirely relinquished the

safety of the trees since its feet retained the ability to grip branches. Its upper skeleton was also well adapted to climbing. We can see equivalent indicators of their lifestyle in the feet of the modern great apes: those associated with the forest have the capacity to grasp whilst those of the open country have a propulsive pad (Figure 1.3).

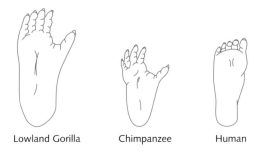

Lowland Gorilla   Chimpanzee   Human

Figure 1.3   The feet of three living Hominoidea. Two retain the capacity to grasp; the other is adapted entirely to walking on flat ground.

Around 3.5 million years ago *Australopithecus* had a brain the size of a chimpanzee's, but stood upright. Bipedalism predated any increase in brain size by over a million years. Perhaps, as Charles Darwin suggested, an upright posture promoted human mental development, by freeing the hands. As they were put to different and more complicated uses, the hands demanded finer control and a larger, more powerful brain. Besides manipulating objects, the hands are also free to carry food and infants, and a family group or band can move together, to feed or escape danger. Bipedal walking is a more efficient way of moving over open ground, at least when we compare modern humans with a chimpanzee using four limbs (Table 1.1).

Being able to see above the grass is likely to confer considerable advantage in finding food and avoiding predators. Standing upright also increases heat loss to the air, and reduces the area directly exposed to the sun, essential on the hot, dry savanna. All human species spent most of their time upright, a shift in posture that requires significant anatomical changes. The pelvis has to locate the legs directly beneath the body and the spine becomes S-shaped to bring the centre of gravity over this line. The muscles in the buttocks are used to pull the spine upright and this is why humans, alone amongst the great apes, have large backsides. The forelimbs now fall along the line of the spine and are swung as a balancing aid in walking. In this vertical position the rib cage has to move up and down, and in modern humans becomes barrel-shaped. Such major changes in posture are not without their consequences, most obviously the back problems many people suffer. These and other conditions demonstrate that we are far from perfectly adapted to an upright life.

We know from their teeth that early *Australopithecus* species were herbivores, but around two million years ago a new genus of humans appeared, with a much larger brain (Figure 1.4). *Homo habilis* was probably a flesh eater, suggested by the crude stone tools dating from the time of its fossils, and found alongside the butchered remains of animals. *Homo*, it seems, had developed new skills. Using the hands to manufacture tools would require a high degree of coordination, and make greater demands

TABLE 1.1 In comparison to chimpanzees, human development is slower, and the infant is dependent on the mother for longer. Humans also have a significant post-reproductive lifespan, possibly because of the role older adults play in educating the young. In contrast to the small growth in the chimp brain, the human brain quadruples its size after birth. Notice also the relative efficiency of bipedal walking.

	Chimpanzee	Modern human
Gestation time	34	38 weeks
Age when		
Head is held erect	2	20 weeks
Walk on all fours	20	40 weeks
Bipedal standing	40	56 weeks
Bones of pelvis fuse	1	7–14 years
Age at adulthood	9	13 years
Years surviving after reproduction has ceased	0	20 years
Brain size		
At birth	300	350 cm³
At adulthood	400	1450 cm³
As a ratio of adult body weight	0.001	0.025
Relative energy use		
Slow walk	1.7	1
Fast walk	1.5	1

on their mental capacity, as would the forethought needed to envisage an end-use for the tool.

Whether these new humans were primarily hunters or scavengers remains a matter of debate, though like many predators, they probably consumed the remains of kills by other species. Perhaps these first tools, simple stone clubs and stone flakes, were used to break open the bones for their marrow or to cut meat from a carcass. For nearly a million years *Australopithecus robustus* and *Homo habilis* lived together in East and Southern Africa, one primarily a fruit eater and the other a herbivore and an

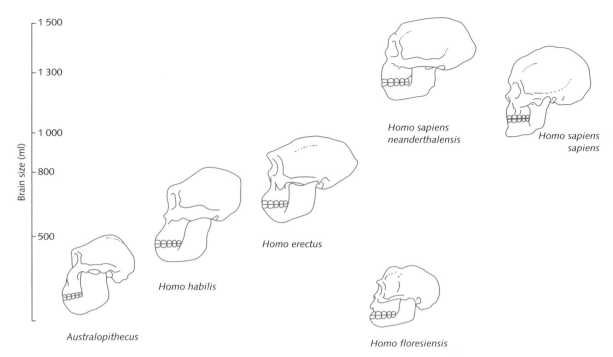

Figure 1.4 Changes in skull shape and the increase in brain size in the hominids. The finer jaws and reduced teeth are possible because of the use of tools to aid feeding. Much of the increase in its size is in areas of the brain responsible for higher mental processes (thinking and memory) and speech. Following the discovery of *Homo floresiensis* in 2003, scientists have to explain how their tiny brain allowed them to use tool technologies previously only associated with larger-brained *Homo*.

opportunistic scavenger. The two species probably shared the same ancestor (*Australopithcus africanus*), but *Homo habilis* had a larger brain and flatter, less protruding face (Figure 1.4).

The continued reduction in the size of the jaw and teeth in subsequent species of *Homo* follows the advent of tools folowing this shift in diet. Smaller teeth would have been no disadvantage if tools were used to butcher and cut meat. The evolution of *Homo* goes hand in hand with the evolution of their tools: over hundreds of thousands of years, tool development, from simple cutting edges through hand-axes to arrow heads, mark the advent of different human species (Figure 1.5).

Within half a million years, the climate had begun to dry again and the first human predator, *Homo erectus*, appeared. With its arrival *Homo habilis* and *Australopithecus robustus* disappeared from the fossil record. Again, *Homo erectus* had a larger brain

and a wider range of tools. Indeed, the arrival of subsequent *Homo* species—generally with larger brains and flatter faces—are marked by their artefacts and the new skills they were practising.

Until 2003, this association of larger brains with refined tools was both consistent and expected. Entirely unexpected was the discovery of a new species of human with a tiny brain, found alongside sophisticated tools and the charred remains of prey (Figure 1.6). From a cave on the Indonesian island of Flores, Mike Morwood, Peter Brown, and colleagues recovered most of the skeleton of an adult human female, just 1 m tall. Remains now collected from a total of nine individuals confirm this was a population of tiny humans, generally accepted as a new species—*Homo floresiensis*.

The remains are thought to represent an example of 'island dwarfing', a phenomenon known for other animal species where the limited resources of an

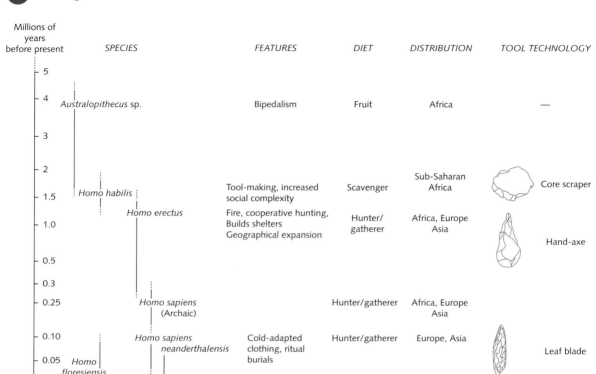

Figure 1.5 The origin of human beings. This diagram shows our immediate ancestors, which are distinct from the other African great apes (chimpanzees and gorillas). The hominids began with *Australopithecus* and its capacity to walk upright, and whose fossils are found as the East African forests were shrinking. With *Homo* came a change in diet and way of life, from a fruit eater to a scavenger/carnivore. Tools used to obtain their food gradually improved and became more finely fashioned. Whether *Homo habilis* or *Homo erectus* was on a direct line leading to modern man is still debated. Equally, there is controversy whether *Homo sapiens sapiens* arose from a single stock of archaic *Homo sapiens* in Africa, or developed simultaneously in several locations in Africa, Europe, and Asia. Remarkably, the remains of *Homo floresiensis* extend the line from *Homo erectus* or *Australopithecus* to virtually the present day . . . and add to this controversy.

island will not support a population of large individuals. *Homo floresiensis* would have been much smaller than any peoples alive today, with adults weighing just 25 kg, but within the dimensions of the earliest hominid, *Australopithecus afarensis* (Figure 1.2). Based on its skull, the Flores female most closely resembles *Homo erectus*, especially those from China and Java. However, other parts of the skeleton point to the more ancient ancestry, and a genus otherwise known only from Africa. Another surprise is the evidence for hunting not previously associated with a human with such a small brain (Case study 1).

Hunting and gathering was a major shift in lifestyle and required a range of new skills and new advances in tool technology. New technologies opened up new habitats: fire, clothing, and the building of shelters provided protection and warmth, allowing early humans to expand northwards into less hospitable environments. Perhaps for the first time, humans were beginning to create or change environments to suit themselves.

Figure 1.6 Far from home. Flores and the location of the Liang Bua cave (yellow arrow) where the soggy skeleton of a tiny female human was discovered in 2003. Deposits indicate that *Homo floresiensis* inhabited the island from 95 000 until just 12 000 years ago, and may have lived alongside modern humans for tens of thousand years. Remains from nine individuals have been recovered but only one complete skull. This is similar in shape to fossil *Homo erectus* skulls from Java and China, the nearest hominin fossils (red arrows). However, the rest of the skeleton suggests links with *Australopithecus afarensis* which was about the same height and had the same size brain. *Australopithecus* is only known from Africa, on the other side of the globe.

Cooperating to hunt, or to ward off predators, would have been essential for these not very fast apes. As with other animals, the requirements and implications of living in cooperative groups would have demanded social organization. At some stage in our evolutionary history the benefit of a precise language became important, aiding the cohesion and coordinating the activity of the band. Using words and putting them together in a meaningful way re-quires conventions accepted by all (Box 1.1). These rules impose a discipline on those trying to make themselves understood, but they bring a major advantage, allowing ideas to be passed from one generation to the next. With a developed syntax and vocabulary, teachers can convey complex and abstract ideas. Tool-making and other skills can then be learnt by instruction, not simply by copying the behaviour of others.

Individuals can rehearse and structure their thoughts using these same rules. Our verbal messages are usually what we wish the listener to hear and we anticipate the reaction our words might provoke. However, like many other species, we also send mes-sages to deceive. While the obligation to share food and to cooperate hold a social group together, there are circumstances, say in competition for a mate, when the advantage is with those who think ahead, who predict outcomes and make trade-offs. Social intrigues and duplicities are known in groups of chimpanzees and some anthropologists suggest that they were key to our own mental evolution, a developed social intelligence that second-guesses social interactions for individual advantage and reproductive success.

Language demands memory and a capacity for imagination and abstract thought, and its increasing importance can be seen in the later fossils. Changes in the braincase of these skulls indicate an expansion of the areas of the brain associated with language, and the anatomy of the mouth and the position of the larynx (voice box) also shift. A larger space developed above the larynx allowing a wider range of sounds to be produced. Unfortunately, it also increases the chance of choking on food, so that articulating speech must have been advantageous indeed to risk this hazard.

The significance of language and mental skills can be seen from looking at other aspects of our biology. Compared to other primates we are born at an early stage of development, before our head grows too large to pass through the mother's pelvic girdle. Relative to our body mass, our brain grows consider-ably after birth. Childhood itself is extended so that many of the anatomical changes associated with an upright posture and mobility occur relatively late in our development (Table 1.1). Even then, the child

Different from the rest

For some thinkers, it is consciousness that makes human beings distinct from the other animals. Although we are often impressed by the computational skills of some animals—such as the navigational skills of migratory birds or butterflies—our species has always enjoyed displaying its own mental powers. Building pyramids or sending men to the moon is as much a celebration of our intellectual prowess as our desire to know more.

This prowess allowed us to expand out of Africa and was the reason we could survive the ice ages of the north, moderating those severe conditions with fire and shelter. We may also have won past competitive battles because our species could use limited resources more efficiently or because we were more aggressive. It is unlikely that we displaced the cold-adapted Neanderthals on the basis of our biology alone.

Whilst it is hedged around with many philosophical problems, our mental capacity, and particularly our sense of self-consciousness, is one trait we use to define our species. Consciousness is the ability to think about our own thinking. It gives our mental machinery the capacity to observe itself, to learn and adapt, to improve from experience. Daniel Dennett argues that humans are uniquely conscious because, unlike other animals, we reflect on our thinking processes. That is, we can isolate concepts, the principles we use to define something, and then treat them as objects. Objects that can be refined, turned over in the mind, much as we might improve a tool through repeated handling. We think about how we think about things, hone our ideas by exploring their properties and their relations. For example, the concept 'bear' may include all the indicative features by which we define bears—their fur, their ears, their snout and so on—but our idea of *bear-ness* can encompass both a cuddly toy and a threatening predator.

We can also combine concepts and play with them—bears roaring, bears sleeping, bears planting a flag on the moon—creating images that we may have never seen. Novel combinations can produce novel solutions, some of which may be adaptive in the same way that novel gene combinations can be. The evidence is that even the brightest of other animals cannot manipulate concepts to create new ideas in this way. Without a fully verbalized language, chimpanzees do not come close. With no prompt to conceptualize—isolating the principle that a word represents—thoughts are not easily manipulated or reviewed.

As Dennett notes, without words, such abstraction is impossible:

> A polar bear is competent vis-à-vis snow in many ways that a lion is not, so in one sense a polar bear has a concept that a lion does not—a concept of snow. But no languageless mammal can have the concept of snow in the way we can, because a languageless mammal has no way of considering snow 'in general' or 'in itself'.

Humans go much, much further. Not only do we use language to extract the principles, we play with the ideas. Words, as symbols, are easily combined. We revel in our lingual skills, toying with the words and enjoying the clash of symbols. Dennett demonstrates his own playfulness a few lines later:

> We can speak of the polar bear's implicit . . . knowledge of snow (the polar bear's snow-how). . . . but then bear in mind that this is not a wieldable concept for the polar bear.

Not only can we manipulate concepts, and occasionally come up with new and viable combinations, our language also provides us with one other key advantage over other animals. We can pass these concepts on to our offspring. Certainly other animals learn from their parents, but they do not deal in concepts and they do not learn to manipulate them. We famously stand on the shoulders of the great thinkers who have gone before us. Mercifully, we do not have to invent the wheel . . . or the word-processor . . . anew with each generation.

remains with the mother and the family for a relatively long period. Along with mother's milk we pass on the tricks of survival, feeding both the body and the mind, inducting offspring into the knowledge and experience held collectively amongst the adults.

Unlike most wild chimpanzees, we also live on for some years after we have ceased reproducing. Drawing on their longer experience, grandparents often play a crucial role in raising the young, and in many cultures they are seen as repositories of a society's wisdom. It is perhaps the lessons we pass on to the young that give our post-reproductive years some adaptive significance.

The complexity of our language, the relative precision of its symbols and structure, puts the greatest distance between ourselves and the apes. Other animals are able to walk upright, some use tools, and most communicate with each other. Some, like chimpanzees, do all these things and live in highly integrated and duplicitous social groups, often with distinctive cultures of behaviour or even technology. But it is our spoken language, and the culture it creates, which distinguishes modern humans. So central is language to human evolution some linguists suggest it is less of an acquired skill and more of an inherited instinct (Box 1.2).

Their success allowed *Homo erectus* to escape from their cradle in the East African Rift Valley, moving through the Middle East into Europe and Asia. Surviving in these different environments would have demanded adaptability and ingenuity, and the ice ages that followed would have further tested human resourcefulness. Yet the hominids continued to spread and to diversify. Our species, *Homo sapiens sapiens*, appeared around 100 000 years ago in East Africa, and moved out through the Middle East. In step with the glaciations, they migrated northwards to join their near relatives, the Neanderthals, in Europe and Asia. Neanderthals had been in the north before the evolution of modern humans and had the anatomy of a cold-adapted people with large chunky bodies. They also had larger brains.

Genetic evidence suggests Neanderthals and modern humans shared an ancestor about half a million years ago, so *Homo sapiens sapiens* was not a descendent of *Homo sapiens neanderthalensis*. Modern humans arrived in Europe around 40 000 years ago but within 10 000 years Neanderthals had disappeared. The two sub-species lived side by side for a long time, though the details of our relations with Neanderthals are not known. It may have been competition for resources that led to the demise of the Neanderthals. Some suggest the Neanderthals were 'absorbed' by interbreeding with modern humans (pointing to an ancient skeleton found in Portugal that shows characteristics of both sub-species) but there is no genetic evidence for this. There may have been more direct conflict, of which our species were the victors. Today we can only imagine what it would have been like to look into Neanderthal eyes, perhaps even to speak with them and learn directly of another species' view of the world. It would be stranger still to meet the more distant *Homo floresiensis*—a very different species, yet one that disappeared only 12 000 years ago. Intriguingly, it may be unwise to rule out this particular meeting of minds just yet.

In this simplified classification, we have described just three or four species of *Homo* (Figure 1.2), those whose remains indicate a major shift in lifestyle or geographical range. A classification based on anatomical or morphological changes, especially in the skull and teeth, would recognize many more, so that some authors count 13 species belonging to our genus alone. How many are distinct species is open to question, but it is certainly true that greater variation followed as our genus moved into new habitats or adopted new ways of life (Sections 2.3, 2.4). Hominids appeared because of a series of environmental changes, to which their biology adapted. Humans diverged from their ape ancestors by expanding the capacity of their brains and it is this mental agility that makes us different. Because of it we have a more flexible range of responses to a changing environment, an adaptability more rapid than any physiological or genetic change. It is the mind, and the development of language and culture, that has made us so successful and allowed us to become manipulators of our world.

We have been describing how one species has developed from another, the sequence of changes that describe human evolutionary history. Compared to chimpanzees and gorillas, we have a relatively rich collection of fossils from our recent ancestry. We continue to search for remains but now we can also look inside our own cells at the history written in our genes. It is changes in this code, slowly selected over millennia, which have produced modern human beings. We now need to describe the mechanism by which changes at this molecular level could give rise to a new species.

BOX 1.2 An ecology of language

> A language, like a species, when extinct, never . . .
> reappears.
>
> Charles Darwin, *The Descent of Man*

A language is the embodiment of a people's culture and, as Charles Darwin observed, has the characteristics of a living species. Languages evolve (and merge, unlike most species) and can also become extinct. Languages carry ways of thinking and of seeing the world, passing ideas from one generation to the next.

Indeed, the cultural and genetic histories of peoples are closely associated. In recent years there have been attempts to trace the origins of our languages, to describe ancestral stock languages from which the roots of all languages can be traced. The most widely accepted is Proto-Indo-European, which provides a root for languages as dispersed as Sanskrit and Gaelic. This can be traced back to a single tribe, which some locate originally in Europe and Southern Russia, and others put in Anatolia. The latter is favoured by the genetic evidence carried in these populations in Turkey today. One suggestion is that the spread of their language was possible because they had a successful agricultural technology which was quickly adopted by other peoples. If this is true, human genes and a language spread on the back of a set of ideas.

Genetic evidence also supports the suggestion that there are three basic families of native languages within the Americas, associated with three waves of colonization from Asia. This scheme groups languages into families, by similarities in names and the way words are used. The largest and oldest is Amerindian which includes most of the languages of the two continents (Figure 1.7). This is followed by Na-Dene, largely confined to the North and North West (but including Navajo and Apache) whilst the extreme North has the Eskimo-Aleut family. Although this grouping is highly contentious, the distribution does match three migrations detected from the archaeological record and also three distinct peoples based on their genetic differences. The first colonizers, associated with the Amerindian family of languages, reached North America from Asia about 30 000 years ago.

When cultures clash, languages often suffer, so subsequent invasions may mean that a language disappears. As Steven Pinker says, languages become extinct for the same reason that species disappear, by loss of habitat. Its speakers

Figure 1.7 The distribution of three families of languages throughout the Americas. The oldest, associated with the first colonizers from Asia, is Amerindian, to which most languages have been allocated. Although these groupings are to some extent subjective, this division does match archaeological evidence and an increasing body of genetic evidence.

are assimilated into other cultures, or in some cases, laws are passed to suppress a language and its culture. Currently there are around 6 500 languages globally, but 96 per cent are spoken by just 4 per cent of the world's population, and 90 per cent face extinction. There may be no more than 300 extant (living) languages in a hundred years' time. Like species, their diversity is greatest in highly divided landscapes with isolated populations—750 languages are spoken in Papua New Guinea alone—but survival depends on the recruitment of new speakers. California had 50 native languages and those still used are spoken only by older citizens. The last speaker of the Northern Pomo language died in 1995 and with her that living tongue. Michael Kraus estimates that a language needs a minimum viable population of several hundred thousand people to persist.

Languages can be reborn from a written record. Cornish died out 200 years ago but today there are 150 speakers. Electronic media and the speed of global communications have contributed to this demise, but also provide the means to preserve languages. As with frozen embryos or stored DNA, we have the prospect of preserving something of their diversity on recorded media.

1.2 Evolution by natural selection

Organisms change as environments change, to better match themselves to the demands of the new conditions. They can adapt physiologically, making short-term, often reversible changes to maintain themselves. However, changes over the longer term, changes in populations and species over generations, offer no advantage to the individual and only become apparent in their offspring. Yet this mechanism, operating through chance, explains the complexity and changeability of all biological systems, at every level, from the latest influenza virus to the human intellect. This is evolution and evolution is, first and foremost, an ecological principle.

The principle of evolution by natural selection says that species change over generations because of the selection of inherited characteristics. Chance and the environment determine which individuals go on to reproduce, and their success means their traits are represented in the next generation. Over many generations, a trait that confers some selective advantage will become dominant in the population. So it was, we believe, that bipedalism became dominant amongst early humans, allowing them to be better fitted to life in open grassland.

The accumulation of a series of small, gradual changes from one generation to the next may eventually lead to a population becoming distinct from its neighbours. When these differences prevent successful matings between populations a new species has arisen (Section 2.6). We can explain this process most easily with an example.

Imagine a field of annual plants, plants that die at the end of the year, but which leave behind seeds to germinate the following spring. The seeds are viable for one year only, so each generation is derived solely from the parents of the previous year. When space or nutrients are scarce the newly germinated seedlings compete for the available resources, and those best able to survive will grow and contribute the most seed to the next generation. Others fail to reproduce because they lose these competitive battles.

Now, although they belong to the same species, we can see differences between the plants. Some are larger, some are dull in colour, and some produce more flowers. One reason for their variation lies in the soil. Perhaps the larger plants are found where nitrogen is abundant. All individuals would be larger if they had the same supply of nitrogen. This is a physiological response to the nutrient and such differences are termed **phenotypic variation**.

On the other hand, some plants produce more flowers than their neighbours, irrespective of soil quality. We shall say this is a trait they have inherited. This difference between the plants reflects information stored in their genes, in this case the code for either single or several blooms. Such inherited differences are known as **genotypic variation**.

Thus there are two possible sources of variation between individuals: genetic differences inherited from the parents and therefore coded in the genes, and phenotypic differences which are not. Phenotypic characteristics may be either reversible, such as wilting at times of water shortage, or irreversible. Irreversible differences are fixed during the development of the organism—for example, once a plant has grown tall on a nitrogen-rich soil it will not shrink again when the nitrogen supply is depleted.

Our main concern here is with genetic differences. The instructions for building a new plant are stored in the **chromosomes** of the fertilized egg, and the code written in their **DNA**. DNA—deoxyribose nucleic acid—is a long molecule, the spine of the chromosome, whose chemical structure carries the information of the genes (Box 1.3).

We shall assume that our flowers are unable to fertilize themselves. Let us say that having multiple blooms gives a better chance of producing seed in very wet years, because their pollinating insects only fly when there is no rain. A flower lasts a short time in wet weather and few will be pollinated before they deteriorate. A plant with a series of flowers is available for pollination for longer and is more likely to produce seed.

In a wet year, the single-flowering plants are unsuccessful. Most die unpollinated and few produce any seed. Those with a sequence of flowers were pollinated in the earlier, drier months and so leave a

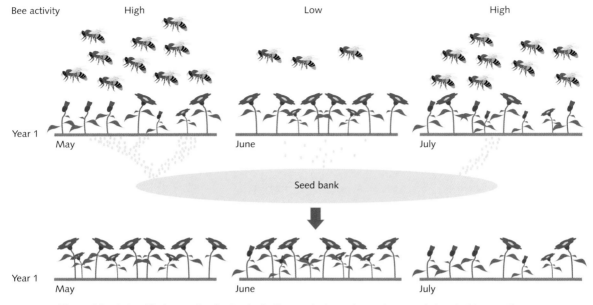

Figure 1.8 A simplified example of natural selection producing a change in a population. In this case, the selective pressure is exerted by bees, responsible for pollinating a flower and thereby ensuring the plant's seed production. In this hypothetical case, the bees only fly during fine weather. In year one this happens only in May, so those plants which flower over a long period, producing several blooms during the season, are pollinated. Those plants which only flower once, perhaps in June or July miss out when the weather is poor and the bees are not active. In this particular year the majority of the seeds in the seed bank contain the gene for multiple blooms. Next year this variety dominates, and the field is covered in flowers in May as well as June.

higher proportion of seed in the soil (Figure 1.8). Next year the field is a carpet of blue for the whole season, since most plants carry the gene for multiple blooms. So, in turn, will their seed. This gene has increased its frequency within the population and our plant population has changed.

Despite being hypothetical, this example demonstrates the inevitability of evolutionary change in a changing world. If a selective pressure favours some individuals over others, and if the selected trait is inherited, then a change in the population, over generations, must follow. Individuals best adapted to a particular habitat will produce more offspring than their competitors. Selection acts on individuals, but it is populations which change—traits that improve reproductive success become dominant within the population.

This is genotypic adaptation and its essential elements are inheritance, variation, and selection. Consider the example again. It makes several important points about the nature of genotypic adaptation:

- *The character selected must be inherited*: to become part of future generations a trait has to be written in the genetic code. Phenotypic adaptations are lost with each generation—and must develop anew each year.

- *An individual carries the traits that were of selective advantage in its parent's environment*: the new generation is dominated by multiple flowers because plants with this trait were more frequently pollinated in the wet conditions prevailing last year. This year conditions may be dry. It was the selection pressures operating upon the parents that determined the nature of the current population. An individual is adapted to its environment only as far as it matches the conditions under which its parents lived.

- *One type dominates only while environmental conditions favour it, or there is no selective pressure against it*: if there was no selective pressure against multiple blooms in the drier years their proportion in the population would not change. Only if they were at some disadvantage would their numbers decline with a series of dry years, for example if their seed were viable for shorter periods or they germinated less readily under dry conditions. Then a series of dry years might shift the balance in favour of single blooms.

- *Individuals need to differ for there to be selection*: without variation, all plants would have the same chance of producing seed. With variation, those less able produce fewer seed. Success—the relative fitness of each type—is measured as the proportion of offspring that each contributes to the next generation.

- *The gene responsible for the adaptation was not induced by the environment*: chance threw up the successful genotype; it was not produced in *response* to a wetter summer. Genetic change is blind to selective pressures operating in the environment. Most major changes in the genetic code are disadvantageous, or confer no advantage. We discuss this further below.

- *The effects of natural selection are seen as changes in the proportions in the population*: a consistent shift to multiple blooms from one year to the next might lead us to suspect this variety was being

selected. This directional change in the population, over a number of generations, might lead to complete dominance or **fixation** of the favoured gene.

We have observed only a change in the preponderance of one variety of the plant, reflecting a change in the frequency of just one gene. But now imagine that the climate turns wetter for an extended period, so that over the generations, fewer and fewer single-flowered plants survive. Eventually the gene for single blooms is lost. If the population remains isolated, not swapping genes with other populations, other genetic changes may follow. An accumulation of differences may lead to reproductive isolation (Section 2.6).

The important difference between the plants is the availability of the flowers because this trait is selected by the environment, in this case by the bees. Yet future generations will only reflect this selective pressure as far as that trait is written into the genes. Only the selection of genetic code can produce inherited change: new species are derived from genotypic variation, not phenotypic variation.

This distinction between the phenotype and genotype is important. To understand its significance we need to recall the facts of life.

1.3 The advantages of sex: the blooms and the bees

The curious might reasonably ask why the plants should flower at all. Rather than hoping for a dry year and the attention of some rather fussy bees, why do they not simply produce some form of runner or tuber that could survive the winter? Then they could do without flowers, gametes, and seeds . . . and without bees.

Given the picture we have painted so far, this might indeed be a good strategy. It would be fine if life in the field never varied. The plant need not change and an over-wintering stage may cost less to produce and also have a greater chance of producing new plants. The species could then avoid all the tribulations and risks of sexual reproduction and a successful genotype would be passed, unaltered, into the next generation.

But if the environment were to change, perhaps with the climate getting wetter, such plants, despite all their runners and tubers, would find it increasingly difficult to survive. It may be that in a wetter soil the tuber is more likely to rot and not produce a new plant next year. Relying solely on asexual reproduction, the capacity of a species to adapt genetically is very limited. Once its phenotypic flexibility has been exceeded there is little prospect of accommodating further change. Sexual reproduction may be a risky, but it does at least produce new combinations of genes and the possibility of increasing the range of tolerance.

Enter the bees. Inadvertently acting as go-betweens, they carry the male genes in the pollen grains to the egg held in the flower. The flower is simply a device

Figure 1.9   Variation. Occasional 'sports' such as this shell from the common garden snail *Cantareus aspersus* demonstrate the variation that can exist within a population. The loose coiling of this shell is almost certainly a mutational change—siblings from the same brood raised under the same conditions produced normal shells. The low prevalence of such sports in wild populations suggest this particular phenotype does not confer any significant selective advantage.

to attract the bee, a feeding station that advertises its presence. The cost to the plant is worth bearing, given all the advantages of sexual reproduction.

Not only are the costs of sexual reproduction high, so too is the risk of failure. Each individual has to meet a partner and two **gametes**, the sperm and the egg, must fuse to form a single cell, the **zygote**. For the female, who may protect and nourish the egg, this means sacrificing half of her genetic code whilst supporting the male genes. The zygote must have a viable combination of genes, yet fertilization always comes with a risk that the new genotype may not support the development of a new individual. In multicellular organisms, the single cell of the zygote has to undergo a series of divisions and transformations to develop all the tissues of the adult. In the process it must, in turn, produce specialist organs for sexual reproduction, such as flowers.

To survive to reproductive age, each offspring, and the new combination of genes it represents, needs to be well matched to the environment in which it finds itself. Producing large numbers of offspring, all slightly different from each other, improves the chance that some will survive. Inevitably there will be wastage and this is a major cost to the parents. Some combinations, some new genotypes will be less fit than their parents. Asexual reproduction avoids all these risks and costs, and is a good bet when the environment is not changing.

The merits of sex are not obvious to biologists and there is a long-running debate about its selective advantages. A variety of animal groups have **parthenogenic** species and many plants are **apomictic** (Section 2.3), where the female produces viable eggs that develop without fertilization. In several cases, the mother's genotype passes unaltered into the next generation, with no role for a male. In others the sperm is used only to stimulate the egg to begin its development but its chromosomes play no part again in the next generation.

However, such species appear to be short-lived in evolutionary time. Clearly, the swapping of genetic information has some major advantages. Sexual reproduction allows for genotypic adaptation by throwing together novel gene combinations and more potential solutions when selective pressures are continually changing. Similarly, sexual reproduction may be favoured because uncommon genotypes allow escape from parasites seeking to adapt to the internal environment of the host (Box 4.5; Section 9.4). Notice, however, that these benefits are indirect, that is, they arise from the chance combinations of genes found in the offspring of the next generation. This variation cannot benefit the individuals bearing the cost of producing the gametes, the parents. Even then, only those genes whose code is passed to reproductively successful offspring have ensured their survival. It must be the genes that promote sexual reproduction—those

which promote this shuffling of the codes—which are favoured in a variable world (Figure 1.9).

For most higher plants and animals, sexual reproduction is the only viable strategy. A changeable environment favours different traits at different times. The variation that comes from shuffling the genes in sexual reproduction creates new genotypes that will, occasionally, translate into new phenotypes better fitted to the environment. Sex also provides the means by which deleterious variations might be lost, at meiosis and at fertilization (Figure 1.12), and via those individuals that fail to reproduce. Variation itself may confer some advantage, primarily for heterozygous individuals that have alternative alleles, perhaps providing them with a greater range of adaptive responses.

1.4 Sources of variation

The infinite variety possible from sexual reproduction provides for change in a changing world. These adaptations follow from alterations in the molecular sequences carrying the genetic code. In Chapter 2 we shall see how the gradual accumulation of such changes over many generations can produce new species. Now we look at how variation between individuals and across generations is created.

Traits may be inherited from either parent, but what is actually transferred by the sperm to the egg? By what mechanism is genetic information passed on? Charles Darwin could only speculate about the nature of inheritance and never knew that Gregor Mendel had begun to describe the mechanism just seven years after he had published *The Origin of Species*. Much later, the **chromosomes** within the **nucleus** of the cell were identified as the site of its genetic information. The code is written as a series of discrete units, the genes, down the length of the chromosome (Figure 1.10).

The code is read sequentially, bearing the instructions either for reading the code (its 'punctuation'), or for building a protein (Figure 1.11). The protein is actually constructed at the **ribosome** from a series of

smaller molecules, **amino acids**. This product is then used in a cellular structure or in cellular metabolism. Genotypic change occurs when part of the code reads differently. We detect this as a phenotypic change when the sequence of amino acids produces a protein with different structural and chemical properties.

In most higher plants and animals, chromosomes come in matching pairs, one from each parent (Box 1.3). A chromosome has to be copied every time the cell divides so that each daughter cell has a copy of the full genetic code (Figure 1.12). Each chromosome is replicated, producing an identical copy, which then separate, one to each cell. Thus, in this process of **mitosis**, the chromosome number is maintained and copies of chromosomes pass unaltered into the daughter cells. This is not true of the gametes, where each cell has only a single set of chromosomes. Sperm and eggs have undergone **meiosis**, a process that firstly duplicates and then separates the paired chromosomes (Figure 1.12). Each parent supplies two of each chromosome to meiosis, but by the end of the process any egg or sperm will only contain one of each pair. When the gametes fuse at fertilization the full chromosome number is restored.

Storing the information to make a living organism

A chromosome is a long strand of **DNA—deoxyribose nucleic acid**—which is supported by protein and held in the **nucleus** of a cell. All higher organisms have several **chromosomes** within a nuclear membrane, though bacteria and their relatives have a single chromosome not confined to a discrete nucleus. Most higher plants and animals are termed **diploid** because they have two sets of chromosomes ($2n$), one of each matched pair derived from either parent. Human beings have 23 pairs ($n = 23$), a total of 46 chromosomes.

The information is actually encoded by a series of **genes**, built up from subunits of the DNA called **nucleotides** (Figure 1.10). There are four chemical forms of nucleotides, distinguished by the type of nitrogenous base in their structure. There are four bases (C, G, T, A) and it is the sequence of these bases along the chromosome which encodes the information. This is rather like a Morse code message, except it has four possible characters rather than two. The code is read in groups of three bases called triplets. As with any language, much of the meaning is derived from the position of each unit of information. Note, however, that genes are defined not by the length of their code but by their function. For this reason the number of nucleotides will vary from one gene to another.

The position of a gene on a chromosome is called its **locus**. A gene is 'read' in a process that leads to the production of a **polypeptide**, a chain of many amino acids joined together (Figure 1.11). Either on its own, or in combination with other polypeptides, this can be folded to produce the three-dimensional structure of a finished protein. The activity of

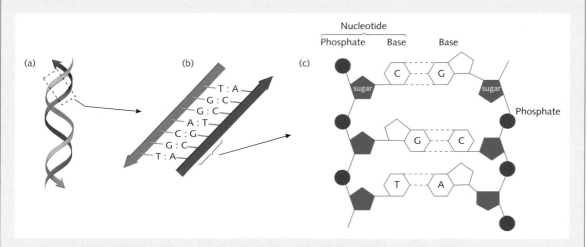

Figure 1.10   The storage of the genetic code in the chromosome. DNA consists of two backbones of sugar-phosphate groups, coiling around each other in a double helix. Each strand consists of a long chain of nucleotides, each of which comprises the sugar-phosphate group attached to a base. The information is actually stored in the sequence of bases (A, G, T, C) that bind the two strands together. The code is read in blocks of three bases. The direction, sequence, and position of these bases give the code its meaning. The bases come in two forms—purines (**adenine** and **guanine**) and pyrimidines (**thymine** and **cytosine**), distinguished by their chemical structure. Notice that only two combinations of bonds will form: A with T and G with C. This means that one strand of the DNA is a complement or 'mirror-image' of the other. To read them, the strands are separated, and the sequence read from the strand carrying the required code. The vast amount of information needed is packed into a small space by winding the double helix around protein bundles, and then packing these 'beads' or nucleosomes in a sequence of loops that eventually comprise one chromosome.

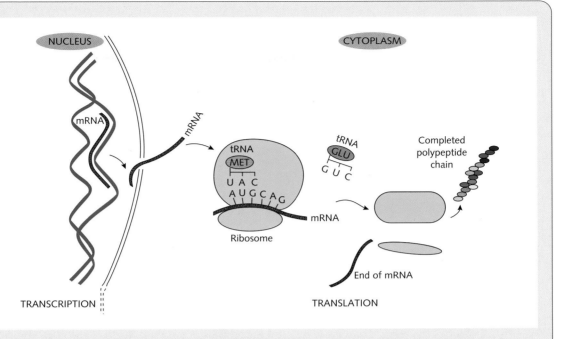

NUCLEUS

CYTOPLASM

mRNA

mRNA

mRNA

tRNA
MET

tRNA
GLU

G U C

U A C
A U G C A G

Completed
polypeptide
chain

mRNA

Ribosome

End of mRNA

TRANSCRIPTION

TRANSLATION

Figure 1.11 The language of the chromosome is translated into the language of proteins. The linear code of bases on the DNA strands has to be translated into the linear sequence of amino acids used to build a polypeptide. There are two stages in this process: **Transcription**, where the two strands of the DNA are parted so that the bases of free nucleotides can pair with their corresponding bases on the strand carrying the code. Note that uracil (U) replaces thymine (T) as the complement of adenine (A) in all RNA. The read sequence is joined together with RNA to form a length of messenger RNA (mRNA) and this then enters the cytoplasm. **Translation** itself takes place in the ribosome. Here the mRNA is read sequentially, with three base sequences matched against their complementary bases attached to a transfer RNA (tRNA). Each tRNA molecule has a specific amino acid attached according to its triplet code and this makes the translation. In this case methionine has been selected and glutamine is about to be. Particular codes also signal the start and finish of the linear sequence.

the protein is determined both by its chemical behaviour and its structure, according to the folding of the polypeptide chain.

Paired chromosomes carry instructions for the same functions and are termed **homologous**. Each gene from the father is matched by an equivalent gene from the mother. When the code is the same on both chromosomes the individual is described as **homozygous** for that character. Otherwise it is **heterozygous**. Alternative forms of the same gene are called **alleles**. Within a population there may be many different alleles for a character. The totality of genes in a population is termed its **gene pool**.

A key distinction is drawn between the genetic information carried by an individual, its **genotype**, and the trait it shows, its **phenotype**. These will differ if an individual is heterozygous for that trait and some of the genetic information is not expressed in the phenotype. For example, a child with brown eyes may actually have the code for both brown and blue eyes, but the action of one gene masks any activity by the other. The brown gene is dominant over its allele. We cannot tell from looking at the phenotype (the brown eyes) whether the child is heterozygous or homozygous. Heterozygotes in the population thus carry some hidden genetic variation with the two alleles at this locus.

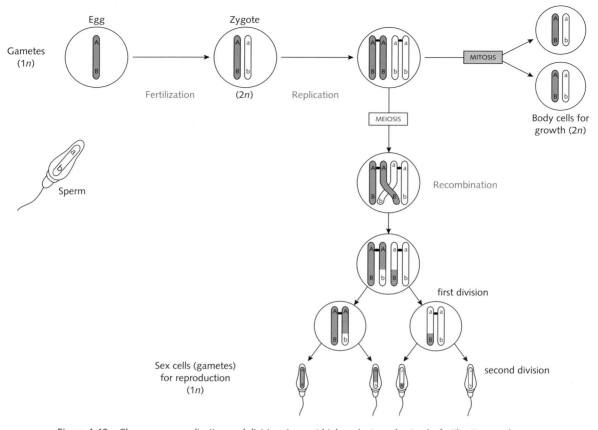

Figure 1.12 Chromosome replication and division. In most higher plants and animals, fertilization produces an individual with matching pairs of chromosomes, one from each parent (2n). In humans there are 23 pairs (n = 23), so we have a total of 46 chromosomes. To simplify matters, here there is only one chromosome pair (n = 1). Chromosomes are replicated for either growth or the production of gametes. **Mitosis** is the replication of the chromosomes to produce two identical daughter cells, each with a full complement of chromosomes (2n). The chromosome number remains the same and the chromosomes themselves remain intact, so that the genetic code in each cell is the same. Mitosis precedes cell division within the non-reproductive cells and has no significance for evolution. However, notice that the 2n cell may contain recessive genes (represented by the lower-case letters) which are not expressed in the phenotype. This is a store of variation retained in the population because it is never exposed to selection. **Meiosis** produces haploid (1n) gametes carrying just one set of chromosomes. When these fuse at fertilization the diploid condition is restored (2n), creating a pair of matching (homologous) chromosomes, one from each partner. Meiosis also involves replication, but now there are two division stages. The first division allows a chromosome and its replicate to remain together (termed sister chromatids) but they are separated from the homologous partner. A second division in each cell then separates the two chromatids, creating four haploid (1n) gametes from the original cell. When homologous chromosomes replicate they may become attached at various points, and pieces of the chromosome may cross over, becoming transposed between non-sister chromatids. This is **recombination** and is an important way in which genetic information is shuffled, producing new combinations and variation. Meiosis does not change gene frequencies in the gene pool, but it does produce new combinations of genes. Variation will also occur between individuals because gametes do not contain equivalent combinations of chromosomes. This is termed **independent assortment**. Gametes differ from each other because of both independent assortment and recombination.

You may look a lot like your parents, but you are not a perfect blend of your mother and your father. Genes carry discrete quanta of information and the proportion of traits you have inherited depends on their shuffling at meiosis and the chance unions at fertilization. There is neither a balancing nor a blending of the traits from each parent.

Chromosomes can also be altered during meiosis. A crossing over of sections of chromosomes may occur between pairs of replicated chromosomes (Figure 1.12), producing new gene combinations on affected chromosomes. In the latter stages, a chromosome and its copy are segregated, one to each gamete. The result is a set of gametes genetically different from each other, and a vast number of possible combinations when sperm fuses with egg at fertilization. Consequently, any individual inherits only some of each parent's genetic code.

Most of the variation within a generation thus comes from the segregation and the shuffling of existing code during sexual reproduction. Entirely novel code requires that the sequence of bases along the chromosome be changed—for mistakes to be made in their replication. Genes are copied and read chemically in a highly reliable process, and cells have mechanisms to try to preserve the fidelity of the copying procedure. But mistakes do occur and these may be passed on in the gametes that produce the next generation.

A mutation is either a change in the code at a single point, or deletions or insertions of lengths of code. A mistake that affects the code for a protein may change the sequence of its amino acids, or the way in which the molecule is folded. Sometimes this has little effect on its function and causes no significant change in the phenotype or its performance. Mutations that cause major change are often deleterious and some are lethal. Very rarely, a mutation improves performance or enables a protein to carry out a new function.

Chromosomal aberrations also result from the duplication and separation processes in meiosis. Large pieces of code can be swapped between non-homologous (non-matching) chromosomes (translocation), or chromosomes are duplicated or lost entirely to a gamete. Again, most chromosomal changes are likely to disrupt the development of the organism after fertilization and so do not survive into the next generation.

Genetic and biochemical analyses show that a large amount of variation is normal in most populations. Much of the variation appears to have little selective advantage, so minor code changes have no apparent impact on reproductive success. This is termed neutral variation. Some biologists believe, however, that a high level of variation may, in itself, contribute to the vigour of a species.

Variation also resides in the code 'hidden' or not expressed in the phenotype. At a heterozygote locus, where one allele dominates another, the recessive gene is not subject to selection (Box 1.3). It is only ever exposed to selection in homozygous individuals, that is when it is expressed in the phenotype. Recessive genes not lost from the population, those hidden in heterozygotes, can be an important source of additional variation at recombination.

In some situations 'ancient' code can re-enter the population. With our hypothetical flowers, we carefully arranged for all seeds from one generation to be derived from plants alive the previous year. If, instead, some seeds were several years old, their code would come from a slightly different gene pool, reflecting the environmental conditions prevailing when their parents were alive. Germination of very old seed (and some seeds can remain viable for centuries) can reintroduce variation into the gene pool.

Variation can also be lost. Over the generations, a directional change in the code may occur as the most fit alleles are selected for a particular trait. Rapid change in the population can lead to a gene becoming 'fixed' at its locus. Then there is little or no variation in this trait. This may happen in small and localized populations where chance and circumstance mean a small gene pool and breeding between a small number of individuals. Isolated from other populations, with no genetic exchange over a number of generations, the population becomes genetically distinct from others. Breeding within a small population, when there is much shared code, means these differences can become very pronounced. Such populations are said to have undergone genetic drift and, with time, may become sufficiently different to create a new species.

1.5 **Rates of evolution**

The idea of species changing over time was not particularly revolutionary amongst the scientists of Darwin's generation. They knew that the fossil record included unknown and strange animals that had died out millions of years previously. Remains of many familiar species appeared in much younger sediments. Giant mastodonts and mammoths recalled living elephants, with similarities suggesting that modern species might be derived from ancient relatives. Their experience of breeding domesticated plants and animals demonstrated the range of variation that can exist within a species and how readily, with human intervention, particular characteristics could be selected. What was needed was a description of the mechanism in nature by which inherited traits were selected and how this could lead to a new species.

The same answer suggested itself to two British naturalists within a few years of each other, both of whom had experienced the diversity of species in tropical forests. Charles Darwin and Alfred Russel Wallace independently originated the idea that the selective force was imposed by the demands of the environment. Given competition for limited resources, and the capacity of all organisms to produce large numbers of offspring, only the fittest, the best suited to that environment, would survive to reproduce. The accumulation of small-scale changes, and selection operating over many generations, would eventually produce the differences by which a new species would become distinguished.

In the middle of the nineteenth century the commonly held belief was that species were ordained by a supernatural power, following a fixed design. Darwin and Wallace showed how generations change naturally under a selective pressure. Not only were species not from a fixed template, the disturbing implication was that humans had evolved from an ancestor shared with the great apes. The human design, once deemed to be perfect, even divine, was now seen as a compromise born out of chance.

When it was first published, there was much that the theory did not explain. Species were not obviously changing from generation to generation so it was thought the process was too slow and the Earth too young to produce the great diversity of plants and animals. We now know much more about the real timescale of life on Earth. Today we can trace life back 3.5 billion years and have evidence of several bursts of speciation from the fossils pressed in the rocks of different eras (Section 9.2).

Several times slates have been wiped clean and new species have evolved to exploit the opportunities as competitors have disappeared. Darwin knew his theory discounted a single day of creation but would have been surprised how frequently and how rapidly species have arisen in the geological past. Massive upheavals in the ecology of the planet have led to both the extinction and the evolution of species. In between times, the fossil record shows long periods with few discernible evolutionary developments.

The study of **speciation** is a branch of evolutionary biology in its own right. A species forms when populations become isolated and genes cease to pass between them. This reproductive isolation can develop for reasons other than just the physical separation of the populations, a subject we take up in Chapter 2. Rapid speciation will follow rapid environmental change, but the mechanism remains the same—small, gradual changes, build on one another over many generations to produce the differences that will separate populations into distinct species.

At the cellular level, there is great uniformity in biological systems: all living organisms share similar mechanisms of metabolism and reproduction, and rely on the same genetic code. These essential features are very conservative in their detail, so that even the flowers and the bees share much of their cellular machinery and a sizeable proportion of their genetic instructions. The differences we see between species result from their interaction with the environment, from their adaptations to their way of life, and the changes to this code. Today we can map differences in the code between individuals and populations and describe how they change with time. In this way we have been able to reconstruct how early humans colonized the

Earth. We can compare geographical distances with the genetic distance between peoples and begin to reconstruct both our evolutionary and cultural histories (Box 1.2).

Chromosomes are immensely complex macro-molecules with elaborate mechanisms for repro-ducing themselves and for reading the signal they carry. Replication and transcription are chemical processes, requiring energy to drive the bonding of the molecules. These processes evolved 3.5 billion years ago and continue to be refined by an environ-ment that selects the most efficient replicators—those that produce the most copies. Life became possible when molecules able to copy themselves were large enough to carry information that aided their own replication. It is the chemical imperative by which genes replicate themselves which drives the whole process of evolution (Box 1.4).

The inevitability of change follows from variations in the sequence of molecules down the strand of DNA and the chances of this code, this combination of molecules, replicating. By such differences indi-viduals are separated, and over many more loci, so are species separated. Famously, we share 98.5 per cent of our genetic information with chimpanzees, but that odd 1 per cent is enough to create a very differ-ent species. It may be that the real difference is the way some genes are switched on and off, especially in the process of building a brain. Some feature of our past ecology selected this trait but, in doing so, pro-vided another adaptable system, giving us a head start on the rest of nature.

BOX 1.4   Units of selection

What does natural selection operate upon? The DNA of the chromosome is not selected directly by the pressures of the environment. It is individuals who reproduce, and it is the relat-ive success of different phenotypes that determines who contributes their genes to the next generation. Yet it is the information coded in these genes which is inherited, not the phenotype. Genes replicate themselves, individuals do not.

Some biologists thus emphasize the role of genetic code in the process of evolutionary change. Richard Dawkins distinguishes between the genes as 'replicators' and the vehicle in which they travel, the individual. Individuals are communal vehicles for a collection of replicators. A gene im-proves its chances of survival by endowing the vehicle, the phenotype, with a trait that better fits it to the environment. Since only this code, this information, survives into the next generation, the selection is only between replicators.

The instruction coded in a gene is, in effect, in competition with its alleles. Actually, Dawkins does not want to identify the gene as the sole unit of selection, but rather any length of DNA, any part of a chromosome, which survives intact into the next generation. Replicators can thus be whole sequences of genes. The length of the replicator depends upon the frequency at which they are fragmented when the gametes are produced after meiosis (Figure 1.12). While individuals and populations change over the generations, the replicator remains the same.

The alternative, and more traditional, view was put by Ernst Mayr, who emphasized that it is individual phenotypes which are adapted to their environment. Each individual represents the expression of thousands of genes and the phenotype is their combined effect. Genes, however de-fined, are, according to Mayr, not the determinants of their own fate. It is the phenotype, the collective expression of the genotype, which succeeds or fails to reproduce. Mayr contended that the effect of a gene and its value to the phenotype can only be understood in the context of that individual in that environment. Under some conditions a gene may be lethal; under others it is advantageous.

This is largely undisputed by the other side. However, they argue that the phenotype, as a collection of genes, is part of the environment in which a replicator finds itself. It succeeds or fails in this context. Combinations of genes that work, mutually compatible replicators are more likely to reproduce and remain together.

For many replicators it is enough not to be selected against. A gene may not be expressed in the phenotype and yet still find its way into the next generation—it has not been selected, but it is has been inherited. Thus far it has been a success simply by not lowering fitness. The traditional view argues that the unit of selection has to be that which deter-mines fitness (the phenotype or individual), but phenotypes are not inherited and genes are—even the quiet ones.

● SUMMARY

The origin of humanity from an ancestral stock shared with the great apes followed a series of changes in the environment of Africa around 5 million years ago. The advent of bipedalism in these apes, prompted by the advancing grasslands, freed the hands for carrying and for using tools. The techniques of making tools and of working within groups would benefit from a precise language, all of which fostered mental development. In time, the problem-solving capacity associated with these increased mental skills made for a more adaptable species.

Such developments had their origins in changes in the genetic code stored in the chromosomes. Natural selection determines which individuals are able to pass their genes onto the next generation. Sexual reproduction, despite its associated risks and costs, produces the genetic variation upon which natural selection can operate, and allows adaptation in a changing environment. A gradual accumulation of genetic changes can lead to a population becoming reproductively isolated from its relatives and, over the generations, evolving into a new species.

● FURTHER READING

Freeman, S. and Herron, J. C. 2004. *Evolutionary Analysis*, 3rd edn. Prentice Hall. A comprehensive and accessible introduction to all aspects of evolution thinking. It includes a detailed summary of human evolution (including many of the species not discussed here) and our fossil record prior to the discovery of *Homo floresiensis*.

Johanson, D. C., Edgar, B., and Brill, D. 1996. From Lucy to Language. Weidenfeld & Nicolson. An excellent introduction to human ancestry. The text and remarkable photographs allow the reader to judge the evidence for themselves, and to review the methodology and history of the subject. Although it doesn't cover the more recent finds, it is an immensely valuable resource.

Cavalli-Sforza, L. L. 2001. Genes, Peoples and Languages Penguin. How genes can be used to retrace the history of our species and our culture. A readable and accessible overview.

● WEB PAGES

General evolution pages:
http://www.ucmp.berkeley.edu/history/evolution.html
http://www.bbc.co.uk/history/historic_figures/darwin_charles.shtml
http://www.nhm.ac.uk

Human evolution pages:
http://www.anth.ucsb.edu/projects/human/

Includes many fossils as 3D images:
http://www.bbc.co.uk/science/cavemen/chronology/

A useful directory of web resources on cell and molecular biology can be found at:
http://www.cellbio.com/

CASE STUDY 1 — Human limitations

If brain size is a mark of being human, what size of brain makes a human being?

In the middle of the last century Louis Leakey famously suggested something over 600 cm³ was an indication that a fossil belonged to our genus, *Homo*. Although we know that brain size alone is not a good predictor of intellectual skills, increasingly large brains are a feature of our recent ancestors. One of the major questions posed by *Homo floresiensis* is whether, with a brain the size of a chimpanzee's, it had the mental capacity to manufacture the tools found with its bones.

These tools, the evidence of their kills and use of fire, suggest that Flores man was a hunter with a relatively advanced technology. Despite the size difference, their skulls resemble those of *Homo erectus*, but whilst *erectus* could have walked into South East Asia, the predecessors of *Homo floresiensis* must have crossed the deep waters isolating Flores (Figure 1.6). Additionally, some of the tools are suggestive of modern humans, rather than *Homo erectus*. Perhaps these finds are not from a new species, or even *Homo erectus* doing something unexpected, but from *Homo sapiens*. Put another way, was its tiny brain within the range of variation we find in *Homo sapiens sapiens* (Section 2.4)?

Brain size scales with body size, but smaller people only have marginally smaller heads. Smaller brains may be an adaptation to a poor or variable diet, for which there is evidence for other hominoids from the same region (Taylor and van Schaik 2007). Different races have different average heights and today there are local tribes on Flores with few people more than 1.4 m high. Across the globe, there is also a condition called microcephaly, a disease leading to much smaller brains. Perhaps the Flores remains were from a race of small but modern human beings, or from a diseased individual.

Dean Falk and a team including the original discoverers (Falk *et al.* 2005a) used scanning tomography to measure the volume and shape of the brain that once sat within the Flores skull. The interior of the skull retains an impression of the brain when it reached its largest size (normally as a young adult) and this technique uses X-rays to map the cranium to create a 'virtual endocast'. This is a three-dimensional computer model of the brain surface from which measurements of its features can be derived. These were compared with those from the skulls of a recent European microcephalic, a pygmy of central Africa, *Homo erectus*, and a chimpanzee.

Falk's team measured the Flores brain size to be around 417 cm³ which, relative to its body weight (around 25 kg), would put this individual closer to the ancient Australopithecines. Based on its overall shape, the endocast resembles those of *Homo erectus* fossils from China and Java (Figure 1.13), and is clearly different from their microcephalic endocast. However, whilst the rest of the skeleton does not fit within the range of proportions for *Homo sapiens*, neither are they a good match for *Homo erectus*. Based on a small number of bones, now from nine individuals, the original discoverers (Morwood *et al.* 2005) note their proportions are within the range of early *Australopithecus* and especially the earliest hominid (Figure 1.2), *Australopithecus afarensis*. Intriguingly, this genus has never been found outside Africa.

The alternative is that the skull is from a diseased modern human. The Falk team compared the endocast with those from a large number of modern and fossil human skulls, including several *Homo erectus* and more ancient species. Using a statistical technique called principal component analysis (PCA) they grouped the endocasts according to measurements summarizing their shape. PCA is commonly used in ecology to quantify the degree of similarity between species or ecosystems, or indeed any entity that can be defined by a series of characteristics. The analysis finds the combination of factors (principal components) which is most important for demarcating groups. Here, the analysis groups the Flores endocast with *Homo erectus* and not *Homo sapiens*, primarily because of its small height to length ratio and change in breadth down its length (Figure 1.13). The team also gives a commentary on the surface features of the endocast and suggest that these are very different from their microcephalic and pygmy endocasts (Falk *et al.* 2005a).

This has not gone unchallenged. Drawing on a larger sample of microcephalics, Weber *et al.* (2005) argue that the possibility of disease cannot be ruled out. Within their sample of 19 microcephalics, brain volumes range from 280 cm³ up to 591 cm³, easily encompassing the Flores skull. They also compare its endocast with that of a microcephalic about the same size and derive similar measurements

(continued overleaf)

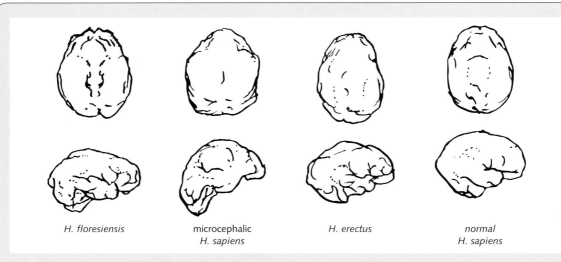

H. floresiensis   microcephalic   *H. erectus*   normal
                          H. sapiens   *H. sapiens*

Figure 1.13   Dorsal (upper line) and lateral views of the virtual endocasts created by Dean Falk and her team. These have been scaled to be the same size for ease of comparison. The overall shape of the Flores brain is suggested to be most similar to *Homo erectus*—both share a relatively low profile and similar dorsal outlines. Additionally, the endocast does not show the surface features known to be associated with the most common forms of microcephaly.

and indices, again suggesting that disease could explain its shape. Endocast dimensions vary widely amongst their microcephalics and Weber's team point to the considerable variation that can exist between modern human individuals. Certainly the Falk analysis would have been stronger had they used more than one microcephalic, but they counter that the Flores endocast shows none of the surface features expected in the most common forms of the disease (Falk *et al.* 2005b).

The two sides argue in some detail over the size and development of different areas of the brain and what this says about the mental capacity of the owner. Microcephalics with brains around half a litre have very limited mental abilities, and the Weber team suggests that the development in one area of the brain may not provide the facility for manufacturing advanced tools . . . if those found with the remains were indeed made by Flores man. Evidence from elsewhere on the island now suggests these tools are not exceptional and that similar tools were being made perhaps 700 000 years before those found with the skeletons (Brumm *et al.* 2006). Falk and her co-workers also point out that the Flores skull is of a female around 30 years old, and modern records show that nearly four out of five microcephalics have died before this age (Falk *et al.* 2005b). Subsequent studies, using other measurements from two other microcephalic

skulls and comparisons of other parts of the skeleton, suggest that the Flores remains are quite distinct, not only from these and from living pygmies, but also from near fossil hominids (Argue *et al.* 2006).

Of course it will need more work and, hopefully, more specimens from Flores to resolve these questions. A larger collection of endocasts showing consistent brain shapes would help to decide their species status. If these people were under immense selective pressure to become smaller, rather than being diseased, we should expect relatively little variation amongst future finds from the same period, and certainly nothing on the scale described by the Weber team for their microcephalics (Weber *et al.* 2005). Whilst they and others contend that *Homo sapiens* remains a possibility on the basis of the single skull, more recent finds of other parts of the skeleton point to a different genus. Perhaps eventually they will be called *Australopithecus floresiensis*.

References

Argue, D., Donlon, D., Groves, C., and Wright, R. 2006. *Homo floresiensis*: microcephalic, pygmoid, *Australopithecus* or *Homo? Journal of Human Evolution 51*, 360–374.

Brumm, A., Aziz, F., van den Berg, G. D., Morwood, M. J., Moore, M. W., Kurniawan, I., Hobbs, D. R., and Fullagar,

R. 2006. Early stone technology on Flores and its implications for *Homo floresiensis*. *Nature 441*, 624–628.

Falk, D., Hildebolt, C., Smith, K., Morwood, M. J., Sutikna, T., Brown, P., Jatmiko, E., Saptomo, E. W., Brunsden, B., and Prior, F. 2005a. The Brain of LB1, *Homo floresiensis*. *Science 308*, 242–245.

Falk, D., Hildebolt, C., Smith, K., Morwood, M. J., Sutikna, T., Jatmiko, E., Saptomo, E. W., Brunsden, B., and Prior, F. 2005b. Response to comment on 'The Brain of LB1, *Homo floresiensis*'. *Science* 310, 236.

Morwood, M. J., Brown, P., Jatmiko, E., Sutikano, T., Saptomo, E. W., Westaway, K. E., Due, R. A., Roberts, R. G., Maeda, T., Wasisto, S., and Djubiantono, T. 2005. Further evidence for small-bodied hominins from the Late Pleistocene of Flores, Indonesia. *Nature 437*, 1012–1017.

Taylor, A. B. and van Schaik, C. P. 2007. Variation in brain size and ecology in *Pongo*. *Journal of Human Evolution 52*, 59–71.

Weber, J., Czarnetzki, A., and Pusch, C. M. 2005. Comment on 'The Brain of LB1, *Homo floresiensis*'. *Science 310*, 236.

EXERCISES

1. Using the list below, decide which of the feeding strategies best describes that used by each of the following species. For each one, point to one feature of its behaviour or anatomy that indicates its adaptation to that way of life:

   lion, cheetah, hyena, leopard, African hunting dog

   Feeding strategies: stalking hunter, cooperative hunter, scavenger, chasing hunter

2. What are the most obvious changes in the jaw lines of hominid skulls shown in Figure 1.4? Suggest why natural selection might favour the finer lines of more recent species.

3. Why can natural selection not work to fit a species to a perfect design?

   Why is no species perfectly adapted to its environment?

4. Classify each of the following as either a cost (risk) or benefit of sexual reproduction. Indicate who benefits and who bears the cost:
   (a) Development from a single cell
   (b) Combining genes with another individual
   (c) Variability amongst offspring
   (d) Change in genotype between parent and offspring
   (e) Large number of potential partners
   (f) Finding a partner

5. Why is 'fitness' relative?

Tutorial/seminar questions

6. Is our adaptability a phenotypic or genotypic trait?

7. There is a possibility that some of the genetic code of *Homo floresiensis* will be extracted from the remains found in Flores. What will be the scientific and philosophical implications of finding high levels of similarity between its code and that of modern humans?

8. Do cultures evolve? Do different cultures evolve at different rates? Do they speciate? When?

Species

'That which we call a rose
By any other name would smell as sweet.'
William Shakespeare: *Romeo and Juliet*

CHAPTER OUTLINE

The binomial system of classification.

- Phylogeny of the major groups of organisms.

- The biological species concept.

- Defining species using morphological, biochemical, and genetic techniques.

- Ecological niche—intraspecific and interspecific competition.

- Types of speciation.

- The potential risks and benefits of genetically modified organisms.

← Mirror orchid (*Ophrys speculum*).

Classification comes naturally to us. We try to compare the unfamiliar with objects or events which are already part of our experience. Recognizing similarities allows us to place something into a category and assume it shares the properties of that group. Establishing categories and describing their characteristic properties is central to the way in which we learn.

This is a good strategy. It means that we can make predictions and perhaps anticipate situations that we have never encountered before. Indeed, pattern recognition is common to all animals with sufficient neural machinery to store the information. Identifying key patterns is so important to the behaviour of many higher animals that some plants have developed means of exploiting it. The male bee looking for a mate is readily fooled by the orchid displaying signals associated with the category 'female'. The flower may have no wings, no legs, and no genitalia, but it does have what a male looks for in a female bee, or at least enough for him to change his behaviour.

One difference between us and the animals is that we use our language to name the categories. Names are labels that imply the properties of a category, a term that serves to summarize a complex description. The general term 'dog' implies all that we take to be characteristic of this category, from being hairy with a leg at each corner, to the fine detail that distinguishes dog from cat. Even then we are sometimes surprised to learn that our category 'domestic dogs' contains everything from a chihuahua to a great dane.

The danger, as Romeo reminds us, is to mistake the name for the category. We need to draw a distinction between the agreed term for an object, its name, and its essential properties. Sometimes our distinctions are not a true reflection of the real world. At other times, we are not sufficiently discerning to pick out the fine detail of nature. This is certainly true of biology, where we have had to repeatedly refine our ideas about the variety and diversity of life.

All sciences begin by creating categories and classifications to help us understand the relations between their important elements. This is invariably a hierarchy, ordered into levels of increasing similarity. In biology this is termed **systematics** and it groups species according to their shared properties. In the process, biologists can pick out the evolutionary lines linking species, more formally called their **phylogenetic relationships**.

In this chapter, we introduce systematics and also the conventions used in describing, naming, and classifying organisms—**taxonomy**. The basic unit in biology is the species, defined earlier as individuals sharing a common gene pool and able to produce viable fertile offspring (Box 2). Although this is a good functional definition, the boundaries between species are often poorly defined, sometimes because we are witnessing the process of speciation in action, sometimes because we are unable to measure the distinguishing traits, or at other times because no real distinctions exist. Here we consider how new species arise and become fitted to their place in an ecosystem by their interactions with other individuals and other species, and with their abiotic environment. We finish by considering the ecological implications of our capacity to change the genotype of species and the variety of genetically modified organisms we have created.

2.1　What's in a name?

The closest we get to a universal language is in the sciences. Even if nations do not use the same names or even the same alphabet we invariably share the same symbols. Chemists speak to each other using one- or two-letter symbols as shorthand for the elements of the periodic table. Biologists have the convention of referring to species with two Latin names, the **binomial system** (Box 2.1). This was developed by Linnaeus in the eighteenth century, who streamlined earlier systems that used several names.

The binomial system establishes a unique combination of two names for each species. When Europeans

BOX 2.1 From the Keys to the Kingdom to the Tree of Life

Modern classification is a hierarchical system based on Aristotle's designation of the Kingdom as the highest category in which living organisms can be grouped. This was further developed into the binomial system by the Swedish botanist Carl Linne, who latinized his own name to Carolus Linneaus (Figure 2.1). His system consisted of seven nested categories ending with the species as the fundamental unit:

- Kingdom (originally either plant or animal)
- Phylum (suffix –phyta for plants)
- Class (suffix –phyceae for plants)
- Order (suffix –ales for plants)
- Family (suffix –aceae for plants and –idae for animals)
- Genus
- Species

Figure 2.1   Carolus Linnaeus (1707–1778).

The hierarchy still works well today, but biologists have found it necessary to extend it considerably, dividing and subdividing these main categories. Nevertheless, it retains an important property—organisms grouped together at the lowest levels have more in common than those grouped further up. Plants or animals belonging to the same genus or family share more characters than others in the same order or class. In effect, the hierarchy helps us pick out the evolutionary history of a species.

By the nineteenth century problems were starting to emerge with the Two-Kingdom system of plants and animals. The development of the microscope revealed an unexpected world of micro-organisms. These clearly did not belong in either grouping and in 1866 Ernst Haeckel proposed a third kingdom, the **Protista**, to include the protozoa, primitive algae, and fungi. These groups were lumped together with the prokaryotic bacteria in what was little more than a biological dustbin for groups that did not fit anywhere else. Eventually this was divided to produce a fourth Kingdom, the **Monera**, containing the bacteria and the cyanobacteria (also known as the blue-green algae). In 1969, Robert H. Whittaker separated the Fungi from the Protista into a fifth kingdom of their own.

Even so, the Five-Kingdom System had its problems. Disparate organisms were sharing the same grouping, all kingdoms had equal status, and the fundamental differences between prokaryotes and eukaryotes (readily distinguished by the organization of their chromosomes) were ignored. Lynn Margulis attempted to address the problem by introducing the concept of **Superkingdoms**, the **Prokaryota** and **Eukaryota**. More recently, advances in genetic analysis have designated three **Domains—the Bacteria**, **Archaea** and **Eukaryota** (Figure 2.2)—to more closely represent the evolutionary links between the major groups of organisms.

There are still problems because some life forms still do not fit readily into any group. Viruses are so varied and so different from anything else that Margulis once half-joked that they were more closely related to their hosts than to each other. To this day, viruses remain outside mainstream classification as little more than replicating molecules that hijack the replication machinery of living cells.

(continued overleaf)

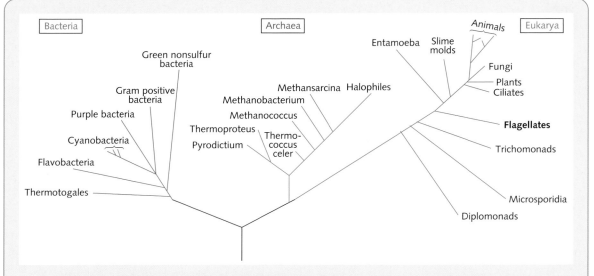

Figure 2.2 The three domains. The comparison of genetic sequences of ribosomal RNA has made it possible to cluster groups of organisms into three domains—Bacteria, Archaea, and Eukaryota—according to their degree of evolutionary relatedness.

Taxonomists have a very different problem with the lichens. Although these may look like individual organisms, they are actually a close association of algal and fungal cells. Lichens are the only true naturally occurring **chimeras**—organisms which have more than one genotype (Section 4.2).

There are relatively few algal species involved in these associations and, in some cases, several different algae occur in a single lichen. However, each 'species' of lichen has its own particular fungal species and so they are therefore classified on the basis of their fungal partners.

and Americans use the term 'blackbird' they are invariably referring to different species with different Latin names. *Turdus merula* is the European blackbird and this belongs in a different genus from North American blackbirds, such as the red-winged blackbird (*Agelaius phoeniceus*). The same is true for plants: the British bluebell is *Hyacinthoides non-scripta*, of the lily family, whilst a bluebell in California is *Phacelia tanacetifolia* (Figure 2.3), from an unrelated family, the Hydrophyllaceae.

- Notice the detail of using a Latin (or scientific) name.

- Each consists of two names (hence the **binomial** system).

- The first name is the **generic name**, the name of the genus (this always begins with a capital letter).

- The second name is the **specific name**, the name of the species (this always begins with a lower-case letter).

- A scientific name is always printed in italics or, if handwritten, underlined.

- It is the combination of the generic and specific names that is unique. So two species may belong to the same genus, and therefore share the same generic name, but will have different specific names; for example, *Mustela erminea*—the stoat—and *Mustela nivalis*—the weasel. Two species belonging to different genera may share the same specific name; for example, *Primula vulgaris*—the primrose—and *Calluna vulgaris*—heather.

- After first being cited in full, the name may subsequently be abbreviated if there is no possibility of confusion; for example, *P. vulgaris* and *C. vulgaris*.

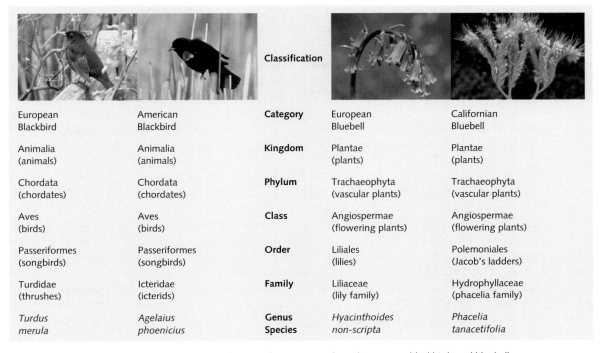

		Classification		
European Blackbird	American Blackbird	**Category**	European Bluebell	Californian Bluebell
Animalia (animals)	Animalia (animals)	**Kingdom**	Plantae (plants)	Plantae (plants)
Chordata (chordates)	Chordata (chordates)	**Phylum**	Trachaeophyta (vascular plants)	Trachaeophyta (vascular plants)
Aves (birds)	Aves (birds)	**Class**	Angiospermae (flowering plants)	Angiospermae (flowering plants)
Passeriformes (songbirds)	Passeriformes (songbirds)	**Order**	Liliales (lilies)	Polemoniales (Jacob's ladders)
Turdidae (thrushes)	Icteridae (icterids)	**Family**	Liliaceae (lily family)	Hydrophyllaceae (phacelia family)
Turdus merula	*Agelaius phoenicius*	**Genus** **Species**	*Hyacinthoides non-scripta*	*Phacelia tanacetifolia*

Figure 2.3 The full hierarchical classification of European and North American blackbirds and bluebells.

- Often the name is descriptive, even for those with just a little knowledge of Latin or Greek. For example, 'vulgaris' means 'common'.

Sometimes the citation also gives the name of the **authority** immediately afterwards; for example, *Cantareus aspersus* (Müller)—the garden snail. The authority is the first person to have described that species in its currently accepted classification and named it. Scientific papers usually cite the authority when the Latin name is used for the first time. There are strict rules about naming and giving authorities for species, with well-known authors having their names abbreviated. In *Homo sapiens* L., here the L refers to Linnaeus, the authority behind our own species.

These rigorous rules make the binomial system universal. Species names are agreed by an international committee which operates the International Codes of Biological Nomenclature. Animals are dealt with by the International Code of Zoological Nomenclature (ICZN), which recognizes organisms down to sub-species level, whilst the International Code of Botanical Nomenclature (ICBN) goes further, recognizing sub-species, varieties, sub-varieties, forms and sub-forms (Box 2.1). The classification of all groups is under constant review, not only as new species are found, but also as we learn more about their evolutionary history.

Even within a well-defined species there can be considerable variation between individuals, and this can make it difficult to decide where one species ends and another begins. We can observe this simply by looking around a room of people, noting the range of sizes and shapes. For many species, much of the variation we observe will be incidental, the phenotypic variation acquired during the life of the individual and which could not be passed on to the next generation. Our problem is to decide which characters are part of the genome, characters that can be inherited, and whether this variation could be important for the species' future.

2.2   Origins of the major groups

In its original form, the binomial classification system used mainly morphological characters to classify around 10 000 plants and animals. Grouping species into this hierarchy, according to their shared characters, made it obvious which species were closely related. Darwin and other naturalists recognized that this gives us important clues to the evolutionary history or **phylogeny** of a species, helping us to pick out the ancestry of multicellular organisms. Originally, the living world was divided between two kingdoms—the plants and animals. Following the improvements in microscopy, this became the Five-Kingdom system with the addition of the bacteria, protista, and fungi (Box 2.1). The advent of the electron microscope revealed small cell organelles and membranes, but these are universal features of cells found across the major groups, so the Five Kingdoms remained the standard classification.

In the last quarter of the twentieth century, other ways of distinguishing life developed and new techniques in molecular biology allowed us to start counting differences in the sequence of genes or the structures of key proteins and nucleic acids. Carl Woese and other molecular biologists have used such differences to reformulate the whole classification of life using the single, albeit ancient, molecule of ribosomal RNA (Figure 1.11). Part of the genetic sequence of rRNA has been used to divide all organisms into just three major domains—**bacteria**, **archaea**, and **eukaryota**.

The three domains help resolve a number of problems arising from the Five-Kingdom system. Three of the original kingdoms (Animalia, Plantae, and Fungi) are now united within the Eukaryota, joining a number of other groups including the protista (single-celled algae and fungi, amongst others). All eukaryotes have their DNA wound around proteins to form linear chromosomes that are held within a nuclear membrane (Figure 1.10). The Monerans, organisms without a distinct nucleus, are now divided between the Bacteria and Archaea, a division that recognizes the latter's shared ancestry with the eukaryotes and a physiology adapted to extreme environments (Box 6.1). However, both bacteria and **archaebacteria** share prokaryotic features, principally a simple looped chromosome with no nuclear membrane.

A key feature of eukaryotic cells is the presence of membrane-bound organelles. In her **endosymbiont theory**, Lynn Margulis argues that organelles such as chloroplasts and mitochondria were derived from prokaryotic cells that combined symbiotically with a host cell. This would explain why both chloroplasts and mitochondria have their own DNA and are similar in size and shape to photosynthetic and chemosynthetic bacteria. However, chloroplasts and mitochondria are synthesized using the DNA of the eukaryotic nucleus, so their code had to somehow find its way into the host chromosomes. Although this is a complication, there are known mechanisms by which this could happen. The **eukaryote** cell is thought to have arisen from of a series of cellular mergers between **prokaryote** cells with all the partners benefiting from this arrangement. Now, after many generations evolving together, they can only reproduce and survive in association with each other.

2.3   The species

Ideally, our classification system would accurately reflect the important differences between organisms. Then, each species would represent a unique collection of traits, the current endpoints of different evolutionary histories. However, the boundaries may be blurred where two species have only recently diverged or if the trait which distinguishes them is not obvious.

Distinctions between orders, families, or genera are easily made. For example, it is easy to distinguish a grasshopper from a butterfly. Though they are both insects, the differences in their anatomy (the enlarged hind legs of the grasshopper and the broad, dusty wings of the butterfly) readily place them in different orders. As we descend down the hierarchy, deciding where a family or genus begins or ends requires closer and closer observation (Figure 2.4). At the level of the species, the difficulty is deciding whether the differences observed between individuals are distinguishing traits. Are the variations in the coloration of the butterfly's wings part of the natural variation within a species, or a character that helps to define a distinct species? Often, variations in colour mark only a different race, a variety confined to a particular area, but whose members are still capable of interbreeding with the rest of the species.

Shared morphology remains the most common means of classification and the basis of most identification keys for higher plants and animals (Figure 2.4). In a key, a series of descriptions are offered at each level in the hierarchy. An unknown specimen is compared against each description and placed in the category with the closest match. This continues, running down a sequence of alternatives, each giving directions to the next stage. By choosing the most appropriate description at each step, the key arrives at an identification and provides a description that could only match the named species.

This works best, of course, with species that are well described, where we know the most reliable characters to observe. Without these, classifications based on appearances alone can be misleading. For example, morphological variation caused Edward Poulton considerable difficulty when he tried to classify the African mocker swallowtail butterfly (*Papilio dardanus*). The males were obvious enough, but there seemed to be no females of the species. Females of the genus were known, but there were at least three different types and these were so variable they neither resembled the males nor each other. Each group of females was therefore classified as a separate species. In fact, the females come in three forms, each of which mimics one of three species of an unpalatable butterfly (*Amauris* species) (Box 4.3). Despite being **polymorphic** in appearance, the different forms share the same internal anatomy. A similar scale of variation in the males would be a major disadvantage because the females need to recognize them to allow them to mate. Equally, females that vary too far from their matching unpalatable species risk being eaten by a discerning predator. Both the males and the females represent cases of **stabilizing selection**, where variation from one or more 'normal' forms may incur greater selective pressure.

Another complication in classifying from morphology alone is that natural selection can produce very similar structures from very different starting points. This is known as **convergent evolution**. One of the most spectacular examples is the thylacine or marsupial wolf. Probably now extinct, this was, to all outward appearances, a member of a dog family, but, as a marsupial, was more closely related to possums and kangaroos (Figure 2.5).

Such structural resemblance is termed **homoplasy** and is particularly common amongst plants. Again, similar selective pressures will tend to produce similar results. Mediterranean-type plant communities are found in five distant regions of the world, but although they comprise very different plant groups, many of their species have evolved thick, tough leaves and a short shrubby stature. These are adaptations to a prolonged summer drought and periodic fires (Section 5.1; Figure 5.2). A more general example of homoplasy are tendrils. Plants with tendrils are able to raise themselves towards the light without the cost of producing substantial stems and trunks. Instead, they exploit the supporting structures of their neighbours. Various families have evolved tendrils, but as modifications of different structures (Figure 2.6) including leaves, stems, petioles (leaf stalks), and stipules (scales).

The biological species

The complications of variation and convergent evolution have led biologists to treat species defined by their morphology with a large measure of caution. Today, most use a more functional definition of a species, termed the **biological species concept**. A biological species comprises those individuals able to interbreed to produce fertile and viable offspring, forming a gene pool and sharing a common set of adaptations. This definition recognizes that species are variable and changeable but by referring to an

© 2005 Zasavica

© Karlheinz Knoch 2005

© Bob Gibbons (www.ardea.com)

© Stephen Dalton (www.nhpa.co.uk)

David Dixon (www.ardea.com)

Figure 2.4   A simple key for some members of the primrose family (Primulaceae).
1. Leaves in basal rosette—go to 2
Leaves in rosette and along stem—water violet (*Hottonia palutris*)
2. Petals flat/turned forwards (*Primula* species)—go to 3
Petals turned backwards—cyclamen (*Cyclamen purpurascens*)
3. Flowers in cluster—go to 4
Flowers solitary—primrose (*Primula vulgaris*)
4. Flowers fragrant—cowslip (*Primula veris*)
Flowers not fragrant—oxlip (*Primula elatior*)

Figure 2.5 The marsupial thylacine is remarkably similar to placental dogs, or perhaps a blend of dog, cat, and hyena. As one of the few major carnivores in Australia during the Pleistocene, it probably filled the role of all three, and shows how evolution can converge on similar adaptive solutions in the case of placental mammals.

Figure 2.6 Clematis uses its petioles as tendrils.

Figure 2.7   An apomictic species: dandelion (*Taraxacum officinalis* agg.).

interbreeding group, it allows the possibility of change in the future. Eventually, some individuals may become excluded from the breeding population and new species may then evolve.

The biological species concept cannot be applied universally. It does not readily encompass the many species that reproduce asexually. More complicated still are groups which reproduce sexually and asexually, and which do so within defined groupings. Although it is given a single binomial, *Taraxacum officinale*, the dandelion (Figure 2.7), is more properly described as an aggregate (*Taraxacum officinalis* agg.), comprising 10 sexual types and 2 000 asexual forms. These latter types are apomictic—they produce seed without pollination—and are consequently genetically isolated. Since the flow of genes from one individual to another has stopped, different apomictic species (also known as micro-species) are often found growing very close together, but remaining distinct from each other—John Richards discovered as many as 100 micro-species of dandelion within a single hectare. Micro-species are difficult to differentiate and we avoid the complication of naming them by adding the abbreviation agg. after the specific name.

A further complication is the freedom with which some organisms exchange genes across species boundaries and thus fail to conform to the definition of the biological species. Sometimes, perhaps under special circumstances (such as cultivation or captivity), the barriers between species may be breached and interbreeding occurs. Any offspring from such a union is called a hybrid. Interspecific hybrids are rare amongst most animal groups. Very occasionally the normally sterile mule (the hybrid from a horse and donkey) may parent its own offspring, but other hybrids are consistently sterile, so that ligers and tigons (crosses between tigers and lions sometimes produced in zoos) are invariably genetic dead ends.

Interestingly, there are examples of fertile hybrid animals which have been classified as different species. Often we have learnt about their true ancestry only after looking at their genetic signature (Box 2.2). An example was the discovery that the red wolf (*Canis rufus*) was actually a fully fertile hybrid between two distinct species, the coyote and the grey wolf (Figure 2.8).

The red wolf underwent dramatic decline after 1900 due to hunting, loss of habitat and increasing hybridization with the coyote, which had been extending

Figure 2.8 The red wolf of North America. Classified as *Canis rufus*, the red wolf is actually a hybrid of the coyote (*Canis latrans*) and the grey wolf (*Canis lupus*).

its range. Robert Wayne and Susan Jenks compared the mitochondrial DNA (mtDNA) of the few red wolves collected together for a captive breeding programme with that of coyotes and grey wolves. They also collected mtDNA from six pelts of red wolves that dated to a time before hybridization was considered to have begun on a large scale. The mtDNA and its cytochrome b sequences revealed the red wolf to be a hybrid of coyotes and grey wolves; its geographical range represents the zone of hybridization between the two parent species. Indeed, Wayne and Jenks suggest that some of the animals they sampled in the wild would have been classified morphologically as grey wolves even though their genotypes were closer to that of the red wolf.

Fertile hybrids are quite common in the plant Kingdom. Many species of orchid of the genus *Dactylorhiza* will readily cross with each other, and whilst some hybrids appear to be a perfect blend of the two parent species, others resemble one or other species. This makes them particularly difficult to identify (Figure 2.9).

The results of crosses between fertile hybrids make it even more difficult to trace phylogenies. For these groups, the idea of a species as a discrete entity is misleading and we have to accept a more fluid classification. We can now at least explore their possible ancestry with molecular biology and genetics (Box 2.2).

Figure 2.9 (a) The southern marsh orchid (*Dactylorhiza praetermissa*) and (b) the common spotted orchid (*Dactylorhiza fuchsii*). These two species naturally hybridize to produce an interspecific hybrid orchid (c).

BOX
2.2 **Reading the molecular code**

The genetic code, the blueprint for a species, may seem the obvious place to look for the real differences between one species and another, but it is only recently that we have been able to read gene sequences. Before that, molecular biologists had to look at the products transcribed from the code (Figure 1.11). This meant comparing the composition of proteins, especially those responsible for key functions, to reconstruct the phylogenetic relationships of a species. The greater the genetic distance between individuals, the less likely they are to share proteins with exactly the same structure.

Even in the absence of a selective pressure, genotypic differences and chance variations will change the amino acid composition or sequence, and such differences in structure are easily quantified in proteins. Immunological methods use the capacity of mammalian immune systems to detect 'non-self' proteins. Antibodies produced against the foreign protein are highly specific, and the degree of binding between an antigen (the protein under test) and its antibody can give us a gross measure of the structural similarities of proteins from different species.

Gel electrophoresis separates proteins according to their electrical charge and molecular weight. The greater the charge across a protein molecule and the smaller its size, the faster it will migrate towards one pole of an electrical field. Two protein samples migrating the same distance through a gel are likely to have the same configuration and amino acid composition. This technique is widely used because it can pick up subtle differences in structure and is particularly good at identifying hybrids that otherwise look the same.

Sequencing their amino acids using a series of enzymic scissors and assays has paved the way for more detailed analysis of proteins. One protein, cytochrome, is of particular interest because it is found in all living organisms and is therefore very ancient. Cytochrome plays an important role in the energy-generating activity of mitochondria and chloroplasts. Analysis of its 104 amino acids has allowed biologists to group species according to these sequences. All share a sequence of 33 amino acids in the same position along the protein chain, but the remainder fall into groups, creating a hierarchy that mirrors the phylogeny of the major groups. All vertebrates, for example, fall into a group of their own, as do the invertebrates, plants, and fungi. The cytochrome of human beings and chimpanzees is identical.

The most direct way of deciding the genetic differences between individuals and species is to read the genetic code itself. Part of the double strand of DNA is unzipped when the genetic code is being read (Figure 1.11). Strands can also be separated by heat treatment and will re-associate as they cool. In DNA hybridization, DNA from two sources is heat treated, mixed together and allowed to cool. Some strands from each source will associate to form hybrid strands. The hybridized DNA is then separated again by heat treatment, but the closer their match, the higher the temperature required to dissociate the strands. Using the temperature scale over which complete dissociation takes place between hybridized and non-hybridized strands, we can calculate the percentage similarity for an entire genome.

More precise information can be gained from gene sequencing. Central to the several techniques available is the polymerase chain reaction (PCR). This allows minute samples of DNA to be copied (amplified) and then sequenced by a process known as the Sanger method. The technique uses a series of dyes to tag each of the bases that make up DNA. PCR is used to build copies of the DNA fragment with the tagged bases which are then separated out using gel electrophoresis and read off using a laser scanner linked to a computer. The result is a series of bases corresponding to the gene sequences within each gene fragment. The more fragments shared between two individuals, the greater their relatedness and the longer their common evolutionary history. These techniques can be used wherever DNA or RNA occurs. Carl Woese used them to rewrite the family tree of the major groups using the sequence for small subunit rRNA (Section 2.2). PCR also allows us to amplify fragments of DNA from the remains of long-dead organisms (including Neanderthal human remains) or identify living individuals through DNA 'fingerprinting'.

It is important to distinguish the DNA of the nucleus from that of the mitochondria or the chloroplast. Mitochondrial DNA (mtDNA) is inherited from the mother with the cytoplasm supplied in the egg cell, making it particularly useful for working out relatedness within parthenogenic species, where females produce young without having to mate. It also helps us sort out the parents of hybrids—for example, Steven Carr was able to unravel the complex mating system of white tailed deer (*Odocoileus virginianus*) and mule deer (*Odocoileus hemionus*) in Texas. He found the mtDNA of hybrids was the same as *O. virginanus*, indicating that crosses between the two species invariably had a mule deer male and a white tail female as their parents.

Classifying species and working out their origins by genes alone can be problematic. Changes in DNA may not accurately reflect evolutionary change and DNA molecules are more changeable in some groups than others. This makes it difficult to construct accurately timed phylogenies on the basis of molecular evolution alone. For example, plants within the sunflower family, the Asteraceae, show little variation compared with other flowering plants. This may mean that sunflowers are a recent evolutionary development and have not yet had time to change or simply that the DNA within their chloroplasts is slow to change.

Defining species by their chemistry

We could also take Romeo's advice and identify a rose by its scent. Flowers are recognized for their ability to produce complex chemical signals used to attract insect pollinators. Some even mimic an animal's sex pheromone, the smell used to attract a potential mate.

For more than three centuries, a group of Panamanian orchids had defied all attempts at classification. Defeated, taxonomists bundled them into a single species, *Cycnoches egertonianum*, and referred to them as the 'Egertonianum complex'. Unable to classify the plants by traditional means, Katherine Gregg looked at the sex lives of the bees that pollinate them. After they have attempted to mate with the orchid, male bees collect the delicate lemon scent and use it as a pheromone to mark out their mating sites. Several species of bees do this and field observations showed that each was extremely selective about its choice of plant. A chemical analysis of the scents revealed they were composed of up to 18 separate compounds, and Gregg found their differences reflected links with different pollinators. By reference to the species of bee and type of scent, she was able to make sense of the Egertonium complex, dividing it into four 'chemotypes', which are now recognized as four separate species.

Although the bees might have got there first, our use of chemistry to classify organisms also has a long history. The standard test for lichen identification was developed by Nylander back in 1866 and grouped lichens into six chemical races. The molecular classification of higher plants is also well developed because of our detailed knowledge of secondary plant metabolites (Section 4.4), the characteristic chemicals produced by some groups of plants, especially those compounds having a medicinal or economic value. Genera such as *Eucalyptus* hybridize freely and it is not always obvious which species were the parents, more so when one of the parents was itself a hybrid. A particular group of these compounds—the terpenoids—have proved invaluable in settling such plant paternity cases and have also been used to work out the origins of hybrids in other aromatic species such as junipers (*Juniperus* spp.) and pines (*Pinus* spp.).

Even though morphological characteristics still remain the basis of most keys to multicellular plants and animals, chemical and biochemical techniques enable us to make finer distinctions. Our goal is to arrive at a classification that reflects the real phylogenetic history of a species, though this can be challenging when the demarcation between species is not clear. The picture of human speciation we painted earlier (Section 1.1) presents the very bare branches of a phylogenetic tree, whereas the current scientific literature shows a much fuller bush. The branch connecting all *Homo*, for example, might contain as many as 13 species, including a separate species for Neanderthals—*Homo neanderthalensis*. Classification is always problematic when we have to rely on an incomplete fossil record. The reduced list we use here refers to those species for which there appear to be distinct ecological differences, but passes over fine anatomical variations. These four species of *Australopithecus* and the three species of *Homo* are those which most authors agree lived in different habitats or adopted distinct ways of life. This same approach is used in microbial ecology, where many micro-organisms are known principally by their actions—by their source of food or the waste they produce. Such a classification uses the **ecological species concept**, recognizing an organism by its role in its environment—that is its ecological niche.

2.4   **Variation within a population**

Variation between individuals is usually fairly obvious. We use these differences to distinguish one person from another and to recall their name—at least most of the time. Under other circumstances, we might wish to know what significance these differences have for an individual's survival and reproductive success.

If we plot some measure of fitness (say growth rate or abundance) against some key environmental factor (such as temperature or water availability), we invariably find a species has an optimum range (Figure 2.10). Usually this produces a characteristic curve, termed a normal distribution. Most individuals are found close to the mean value and numbers rapidly tail off with distance away from the mean.

This tells us something about the species' adaptation to that parameter. Away from the optimum,

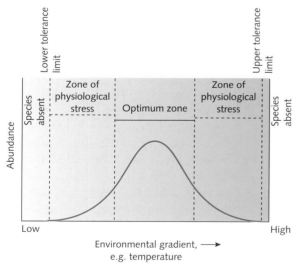

Figure 2.10 A plot of abundance of a species against some environmental parameter, in this case, temperature. This shows the range to which a species is best adapted, and the same sort of pattern would be found for any important parameter, such as water availability, nutrient level, salinity, and so on. Outside the optimum range, numbers decline rapidly because individuals have to expend energy or other resources to adapt physiologically to the poor conditions. At extreme temperatures, conditions are too harsh and the species is not found. We could create the same plot for any other measure of a species' performance, such as reproductive success or growth rate.

conditions would be less favourable and individuals will incur physiological costs to maintain themselves. For a warm-blooded animal (or endotherm), for example, this may mean using energy to keep warm or to keep cool. If the costs are high, or have to be sustained for a long time, an individual will have fewer resources to devote to growth or reproduction (Section 4.1). Those forced to live at the margins of their physiological range may thus fail to reproduce or grow.

Over time, individuals may arise which are better adapted to living outside the optimum range of the larger population. Some change in their genotype allows them to flourish where most struggle. This is the nature of variation within a species—not every individual responds in the same way to a selective pressure, and there will be some who favour conditions away from the population optimum.

A dramatic example is seen in human populations' resistance to disease. When challenged by unfamiliar diseases, those able to resist the pathogen, either through variation in their immune system or some other adaptation, have a greater chance of survival. The landing of Columbus and his successors sealed the fate of many millions in America because the native peoples had no immunity to the diseases carried by the conquistadors. It is estimated that the indigenous population of Mexico fell from 30 million to 3 million between 1519 and 1568 following their first exposure to smallpox and measles. Genetic variability within the native population meant some developed resistance and it is their descendants we see today. Isolated Amazonian tribes have more recently succumbed to the diseases of invaders, though on this occasion carried by lumbermen and miners (Box 2.3).

Much of the variation we observe between individuals does not have an adaptive significance, and individuals then differ because of the chance combination of genes they were dealt at fertilization. This is termed neutral variation (Section 1.3). Proteins have variations in structure (Box 2.2) that appear not to be adaptive and impart no selective advantage

BOX 2.3   **The shock of the new**

The native Americans arrived in the New World around 13 000 years ago when they crossed the Bering Strait between Alaska and Siberia. Their genes, and perhaps their language (Box 1.2), record the history of their walk through the continent. The early colonists had to survive the harsh and unpredictable conditions of the Arctic Circle, to pass through what has been called the 'arctic filter'. From the evidence of today's peoples the severe conditions in the north favoured particular genotypes, particularly those able to survive periods of famine.

Life must have become easier as they moved south into the temperate regions, as a thriving and expanding population quickly established itself. However, what started as a relatively rapid spread then seems to have slowed. Archaeologists and anthropologists have repeatedly found evidence that colonization was checked further south. Some new challenge must have faced those moving towards Central and South America. One suggestion is that these peoples were confronted with a range of diseases for which they had no immunity: diseases prevalent in the tropical forest and resident in the mammals of these regions. Human populations had to adapt and acquire immunity before they could move on. Again, only those genotypes able to pass through the tropical filter would enter South America.

This second genetic bottleneck is again suggested by the genetic make-up of native South American peoples. By the same token, their geographical and genetic isolation became obvious when they were exposed to the diseases of later European colonists. Many millions fell to the diseases of the conquistadors and those who followed.

Ironically, their geographical and cultural isolation saved some natives from the worst of these epidemics, at least in the first waves of European colonization. The Yanomamo Indians number 20 000 individuals spread across 150–200 villages in remote forest areas on the border between Brazil and Venezuela. They survive through a regime of hunting, gathering, and cropping, but this requires large areas of forest so their villages tend to be widely dispersed. Because there is little contact between the villages, individuals tend to marry within their own community and other pairings occur only rarely. The small size of communities and their reproductive behaviour has had important consequences for the disease resistance of the Yanomamo.

For a viral disease like measles or influenza to remain within a population there has to be a sufficient number of individuals to maintain a reservoir of infection. Measles, for example, is only virulent within an individual for 14 days, after which it needs to move on to someone else. Epidemiologists estimate that measles will remain active within a population of more than 200 000 individuals. In smaller, isolated populations the disease is easily lost, and with time, so is the immune response of the people. When the disease returns, it can devastate such a population.

This is exactly what happened when the Yanomamo people were 'discovered' by the outside world in the latter half of the twentieth century. Their first encounters with anthropologists, explorers, and missionaries in the 1960s led to a series of epidemics amongst the natives in which the 'diseases of civilization' such as measles, mumps, and influenza swept through their villages. With our frequent exposure to such diseases, most of us take these infections in our stride—our regular contacts with others helps to build and maintain our immune response.

The Yanomamo did not have this frequency of contact with other people and were further devastated when these and other diseases arrived with later colonists. The building of the Northern Circumferential Highway in the 1970s brought construction workers and, more recently, an influx of loggers, ranchers, and miners. Now tuberculosis and HIV/AIDS are the major killers of the Indians and a whole way life is under threat. Whilst genotypic variation is important for disease resistance, so is exposure to low levels of infection from childhood, without which the immune system cannot adapt. Globally, the human population is threatened in the same way by a new pathogen or a variation on the influenza virus at regular intervals. Today, the avian H5N1 virus could result in the next flu pandemic, should it cross the species barrier from birds to humans.

Isolation only offers protection as long as it lasts . . . thereafter, the shock of the new can prove fatal.

or cost. Even so, in a new or changing environment, such variability may equip, again by chance, some individuals to survive and reproduce with a higher frequency than their neighbours. Then the amount of variation in some critical trait may allow a species to adapt and evolve quickly. A population lacking such variation has fewer alleles from which to select, and less chance of adapting to change in its habitat.

2.5 Ecological niche

If a species is defined in nature by its collection of adaptations, which selective pressures have been most important in shaping it? And what selective pressures have led to it becoming separated from its close relatives?

Major morphological adaptations, such as those needed for flight, have a long evolutionary history and will separate animals high up in the hierarchy, at the Class level (Class Aves, the birds) or Order level (Order Chiroptera, the bats). Smaller adaptive changes—variations in colour, for example—may not require major physiological or structural changes. They distinguish closely related species which have diverged relatively recently. Even more recent are the distinctions between varieties of plants or sub-species of animals that are perhaps only separated geographically rather than genetically.

But what about closely related species that share the same geographical range, but apparently remain distinct? Often, it is the details of their life history, or the habitat they use, which keeps them separate. For example, two micro-moths, *Lampronia rubiella* and *Lampronia praeletella*, share a similar range, and as adults, have similar markings. Their caterpillars are leaf miners, but one attacks strawberry leaves (*L. praeletella*) and the other raspberry fruit stalks (*L. rubiella*). They have evolved to use different resources and, indeed, their larvae live in different habitats. This separation at one stage in their life cycle means they are not in direct competition with each other. The larval food plant is one of a large range of factors to which each species has to adapt, any one of which might serve to distinguish them from other species in the same genus.

For any environmental factor, whether biotic or abiotic, we can define the optimum range for a

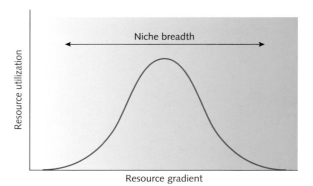

Figure 2.11   Niche, defined by one environmental gradient—in this case, the resource being used. For every environmental parameter we can define an optimum (Figure 2.10), but the full range over which a species is found is its niche breadth. For example, plants may respond to a gradient of light at different intensities; for an insectivore it might be the size of insects on which it feeds.

species and measure its performance over this range (Figure 2.11). The distribution of the species will reflect its adaptation to that factor and, more generally, to the other myriad factors to which it responds. This totality of factors to which a species has adapted is called its ecological niche.

Niche is a complex idea that tries to describe the fit between a species and its environment. An ecological niche is not a place in the normal sense of the word and is much more than a species' habitat. It represents the interaction between a species and its habitat, describing where a species lives and how it lives there. Niche is sometimes described as a species' role in its community. That role depends on its interaction with other members of the community and so can include everything from being, say, a strawberry leaf miner to a prey item for a bird, or a host to a parasite.

Although niche is difficult to define, it does have a very real ecological meaning. We can see this in convergent evolution, where species filling equivalent niches in different areas arrive at the same adaptive solution. The thylacine and the dog are one example (Figure 2.5). Another is the anteaters found on the three southern continents, all of which have elongated snouts and long probing tongues—equivalent adaptations to the same selective pressure. Like the thylacine, one of these (*Myrmecobius*) is a marsupial, another (the echidna) is an egg-laying mammal, while the pangolins of Africa or the tamandua of South America are placental mammals. These fundamental differences in reproductive physiology place their separation high up in the hierarchy (at the level of the sub-class), indicating a large phylogenetic distance between them. Yet each has evolved equivalent adaptations to collect ants and termites.

These similarities emphasize that a particular niche will apply particular selective pressures which may produce comparable adaptive solutions. But they are not perfect matches and differences between most convergent species are obvious on even the most casual inspection. A niche is defined by the organism and its fit with its living and non-living environment. Rather than simply being a role that might be played by a number of species, a full description of a niche must include the detail of the species which occupies it. So, part of the niche occupied by *L. rubiella* is defined by the raspberry stalk it uses when a caterpillar.

Clearly, there are practical difficulties in measuring every significant factor in a species' niche. Usually, studies have concentrated on those which seem to be important in separating species and which are also easily measured. **Niche breadth** is the range of a factor over which a species is found (Figure 2.11). This could be an abiotic factor like temperature, or (more often) the range of a resource the species uses. An insect that feeds on the leaves of a variety of plants is said to have a larger niche breadth than one which feeds on only a single species. A large niche breadth implies a **generalist species**, able to exploit a wide range of resources, whereas specialists are adapted to a narrow range of resources. Being a generalist is a good strategy in an unpredictable habitat, when switching to a different resource is advantageous during times

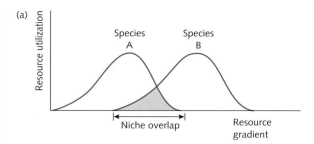

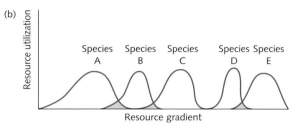

Figure 2.12   (a) Two species that exploit the same part of a resource gradient (or 'spectrum') are said to show niche overlap. In some cases, this may indicate competition between them for that resource. (b) A collection of species exploiting the same resource can only coexist if they exploit different parts of the gradient.

of shortage. Being a **specialist** is only viable when the resource supply is reliable, in constant or predictable habitats (Section 3.5). In these situations a species can become highly adapted to a resource, out-competing its neighbours by being more efficient.

Niche overlap is the part of a resource spectrum that two species occupy (Figure 2.12a). Sometimes, overlap indicates that two species are competing with each other for the resource. At other times it can mean the opposite.

Niche and competition

Different species invariably concentrate on different parts of the range of a resource—they are spread out along the resource spectrum (Figure 2.12b). A large zone of overlap may indicate where two species are competing for a resource, say the same prey or the same soil nutrients, or some other limited resource. If, however, this resource is abundant, the two species may coexist and the overlap may persist over many generations. In this case there is little competition because neither species is limited by this particular resource.

By the same token, a lack of overlap between two species only indicates there is no current competition for this particular resource. It tells us little about their past battles or their ongoing fight for other resources. We may, in fact, be observing the outcome of a competitive battle fought long ago, after each has become adapted to a different range.

Or we may be witnessing a competitive battle in progress. Where a resource checks the growth or reproduction of both species, utilization by one species depletes the supply for the other. **Interspecific competition** (competition between species—Section 4.3) means either species would perform better in the absence of its competitor. Then niche overlap does indeed denote competition.

Under these circumstances, ecologists distinguish two types of niche. The range a species could occupy in the absence of interference from other species is its **fundamental niche**. The range to which it is confined by competitors or predators is its **realized niche**. Under severe competition, a species may only use a very narrow part of a resource spectrum and have a small realized niche. Then selection will be intense, favouring those individuals able to make best use of what is available. These will be the most successful reproducers and will soon dominate the population and gene pool. In this way, a species becomes highly specialized, often showing distinct morphological or other changes that adapt it to use a resource most effectively. This is known as character displacement and becomes most obvious when two closely related species or races begin to diverge.

Competition between individuals of the same species is termed **intraspecific competition** (Section 4.3) and is one of the prime forces driving natural selection. Those most able to secure resources—food, space, or shelter—are better able to reproduce and ensure their genetic code enters the gene pool of the next generation. Ordinarily, we would expect those individuals close to their optimum range to have the greatest reproductive success. However, genetic change might adapt some individuals to a different part of this environmental gradient. Perhaps a change at one locus means a key enzyme is more effective at higher temperatures. The new genotype thrives at a different optimum and its numbers increase in the absence of competition from its neighbours.

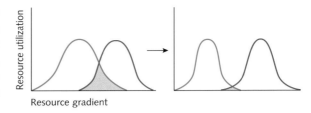

Figure 2.13   Where overlap represents a high competitive pressure on each species, one or both of them will, through natural selection, gradually change to reduce the overlap, shifting their niche. This is termed character displacement and may be represented by some change in physiology or morphology.

Mutation has produced a variety that might eventually lead to a new species, though at the moment, it can still breed with the rest of the population. Over time, the entire population may become dominated by the new genotype and its optimum range shifts (Figure 2.13).

Dolph Schluter has shown how rapidly intraspecific competition can lead to significant character displacement. Schluter and his colleague Don McPhail work on the three-spined stickleback (*Gasterosteus aculeatus* complex), a variable group of 'species' whose taxonomy has yet to be fully worked out. These fish have been isolated in a series of small coastal lakes in British Columbia since the retreat of the ice sheet 10 000–13 000 years ago. In some lakes, two different 'species' are found—a larger and rounder fish that feeds on invertebrates on the lake bottom ('benthic species') and a smaller slender 'species' that feeds on the zooplankton closer to the surface ('limnetic species') (Figure 2.14). In

Figure 2.14   Benthic (above) and limnetic (below) forms of the three-spined sticklebacks (*Gasterosteus aculeatus*).

lakes where there is only one 'species', an intermediate form is found that is able to exploit both food sources. Although limnetics and benthics do not readily interbreed with each other in the wild, they will breed in captivity.

This has enabled Schluter to run experiments to measure the pressure for character displacement when two forms share similar morphologies and the same diet. In one experiment he bred three intermediate forms: one close to the limnetic form, another close to the benthic from, and the third a true intermediate form. These were then grown in experimental ponds in the presence or absence of the wholly limnetic 'species'. By measuring the performance of each form (in this case by measuring their growth rate over a season) Schluter was able to work out the response of the different intermediate forms to the presence of a limnetic (Figure 2.15).

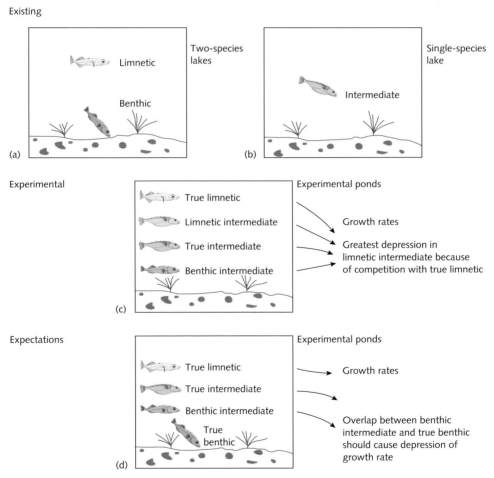

Figure 2.15 Schluter's experiments with three-spined sticklebacks (*Gasterosteus aculeatus* complex) demonstrating character displacement. Lakes in British Columbia have either two species of stickleback with different morphologies or one which is intermediate between the two. (a) One form (the benthic 'species') feeds on bottom-dwelling invertebrates and has a large mouth and rounded body; the other feeds on zooplankton in open water and has a smaller mouth and slender body (the limnetic 'species'). (b) Intermediates feed in both habitats. (c) Schluter has shown that intermediates with forms closest to the limnetic species have the greatest growth depression. (d) Similarly, we would expect benthic intermediates in closest competition with benthic species to grow less well when competing with benthics instead of intermediates.

As expected, the intermediates closest to the limnetics grew less well in the presence of limnetics. In control experiments (in the absence of a limnetic) the same intermediates showed no depression in their growth rate. Thus the greater the similarity between an intermediate form and limnetic, the greater the niche overlap and the greater the reduction in the intermediate's growth rate. Eventually this might result in the loss of intermediates since they would not compete with the efficient limnetics. Over a number of generations the intermediates would be expected to show character displacement, presumably approximating to the larger, rounder form of benthic sticklebacks, shifting to a niche not occupied by the limnetics.

It seems that the niche separations seen in some lakes today are the result of past displacements. Perhaps, with time, the intermediate forms found in single species lakes would begin to separate into two distinct types, and partition the resource spectrum. In further experiments, fish from single-population lakes (either all benthic-like or limnetic-like) were introduced to lakes where both forms occur. Given the option, females tended to select male partners that were closer to their own type—a preference known as **assortative mating**. Schluter's team suggest that lowered hybrid fitness will favour such selection, reinforcing the process of character displacement.

The alternative to moving to a different part of the resource spectrum may be oblivion. If the competition for a resource between two species is very intense, one may lose out completely. This is the competitive exclusion principle. Simply stated, it says that two species may not occupy the same niche at the same time in the same place (Section 4.3). Competitive exclusion means there are probably limits to the number of species that can be packed along a resource spectrum, depending on how different they have to be to avoid excessive competition. This has proved very difficult to measure, partly because a separation along one gradient is often tempered by the interactions between the species on other gradients.

However, not every niche has to be filled. Sometimes an alien species manages to establish itself in a new habitat and find an unoccupied niche. Sometimes an invader may squeeze between two resident species, by being better adapted to a resource space neither fully occupies. For any newcomer to persist, it needs to be sufficiently different from its new neighbours to carve a niche for itself (Box 5.3). There are numerous examples of the alternative outcome, where an introduced species out-competes the native niche-holder, and the resident is lost (Section 4.3).

Niche theory is powerful because it links the evolution of species to their use of resources and the number of species that a community can contain (Sections 5.2 and 9.1). It also links the range of resources used by a species to the constancy of its habitat and explains why we might expect more species and more specialist species in constant or predictable habitats, a subject we explore in some detail in Chapter 9.

2.6 Speciation

Whilst Dolph Schluter's sticklebacks are in the process of becoming new species, they are not yet perfectly separated. The reproductive barriers that would define them as distinct species have not yet formed, but they are underway: hybrids produced in the wild have reduced fitness and tend to die off by the next breeding season. If two populations remain isolated for long enough, genetic differences emerge which eventually prevent them from mating at all. Their gene pools are then separated and any changes that develop in one pool are not transmitted to the other. Schluter and his colleagues are observing the development of such reproductive isolation between benthics and limnetics through the behavioural isolation of assortative mating (Table 2.1).

Separation in its various forms is an important part of speciation. Consider the allele we described earlier, coding for an enzyme that had a different temperature optimum. Under character displacement, individuals with different alleles flourish in different places in which the temperature regime is most suitable for each type. This means they are physically separated—perhaps in space, say by living on different sides of a hill, or perhaps in time, if they are active at different times of day. Either way, the effect may be the same: the two populations rarely meet or interbreed with each other and genes are only exchanged between individuals active at the same time or in the same place. Then differences between separated populations are never diluted by gene flow. After a long period of isolation their accumulated genetic differences may preclude mating, even if they are reunited, a process that appears to be underway with Schluter's sticklebacks.

Divergence following a separation in space is termed **allopatric speciation**. With sympatric speciation, populations may live together, but gene flow is initially restricted because of some genetic change.

Allopatric speciation

Spatial separation can occur when a population becomes divided into isolated fragments, perhaps as some become adapted to more marginal environments. This happens most readily in highly fragmented habitats, where very different conditions are found within short distances, selecting different traits in their resident populations. For example, adjacent valleys can differ in the amount of sunlight they receive or in their geology and soil type. This is a major cause of the rapid speciation and high plant diversity in mediterranean-type ecosystems around the world (Section 5.3).

Adaptations needed to survive in one pond may differ from a second pond perhaps only a few kilometres away. Then an individual moving to the next pond or the next valley has a lower chance of survival in its new habitat, where it is less well adapted. The same sort of separation can occur temporally, where populations become specialized for flowering or feeding at different times of the day or year.

Sometimes, a small section of a population is cut off by a major disturbance, such as a flood or a forest fire, and gene flow with the parent population is reduced. In the same way, individuals invading a new

habitat (say an island) may only breed with those present. With just a small number of breeding individuals, either as colonists or survivors, any novel code is found in most of the offspring within a few generations. Only a small number of individuals may show any adaptive change, but if this proves successful their numbers quickly build as they benefit from a lack of competition. Their separation means they mate primarily with each other, rather than with the main population. As a result, their adaptive traits soon become fixed within their sub-population.

These few individuals will have only some fraction of the genetic variation of the parent population, with two important consequences for future genetic change. First, the few individuals establishing the sub-population may be unrepresentative of the larger population, and traits that might occur at a low frequency in the main population could, by chance, be common here. Within the small breeding population this code is preserved and differences from the main population become exaggerated. This is known as the **founder effect**, and allows small gene pools to diverge rapidly from the parent population.

Second, rare alleles can easily be lost in small populations. In large populations rare code may persist because, even with a low incidence, a significant number of individuals will carry the gene. The same proportion in a small population implies very few individuals and a small chance of the code passing to the next generation. The frequencies of some alleles change rapidly because rare code has a high chance of being lost, a process called **genetic drift** (Section 3.6). Again, with no exchange with the parent population, the two gene pools quickly diverge, perhaps reaching a point when they can no longer interbreed.

Reproductive isolation and genetic divergence underpin the concept of the **evolutionary significant unit** (ESU). Ollie Ryder coined this term to describe populations considered to be on a different evolutionary trajectory from their counterparts. Conservation strategies aim to identify populations requiring particular management regimes to maintain their long-term genetic integrity, and in some cases this recognizes the distinct genetic identities of some local populations. ESUs have been used extensively to manage fishery stocks, in particular the Pacific salmon (*Oncorhynchus* spp.), where sub-species have been given the legal protection more usually accorded to species. For example, one species, the Chinook (*O. tshawytscha*), has six ESUs, some confined to a single river system where they are reproductively isolated from other populations. Protecting such ecotypes, however, raises concerns that preserving variation may negate the evolutionary process itself, perhaps deflecting or halting the very trajectory that makes the population evolutionarily significant.

Reproductive barriers

Of course, populations that have become physically separated can be reunited again, when gene flow may be resumed. To produce a new species, their physical separation must last long enough for the two populations to diverge. Then any subsequent mating between them is either impossible or unproductive.

Reproductive barriers take one of two forms according to whether they operate before or after fertilization—pre- and post-zygotic barriers (Table 2.1). **Pre-zygotic** barriers are mechanical, physiological, or behavioural. For example, many insects have intricately sculptured genitalia, so that the male fits the female rather like a key inside a lock. This is one (mechanical) way in which the female ensures she is fertilized by the correct male.

An equivalent situation exists in the apple (*Malus × domestica*) but in this case it prevents self-fertilization using a biochemically based pre-zygotic barrier (Figure 2.16). Apple growers know that isolated trees seldom self-pollinate and that another variety is needed nearby to ensure a good harvest. This is due to an 'anti-selfing mechanism' known as genetic self-incompatibility (GSI)—a process controlled by a single gene at the S locus. The S gene is highly polymorphic, so each variety of apple produces a slightly different S protein. The stigma (the female part of the flower) is able to differentiate its own S proteins from those of other varieties. Fertilization proceeds only when non-self S proteins are detected.

Even when pollen is introduced to the ova or sperm are introduced to the egg, fertilization may still not happen. Often there are physiological barriers between the gametes of different species, and at

TABLE 2.1 Reproductive barriers

Pre-fertilization (pre-zygotic barriers)

Ecological isolation

Populations are separated by distance or barriers (such as mountains or water bodies)

Temporal isolation

Populations may be reproductively active at different times; they may flower at different times or have different breeding seasons

Behavioural isolation

Without the correct signals to initiate reproductive activity, males and females of different populations may never interbreed

Mechanical isolation

Reproductive organs need to complement each other for the exchange of gametes

Anatomical differences can thus prevent fertilization

Gametic isolation

Unless the sperm and the egg recognize each other, fertilization may be prevented by their failure to fuse

Post-fertilization (post-zygotic barriers)

Hybrid inviability

Embryonic development may be impaired so a hybrid never reaches the adult stage

Hybrid sterility

Offspring are produced but they are infertile, producing either dysfunctional gametes or no gametes at all

Hybrid breakdown

Although the offspring are fertile and may reproduce, their young fail to develop properly, cannot reproduce, or are poorly adapted to new habitat

the cellular level, the two cells have to recognize each other. The failure of an egg cell to recognize a sperm (or *vice versa*) prevents fertilization between species, and sometimes within species too.

Compared to the intricacies of cellular recognition, behavioural barriers to reproduction are easily observed since they frequently involve spectacular courtship rituals. In a variety of animals, females

Figure 2.16   Apple (*Malus* × *domestica*), in this case Kidd's Red Orange, is a mid-season variety that is not self-fertile. The stigma of the flowers distinguish between self and non-self pollen on the basis of a protein coded by a single gene.

select between males according to some indication of the quality of their genotype—the so-called **good gene hypothesis**. The males boast of their prowess by their appearance, or by their song or dance routine. Fanciful plumage or the capacity to build a complex nest is an indication of the fitness of the male—those with poor colours or offering a poorly built nest may not have genes that will confer an advantage to any offspring. The female selects between competing males, using her judgment of their fitness. This can be a potent selective pressure that helps explain the elaborate displays and behaviour of many males, particularly amongst birds, such as peacocks or the birds of paradise.

This is termed **sexual selection** because the success of the male is determined by the selectivity of the female (or *vice versa*). In some species, males select females in the same way, especially if the male makes a large contribution to the reproductive effort, perhaps by helping to feed or protect the young.

Sexual selection has been a key driver of evolutionary change. It is one reason why there are visible differences between the sexes (**sexual dimorphism**) in some species, including ourselves. Sexual selection may also be responsible for the diversity within a population. The jumping spider *Habronattus pugillis* lives on isolated Arizona mountain tops, which limits gene flow between populations. Females choose mates according to their capacity to drum their legs and wave their palps. Damian Elias and his colleagues found the females consistently preferred males that could drum to a different beat rather than hammer out the familiar rhythms. This may be a means of ensuring outbreeding, by selecting partners with very different genes, or simply those which are adaptable and more inventive.

These displays or signals can be important in ensuring that mating does not take place between closely related species. Part of the massive variety of fruit flies (*Drosophila*) in Hawai'i is due to sexual selection. Males of some species are required to dance for the right to mate with a female. By their dance so they are known, and getting the steps right signals to the female that the male belongs to the correct species. In this way, different species have developed on different islands and have remained separate. If the signals are not recognized by the female or the male fails to impress, he may never get to pass on his genes.

Other reproductive barriers operate after fertilization has occurred. These post-zygotic barriers operate at various stages in the development of the zygote. A mismatch between the number of chromosomes in the sperm and the egg means that the development of the offspring is not likely to proceed very far. Even amongst closely related species, matching (homologous) chromosomes may differ according to their gene sequence and such incompatibilities usually cause meiosis to fail.

Sterile hybrids unable to produce effective and viable gametes are often the result of interspecific crosses. Even where a hybrid develops to full maturity, there are a number of reasons it may not breed, collectively called hybrid breakdown. Many hybrids are poorly adapted to the habitat. One example is the intermediate forms of *Papilio dardanus*. When females are a poor match to any of the distasteful butterflies mimicked by others, they are readily taken by predators. In the same way, hybrids are likely to lose competitive battles with a parent population closely adapted to a specific niche.

Reproductive barriers preserve species differences and, for animal species at least, make them the most well-defined taxonomic unit. Whereas the rest of the hierarchy is our classificatory convenience, many species are distinct and operate as functional units. Reproductive barriers prevent closely related species from blending into each other. The faster these reproductive barriers are erected the quicker the identity of a species becomes established. The longer the barriers have been in place, the less likely it is that hybrids will form.

Sympatric speciation

With sympatric speciation, species are formed by becoming reproductively isolated through genetic change, even though they are living side by side. Gene flow is halted not by a physical barrier but by individuals becoming separated by their genetic differences.

One mechanism by which sympatric speciation can occur is through polyploidy. This happens when normal diploid ($2n$) individuals produce gametes with multiple copies of their chromosomes. A gamete that fails to undergo meiosis will remain diploid. When it unites with a standard haploid gamete, a triploid ($3n$) zygote is formed. Tetraploids ($4n$) form when two diploid gametes combine. Polyploidy rarely results in viable or fertile offspring in animals and is much more common in plants. This may be because plants tend not to have sex chromosomes (save for a few exceptions such as hops (*Humulus lupulus*), cannabis (*Cannabis sativa*), and white campion (*Silene latifolia*)—Figure 2.17, most are hermaphrodite and any polyploidy offspring grow bigger and may outcompete their diploid counterparts). Polyploidy has been an important element in the speciation of several crop plants, especially cereals (Box 2.5).

Sympatric speciation may follow relatively small changes in the environment. Character displacement is seen in Schluter's sticklebacks as individuals begin to feed in one part of the resource spectrum, reinforced by the genetic isolation created by assortative mating. Other evidence that speciation can follow genetic change alone comes from other freshwater fish—the remarkable cichlids of Africa.

The ribbon of lakes within the African Rift Valley includes both fresh and saline waters. Water levels change dramatically from one year to the next, and over thousands of years the lakes have expanded and contracted considerably. Some lakes have been cut off from each other for thousands of years, during which time their fish communities have speciated into a large number of forms. Even within a single lake, differences in habitat type as well as the range of available habitats created by changing water levels have promoted speciation. Lake Victoria, for example, has over 300 species of cichlid fish that have evolved in the last 750 000 years, largely differentiated by their food source and method of feeding. For example, there are several species that feed primarily on molluscs but are differentiated by their feeding methods. The similarities in their genetic code and their mitochondrial DNA suggest the cichlids in Lake Victoria have evolved from a single ancestral species.

Outside the Rift Valley there are some lakes which have never been connected to other water courses.

Figure 2.17 White campion (*Silene latifolia*), one of the few plant species to have sex chromosomes.

Two volcanic crater lakes in Cameroon studied by Ulrich Schliewen and his co-workers also have endemic cichlids unique to each lake (11 and 9 species respectively), despite being very small and with little habitat differentiation. Again, their molecular biology suggests each group was probably derived from a single colonization event in each lake. Since the lakes are very uniform, with no effective inflow from surrounding rivers, the speciation within each was almost certainly sympatric. Indeed, the phylogenetic tree for the cichlids of each lake suggests that their divergence was prompted by niche differentiation driven again by their feeding behaviour.

Some ecologists also recognize another form of speciation termed **parapatric speciation**, which is a feature of small, isolated, and rapidly reproducing populations. This is a particular form of sympatric speciation in which gene exchange is confined to individuals occupying a small area. The resultant inbreeding leads to highly adapted local populations, found in distinct and discrete habitats. The immobility of plants with especially restricted gene flow can produce local races or **ecotypes**. A good example of this is the metal-tolerant ecotypes of various grasses (Box 2.4).

In this case, some individuals have colonized a marginal habitat poisonous to most members of the parent population. Here they grow in the absence of competition. We often find that highly adapted ecotypes are less competitive forms that would lose any intraspecific battles with the normal types. It seems that the costs of withstanding the high levels of stress (the toxic metals) put them at a disadvantage in normal soils.

However, the benefits can far outweigh the costs when growing unhindered on contaminated soil. Tom McNeilly and his colleagues studied a colony of common bent grass (*Agrostis capillaris*) growing on an abandoned copper mine in North Wales. The metal-tolerant ecotypes were surrounded by non-tolerant individuals on uncontaminated pasture but with the wind carrying pollen on and off the contaminated site. McNeilly found metal-tolerant individuals growing 1 m upwind and 180 m downwind of the mine, so tolerant genes did move off the site (Figure 2.19). Similarly pollen from non-tolerant plants arrived on the site, but, not surprisingly, seed containing non-tolerant genes failed to grow on the toxic spoil. Because of their competitive disadvantage,

BOX 2.4   **The evolution of the metal tolerance in plants**

Large-scale extraction of metal ores has created a series of distinctive habitats in which most plant species cannot survive. Yet some plants have colonized these sites, providing evidence of the speed of adaptive change. In Britain, sites first exploited 200 years ago now have varieties of grasses tolerant of toxic metals, several of which appear to have evolved from local populations where no tolerance is evident. These grasses have proved useful in restoring spoil heaps, where little else will grow (Section 7.3).

Tolerance seems to evolve fairly readily in some grasses. Adaptations to lead, copper, zinc, and others have arisen in different species and some varieties are tolerant to combinations of metals. Sowing normal grass seed onto a toxic spoil will usually produce one or two seedlings that are able to survive. This is an indication that the genetic information for tolerance occurs as part of the background variation in the population. Interestingly, while a range of plants are known to have such alleles, albeit at a low frequency, others never show this capacity. Locally adapted varieties are known as ecotypes. Metal-tolerant ecotypes are also found among some animal groups, especially soil-dwelling invertebrates.

The genetic change needed to create an ecotype may not be that large. Some plants produce special proteins that bind toxic metals and prevent them passing from the roots to more sensitive parts of the plant. The most important adaptive change is simply an increase in the amount of this protein produced. Others actually accumulate the metal and store it in parts of the cell where it will do the least damage. This includes vacuoles within the cell or storage in the cell wall. Several plants are so effective at accumulating metals in this way that they are being developed as a means of concentrating precious metals from wastes (Section 7.3).

Strangely, the nature of the tolerance mechanism in grasses may have been revealed by studies of non-tolerant ecotypes of Yorkshire fog (*Holcus lanatus*—Figure 2.18). These will grow on arsenic-contaminated soil provided they are supplied with additional phosphate. Mark MacNair and Quinton Cumbes found *Holcus* was unable to distinguish between the poison and the nutrient, simply because the size and shape of their two ions are so similar. By swamping the soil with excess phosphate the uptake of arsenate could be reduced to a level low enough for the plant to survive. Tolerant ecotypes of *Holcus* can make the distinction

between the nutrient and the poison. Not only does this tell us how this tolerance mechanism functions but it also shows how non-tolerant plants may become poisoned.

Very often, we find that tolerant ecotypes pay a price for their capacity to live in marginal habitats. In normal conditions—say growing in an unpolluted soil alongside normal plants—the ecotypes grow less vigorously than their non-tolerant neighbours. It seems the costs they incur in being tolerant, perhaps the costs of producing metal-binding proteins, mean they are less well adapted to this habitat. This is possibly the reason their genotype is only found at a low frequency in unpolluted habitats.

Figure 2.18   Yorkshire fog grass (*Holcus lanatus*). Varieties of this species have evolved a tolerance to toxic metals.

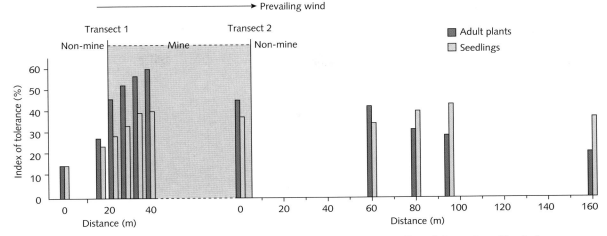

Figure 2.19 Copper tolerance in adult plants (darker coloured bars) and seedlings (lighter coloured bars) of common bent grass (*Agrostis capillaris*). Note the polarization of gene flow, where non-tolerant pollen fails to dilute the tolerant population and the plants growing on the mine spoil. Downwind of the mine, seeds containing genes for metal tolerance fail to develop into adult plants.

metal-tolerant ecotypes were, in the same way, unable to establish on uncontaminated soil.

McNeilly and Janis Antonovics found the two ecotypes had different flowering times, with metal-tolerant individuals producing their pollen much earlier than their non-tolerant counterparts. This minimized the chance of picking up non-tolerant genes and maximized the sharing of genes from other tolerant individuals. McNeilly and Antonovics were able to show that early flowering was an inherited characteristic and that a reproductive barrier was forming between the two ecotypes.

Man-made species

Over the past 10 000 years or so, humans have been the major selective pressure in the lives of thousands of plant and animal species. By means of domestication and cultivation, we have selected those varieties most suited to our purpose (Box 2.5).

The seeds of civilization

Around 12 000 years ago, humans began the move away from hunter-gathering towards a more settled life dominated by agriculture. Wheat, as one of the first plants to be cultivated, was central to this change and remains one of the staple foods of many peoples today. Indeed, the evolution and spread of wheat and other cereals closely follow those of a number of civilizations (Box 5.1, Table 5.2).

Archaeological evidence from the Near East suggests that wheat was widely grown in the Jordan Valley, Jericho, and Damascus around 10 000 years ago, though its first use may have been a thousand years before, during a cooling in the climate known as the Younger Dryas. Perhaps then the grain was simply collected from the wild, but eventually some seed was saved and sown to ensure the size of the harvest the following year. In so doing, human beings began its cultivation and in the process, its domestication, selecting and sowing seed from plants with the most useful characteristics.

The cultivation of wheat probably began in the 'Fertile Crescent'—an area stretching across the Near East through to Mesopotamia (the land between the rivers Tigris and Euphrates). The origins of wild wheat have been revealed

(continued overleaf)

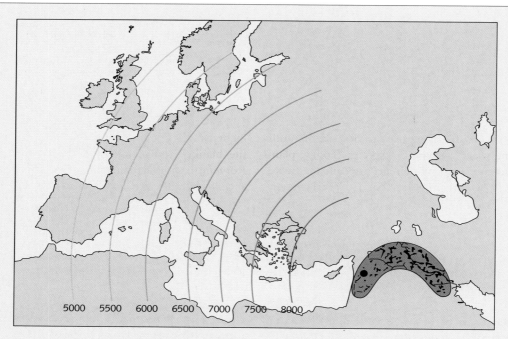

Figure 2.20   The rate of spread of modern wheat (*Triticum aestivum*) and barley (*Hordeum distichon*) from their origins in the 'Fertile Crescent' (blue area). Lines show the approximate date of arrival in years before present.

by genetic sequencing using amplified fragment length polymorphisms (AFLPs), a genome-wide measure of genetic similarity (Box 2.2). The technique identifies einkorn (*Triticum boeoticum*), a grass growing in the western foothills of the Karacadag mountains of southeastern Turkey, as the first cultivated wheat.

By 12 500 years ago, the wild form had been domesticated and was being cultivated in the Fertile Crescent as a new sub-species of cultivated einkorn (*T. monococcum*, ssp. *monococcum*). Cultivated einkorn quickly spread throughout Europe, though by the time of the Bronze Age it too had been superseded by new varieties.

Unlike the diploid einkorn, more recent wheats are polyploid and are derived from wild emmer wheat (*T. dioccoides*). Emmer enjoys several key advantages from being a tetraploid (4*n*)—it is larger, grows more vigorously, and is also free-threshing, allowing the seeds to be readily separated from their ears. Emmer was the product of a hybridization event between another wild wheat (*T. urartu*) and goat grass (*Aegilops tauscii*). By 9 000 years ago, cultivated emmer wheat (*T. dicoccum*) had supplanted einkorn

to become the single most dominant wheat in cultivation. Within 2 000 years it had spread as far as Europe, Ethiopia, and India, and along with barley (*Hordeum distichon*), became the staple cereal of the Neolithic Period (Figure 2.20).

Around this time, another new species appeared that would eventually come to dominate world agriculture. This was bread wheat (*T. aestivum*), a hexaploid with six sets of chromosomes (6*n*). Again, there seems to have been an intergeneric hybridization with goat grass (*A. tauscii*), this time with the tetraploid cultivated emmer (*T. dicoccum*) somewhere in the Southern Caspian Basin.

Although bread wheat has become the most widely grown of all wheats, with hundreds of different cultivars adapted to different soil types and climates, other wheats are still cultivated. Chief among these is durum or 'macaroni' wheat (*T. durum*). Durum, a descendent of the tetraploid emmers, has a high protein content and is also rich in the pigment beta-carotene, and these give pasta its firm texture and golden colour. This has ensured its continued use in traditional agriculture and cookery and, along with the other cereals, it has had an important role in civilizing humanity.

There are thought to be around 70 000 cultivated plant hybrids derived from a mere 1 100 wild species, though some have an ancestry that includes as many as 35 species. Not surprisingly, this makes them particularly difficult to classify. In the case of hybrids, we simply list the most recent parents using an '×' to indicate the cross.

In the past, cultivated plants were known as varieties, but this term is now used to describe naturally occurring variation within a species. Today, plants bred and cultivated by humans are called **cultivars**. According to the International Code of Nomenclature of Cultivated Plants, a plant must be listed according to its genus and species followed by its cultivar name (usually abbreviated to cv.). For example, Red Ace, a cultivar of shrubby cinquefoil, is usually written as *Potentilla fruticosa* cv. Red Ace.

Hierarchical classification tends to break down at the level of hybrids and cultivars, primarily because we have so often blended very different genotypes to create new varieties. Even so, understanding the evolutionary history of domesticated plants can be very important. Their wild-type relatives often hold genes which make them tougher and more resilient, and we use this genetic diversity to produce more resistant cultivars. For example, potatoes are plagued by aphids and the cost of insecticides to protect them is considerable. The wild hairy potato of Central America has been used to produce aphid-resistant cultivars that impale aphid attackers on short, sharp hairs.

Unfortunately, pests evolve too and our pest-resistant cultivars often have only a limited useful life. Most strains of wheat resistant to fungal attack ('rust') last little more than 5 years before the fungi itself adapts to the change in its niche. Within a decade, the build up of pests and diseases associated with a cultivar may make it uneconomic to grow. Crop-breeding programmes continue the evolutionary battle between our domesticated plants and their pests (Section 4.5).

Genetic engineering has taken some of the guesswork out of crop and stock improvement. By incorporating selected genes from other organisms into the genome of economically-important species, we can enhance their productivity or confer protection against pests and disease. Gene technology offers the opportunity to move genetic code across species boundaries and even between phyla. Today, we can practise evolution by intelligent design.

The first commercial applications of genetic engineering sought to extend the shelf life of tomatoes by manipulating the genes that control the ripening process. Since then, herbicide resistance genes have been incorporated into the genome of soybeans to allow the blanket application of herbicide (in this case *Roundup*) without damage to the crop. However, such techniques are highly controversial. First, it seems contrary to our increasing emphasis on sustainable and 'green' agricultural practices, relying as it does on biocides. Second, the technology is seen by some as tying farmers into buying both seed and pesticide from the same biotechnology company. Finally, there is the possibility of the introduced genes moving into less desirable species, perhaps creating resistant weed species.

The latter is a critical issue, given the ease with which most crop plants will hybridize with their wild relatives (Box 2.5). At least 13 major crop species are known to do this (Table 2.2). Plants of the cabbage family, the Brassicaceae represent a particular risk for transgene escape. Wild cabbage (*Brassica oleracea*) has been domesticated to produce varieties which include cauliflower, broccoli, spring greens, and sprouts. Rapeseed (*B. napus* ssp. *oleifera*), a major oilseed crop throughout the world, arose out of a chance hybridization between *B. oleracea* and wild turnip (*B. rapa*) in the western Mediterranean. Both remain highly inter-fertile with other species within the Brassicaceae.

The potential risks of transgene escape led Mike Wilkinson and his colleagues to measure the probability of hybridization between rapeseed and their wild relatives. In the United Kingdom, wild turnip occurs around crops and alongside watercourses. Wilkinson and his team used a series of techniques—satellite imagery, biological databases, survey work, and genetic analysis—to locate areas of overlap and to measure its rate of hybridization with rapeseed. By identifying gene sequences unique to these plants, they could distinguish hybrids from the diploid wild turnip and tetraploid rapeseed. Mixed parentage

TABLE 2.2	Plant families known to hybridize with wild relatives
Plant family	**GM species of potential risk**
Asteraceae (formerly Compositae)	Lettuce, sunflower
Apiaceae (formerly Umbelliferae)	Carrot
Brassicaceae	Broccoli, cabbage, rapeseed (canola)
Chenopodiaceae	Sugar beet
Ericace	Blueberry
Juglandacea	Walnut
Leguminosae	Alfalfa, soybean, peanut
Malvaceae	Cotton
Poaceae (formerly Graminae)	Creeping bentgrass, maize (corn), rice, sorghum, wheat
Rosaceae	Apple, strawberry
Salicaceae	Poplar
Solonaceae	Aubergine (egg plant), potato, tobacco, tomato
Vitaceae	Grape

TABLE 2.3	Potential sources of foreign genes or gene products of genetically modified organisms
Organism/product	**Sources**
Animals	Carcasses
	Faeces
	Urine
Plants	Biomass
	Food chain effects
	Pollen
	Root exudates
Microbes	Fermentor malfunction
	Waste media
	Waste micro-organisms

which included *B. rapa* was found in 1.46 per cent of the samples, one measure of their likely hybridization.

Transgene escape might be avoided by incorporating the engineered genes into chloroplast DNA rather than the nuclear genome. This would ensure that genes do not escape within the pollen, the male gamete, since chloroplast DNA, like mitochondrial DNA, is inherited only through the maternal line. Moving a gene from one species to another is an obvious genetic transformation, but how do we regard copying an existing gene and placing multiple copies in the original owner? No foreign code has been incorporated into the genome, but some form of genetic modification has taken place. Joachim

Messing and Jinsheng Lai have modified the genome of maize to increase its content of the amino acid methionine by altering the genes controlling its production. They suggest this might be regarded as no more than a special, accelerated form of plant breeding.

In contrast, foreign genes can pose practical and ethical dilemmas to those that consume them. Engineering high-methionine soybean by incorporating a gene from the Brazil nut resulted in beans with a protein capable of triggering nut allergies in susceptible consumers. The risks to human health were considered so great that the project was eventually abandoned. A clear identification of the genetic composition of transgenic material and its consequences is therefore needed to protect those who avoid contact with a species for reasons of health, custom, or ethical concerns.

There is also the question of specificity, and whether non-target species can be affected. A gene derived from the bacteria *Bacillus thuringiensis* produces δ-endotoxin, a protein that could protect plants from insect pests. Unfortunately, plant material from transgenic maize containing the endotoxin gene has been implicated in poisoning the larvae of the Monarch butterfly (*Danaus plexippus*). Other non-target species may be at risk because δ-endotoxin will leach out of the host plant and persist in the environment for up to 180 days. Traces have been found in earthworms, and this has raised concerns that it might pass further along the food chain.

In trying to escape the problems of conventional biocides—such as toxicity and persistence—we have created a new generation of pest control technologies with some of the faults of the old and the fears of the new (Table 2.3). Gene technology offers the potential to combat disease, hunger, and pollution but we are only just beginning to understand its ecological implications.

● SUMMARY

The binomial system for classifying and naming living organisms provides an internationally accepted convention for naming species, and its hierarchical structure is indicative of their phylogenetic relations. Most classification has traditionally been based on morphology, but we now measure genetic differences and arrange phylogenies using molecular biology. The biological species concept is useful for many higher plants and animals but is not readily applicable to those groups that hybridize freely.

A species' ecological niche is the totality of factors, biotic and abiotic, to which it has adapted. Species occupying a narrow niche are specialized for a particular part of a resource spectrum whilst generalist species have a broad niche. Two species may not occupy the same niche. Intense competition for a resource will lead to character displacement, and within a species, some individuals adapt to exploit a different part of a resource spectrum.

Speciation occurs where gene flow ceases between two populations. This can occur because the populations become physically separated (allopatric speciation) or isolated from each other by a genetic change (sympatric speciation). Barriers to gene flow can occur before or after fertilization (pre- and post-zygotic barriers). Speciation follows not only from natural selection but also occurs as a consequence of selective pressures applied by humans in their breeding of plants and animals. Genetic engineering offers new opportunities but also presents new challenges in environmental protection.

● FURTHER READING

Jeffrey, C. 1977. *Biological Nomenclature*. Edward Arnold, London. A useful introduction to the rules of naming.

Jones, S. 1999. *Almost like a Whale*. Anchor, Doubleday, London. A highly readable review of current evidence and understanding of evolution by natural selection.

Price, P. W. 1996. *Biological Evolution*. Saunders, Fort Worth. A systematic and comprehensive introduction to current thinking on evolution.

Weiner, J. 1994. *The Beak of the Finch*. Knopf, New York. An excellent account of detailed research into the Galapagos finches originally studied by Darwin and other research, including Schluter's work on sticklebacks.

Zohary, D. and Hopf, M. 2000. *Domestication of Plants in the Old World*. Oxford University Press, Oxford. A useful reference source.

● WEB PAGES

The Tree of Life site provides a wide range of information and links on phylogeny and biodiversity:
http://tolweb.org/tree/phylogeny.html

The following is a web directory to several aspects covered here, including biological nomenclature:
http://www.biologybrowser.org

More about Dolph Schluter and his sticklebacks can be found at:
http://www.zoology.ubc.ca/~schluter/

The Sanger Institute is a key participant in the Human Genome Project. Its website has general resources and specific information on gene sequencing techniques:
http://www.sanger.ac.uk/

How to produce a new palm

The Kentia palm (*Howea forsteriana*) is a familiar sight in homes and offices across the world, yet it is far away from its evolutionary home. It grows naturally on Lord Howe Island in the South Pacific, alongside its close relative, the curly palm (*H. belmoreana*) (Figure 2.21). How did the two species arise and did this result from their proximity on Howe?

How indeed? Plant populations on the road to sympatric speciation often reinforce the process by shifting their flowering times. Vincent Savolainen of the Royal Botanic Gardens, Kew investigated the evolutionary history of the two palms that were known to have reproductive phases approximately six weeks apart (Savolainen *et al.* 2006; Figure 2.22).

Savolainen and his team used amplified fragment length polymorphism (AFPL) to measure how much of their genome is shared by the two species. This was then compared against a DNA-based phylogenetic tree for all Indo-Pacific palms (Dransfield *et al.* 2005) to place the *Howea* species on this evolutionary time-line. *Howea* appears to have split into two species around 1.92 million years ago, having evolved from the Australian palm *Laccospadix*.

The pattern of genetic differences within their genome also indicated the time since their separation occurred. With allopatric speciation, differences caused by genetic drift tend to be spread evenly across the genome, across many different loci, and these will accumulate as the period of genetic separation increases (Via 2001). In contrast, the differences found between the *Howea* were restricted to a small part of

the genome, just four loci, indicative of a relatively recent separation.

H. belmoreana grows in acid and neutral soils whereas *H. forsteriana* tends to favour more alkaline conditions (Figure 2.23). A common ancestor would have had to colonize the nutrient-poor, acidic volcanic ash in the early days of the island. Over time more alkaline soils would have developed, especially in lowland areas where organic material accumulated and calcareous deposits blew in from the ocean (Figure 2.24). This created a new abiotic factor in relation to which the plants could adapt and differentiate themselves, a new niche associated with the new soil conditions. However, since individuals would be growing near enough to cross-pollinate, it would need a shift in flowering time, rather like in the case of the metal-tolerant grasses (Section 2.6), to produce the divergence. A recent sympatric speciation therefore seems the most likely mechanism.

Savolainen's findings are supported by other evidence, in particular the timing of the alkaline coastal deposits, which coincides with the genetic splitting of the species (Brooke *et al.* 2003). As with the grasses, the advantages of adapting to a different soil were best preserved by keeping their genes to themselves rather than swapping them with their neighbours, so a difference in flowering time may once again have been the mechanism of sympatric speciation in the palms. Molecular changes in their genotypes correspond to the changes in the soils of Howe and the opportunities for a new niche that the island offered.

Figure 2.21   (a) Lord Howe Island, a small volcanic island less than 12 km². Australia is the closest land mass, 580 km to the west. (b) Kentia palm (*Howea forsteriana*), an endemic of Lord Howe Island but known the world over as an ornamental plant. (c) The curly palm (*H. belmoreana*) the other endemic *Howea* of the island.

(continued overleaf)

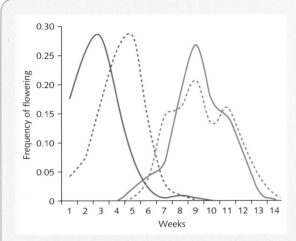

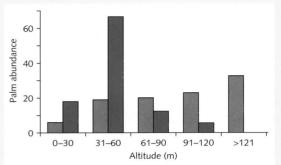

Figure 2.24   Distribution of *H. forsteriana* (black) and *H. belmoreana* (grey) according to altitude: with *H. belmoreana* tending to favour the higher ground and *H. forsteriana* lowland areas.

Figure 2.22   The flowering times of the two *Howea* species. *H. forsteriana* in blue and *H. belmoreana* in red (solid line male flowers, dotted line female flowers). Note the strong separation between the peak flowering period of the two species.

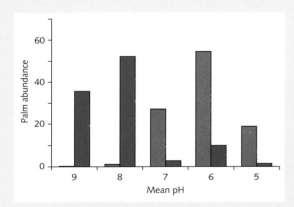

Figure 2.23   Distribution of *H. forsteriana* (blue) and *H. belmoreana* (red) according to mean pH. The two species show distinct preferences for acidic (*H. belmoreana*) or alkaline soils (*H. forsteriana*).

References

Brooke, B. R., Woodroffe, C. D., Murray-Wallace, C. V., Heijinis, H., and Jones, B. G. 2003. Quaternary calcarenite stratigraphy on Lord Howe Island, southern Pacific Ocean and the record of coastal carbonate deposition. *Quaternary Science Review 22*, 859–880.

Dransfield, J., Uhl, N. W., Asmussen, C. B., Baker, W. J., Harley, M. M., and Lewis, C. E. 2005. A new phylogenetic classification of the palm family, *Arecacea. Kew Bulletin 60*, 559–569.

Savolainen, V., Anstett, M.-C., Lexer, C., Huton, I., Clarkson, J., Norup, M. V., Powell, M. P., Springate, D., Salamim N., and Baker, W. J. 2006. Sympatric speciation in palms on an oceanic island. *Nature 44*, 210–213.

Via, S. 2001. Sympatric speciation in animals: the ugly duckling grows up. *Trends in Ecology & Evolution 16*, 381–390.

● EXERCISES

1. Using Figure 2.4 as an example, devise a key to classify the following insects according to the characteristics listed in the table below. Seek to use the minimum number of steps.

Characteristic	Insect type					
	Wasp	Beetle	Butterfly	Fly	Grasshopper	Ant
Wings present	Y	Y	Y	Y	Y	N
Number of wings	4	4	4	2	4	0
Hard wing case	N	Y	N	N	N	N
Large dusty wings	N	N	Y	N	N	N
Large jumping legs	N	N	N	N	Y	N

2. Match the following modes of speciation with their correct descriptions:
    (a) Allopatric
    (b) Parapatric
    (c) Sympatric

    (i)   Divergence within populations which share the same range but become adapted to local conditions.
    (ii) Geographical isolation within a population leading to the formation of new species.
    (iii) Speciation arising within a population as a result of a genetic change that restricts gene flow.

3. What are the advantages of using mitochondrial DNA in genetic analysis?

4. Match the following types of reproductive barriers with their appropriate description. Then group them into either pre- or post-zygotic barriers.
    (a) Behavioural isolation
    (b) Ecological isolation
    (c) Gametic isolation
    (d) Hybrid breakdown
    (e) Hybrid inviability
    (f) Hybrid sterility
    (g) Mechanical isolation
    (h) Temporal isolation

    (i)   Anatomical differences can prevent fertilization as reproductive organs need to complement each other for the exchange of gametes.
    (ii) Although the offspring are fertile and may reproduce, their young fail to develop properly, cannot reproduce or are poorly adapted to new habitat.
    (iii) Embryonic development may be impaired so a hybrid never reaches the adult stage.
    (iv) Offspring are produced but they are infertile, producing either dysfunctional gametes or no gametes at all.
    (v)   Populations are separated by distance or barriers (such as mountains or water bodies).

(vi)   Populations may be reproductively active at different times; they may flower at different breeding seasons.

(vii) Unless the sperm and the egg recognize each other fertilization may be prevented by their failure to fuse.

(viii) Without the correct signals to initiate reproductive activity, males and females of different populations may never interbreed.

5.   The following measurements of niche breath were made for four species along a resource gradient and its overlap with neighbouring species:

(Note—1.0 indicates maximum niche breadth across the entire resource spectrum. 1.0 also indicates complete niche overlap with other species.)

	Niche breadth	Niche overlap
Species A	0.7	0.8
Species B	0.2	0.1
Species C	0.2	0.8
Species D	0.7	0.1

Identify:

(a) Specialist species likely to be suffering intense competition
(b) Generalist species likely to be suffering intense competition
(c) Specialist species likely to be little competition
(d) Generalist species likely to be little competition

Which species may have a realized niche almost as large as its fundamental niche?

6.   Explain what is meant by:
(a) the morphological species concept
(b) the biological species concept
(c) the ecological species concept

In your answers indicate the limitations of each concept.

Tutorial/seminar questions

7.   Does genetic variability have a conservation value? How should we measure it and would it be comparable between species we are trying to protect?

8.   Given that we have bred domesticated plants and animals for many thousands of years, why should we not regard genetic manipulation as another form of artificial selection?

Populations

'Everything should be made as simple as possible, but not simpler.'

Albert Einstein

CHAPTER OUTLINE

- Regularity in ecological systems and the use of models to predict change.

- Simple models of population growth.

- The use of population models in applied ecology. Fisheries models.

- Metapopulations. Population genetics and life history strategies.

- The genetics of small populations and the conservation of endangered species.

- The social and economic factors in large mammal conservation.

- New techniques in single species conservation.

← Part of today's catch is discharged into the hold.

Part of the attraction of ecology is its appeal to our sense of order and balance. Many people see in ecology a set of principles that explain a constancy and regularity in living systems: the cycle of life and death, of eating and being eaten, of continuity between generations. The natural world seems to be ordered by simple rules that maintain some sort of 'balance of nature'. We capture these ideas in familiar phrases such as ashes to ashes, big fish eat little fish, and the cycle of life.

Perhaps there is a natural order out there, but few ecologists today would try to write out the rules of the game. It is difficult to show that natural populations and communities are actually stable or unchanging for long periods, regulated by one or two environmental factors. Various ecosystems may appear to be organized and controlled according to the same principles, but differences start to emerge when we look closely at each one in detail. Close up, chance and variation blur the sharp outline of these principles. To use another common phrase, the devil is in the detail. Our ideas might make intuitive sense, but their simple logic often fails to capture the complexity of the living world. Ecology the science often muddies the waters, showing that nature does not always come as clean as we might think it.

Nevertheless, most of us have some conception of an order in our environment, a commonsense view of the economy of nature. We see that nature works by cycling nutrients, that food and the energy it provides fuels living systems and governs the growth of individuals and populations. Such ideas are part of the knowledge and concepts we all need to survive and to exploit our various environments. We are all ecologists at heart.

A simple example is relevant to this chapter. A fisherman sitting by a pond knows there are a limited number of fish in the water. With no stream bringing new individuals, the fish caught can only be replaced by the reproductive efforts of those remaining in the pond. The frugal fisherman understands that the population must be allowed time to recover.

Consider the elements in this simple system. The pond has a population of a limited size and the fish reproduce at a limited rate. The population is defined solely by the boundaries of the pond, with no stream allowing fish to enter or leave. To maintain a constant population, the fisherman can remove fish only at the rate at which they are replaced by reproduction.

The common sense of this example underpins the first part of this chapter. Here we use simple models to examine population growth and the limits on population size, as well as to estimate how many individuals might be caught without damaging a population's reproductive potential. We shall find the simplicity of these models is clouded by the detail of real fisheries, when data on species' life histories are needed to manage their exploitation sustainably.

Later, the same models are used to examine why some populations and species face extinction. The loss of a population is inevitable when its reproduction fails to keep pace with its death rate over the long term, but to save a species from extinction we first need to understand the detail of its ecology, but we shall find this detail again tempers the use of general rules in conservation.

3.1 Modelling

Models are used extensively in ecology and in this chapter, so it is worth saying something briefly about their nature and their purpose.

Science uses models to do either of two things. First, they can simplify a complicated system by removing non-essential information or 'noise'. Think of a road map. This is a model in which most of the background detail of the real world has been removed to highlight the important information—the pattern and direction of roads. We model ecological systems in the same way by isolating the key factors in a study. The model helps clarify our understanding of a system by defining its important elements and their relationships. For this reason it is termed an **analytical model**.

A second type of model simulates the behaviour of a system. Again, much of the spurious information of the real world is pared away, but now sufficient detail remains to produce accurate predictions. A flight simulator is a physical model (or a computer program) for training pilots, providing enough information to give a realistic impression of flying an aircraft. A **simulation model** attempts to make realistic predictions about a system, rather than analyse its workings. Ecological simulations make predictions about populations, communities, and ecosystems but demand large amounts of data. Like flight simulators they also allow us to try out different manoeuvres, with no risk of destroying the real thing.

We could also divide models by the way in which they process data. **Deterministic models** describe the relationships between their components using fixed values or fixed equations, such as the changing weight of a fish in relation to body length. With these models, a given set of inputs produces a small number of possible outputs. In contrast, **stochastic models** allow for variation or chance factors in these calculations, accepting that not all relations are consistent and many show variability. This produces a wider range of possible outputs, mimicking the variability of the real world. With repeated runs, such models can tell us the most likely outcomes. Both types can be used to create simulations, but generally, analytical models use deterministic methods.

Models abbreviate reality, enabling us to analyse the workings of a system, or to make predictions about its behaviour. Here we begin by making simple analytical and deterministic models of population growth, using the minimum of information. Later, we add details that improve our predictions and help us to manage both exploited populations and those in danger of extinction.

3.2 Simple models of population growth

The number of fish hatched in the pond at the end of the breeding season will depend on the number of adults breeding there at the start. In describing the relationship between these two numbers we create a simple population model. This has just two elements and one function: to make predictions about population growth in future years.

Our population is defined by the boundaries of the pond (Box 3.1). To simplify matters, we shall assume there is no migration in or out and no limits on the number of fish in the pond. Another unrealistic assumption is the absence of deaths during the breeding season. This model is for a fish with a single breeding season each year, and this is at least true of several species in temperate waters.

We make two other simplifying assumptions— that all fish live for just one year, with the adults dying after spawning, and that all fish have an equal chance of reproducing. If there are 100 adults, each of which, on average, gives rise to 2 adults next year, the population in the next generation would be 200 fish. This is the **reproductive rate**:

$$\text{Reproductive rate } (R_0) = \frac{N_g}{N_0}$$

The number of adults at the start (N_0) was 100. The number of offspring reaching reproductive age in the next generation (N_g) is 200. The reproductive rate (R_0) equals 2.0.

Note that R_0 is the *average* reproductive rate per individual in the population; each adult does not produce exactly two offspring. If the reproductive rate remained fixed, we could easily predict the population size in each of the following years.

In this simplest of models, the generations do not overlap and the adults do not survive long after reproducing. More realistically, fish populations include individuals that survive for several years and take part in several breeding seasons. Keeping a count is then more difficult, requiring us to note the age of reproduction and of death for each individual in the wild. Instead, we can measure the reproductive rate by taking a census at certain times. Counting the number alive before and after a period of time allows

BOX 3.1 Individuals, populations, and metapopulations

Ecologists counting individuals and treating them as a population need to define their terms very carefully. Criteria that most people would use to define the individual become problematic when we look closely at different species.

Consider organisms that increase their numbers by budding or some other form of asexual reproduction. Many plants and several groups of colonial animals grow by forming modules that may separate and grow independently. For these organisms we have to distinguish between genetically identical, asexual clones termed **ramets**, and genetically distinct individuals (having arisen from different zygotes) or **genets**. In some species (e.g. potatoes) a count of individuals will include both genets (seedlings) and ramets (tubers).

We then have to decide whether we include all stages in the life cycle in the count: for example, do we include tadpoles or only adult frogs; seedlings as well as overwintering tubers? Our decision will depend upon the organism and the purpose of the study.

Even then, defining a **population** is not always straightforward. A good functional definition is a group of individuals of the same species occupying a particular area at a particular time. Notice we define notional boundaries that delimit individuals who can actually breed with each other—

for example, the elephants of the Serengeti in 2007. Elephants in the Serengeti would not normally mate with those many miles to the south. Nor, very obviously, could they breed with the elephants in the Serengeti in 1907. Even so, these spatial and temporal boundaries are only truly distinct in the mind of the ecologist, as a necessary convenience to define the terms of a study.

Most species have a collection of populations alive at the same time, separated spatially. Within a well-defined region, this population of populations is called a **metapopulation** (Box 3.4) where relatively distinct populations occasionally breed with each other and therefore swap genetic material. While some populations may interbreed fairly regularly, others may only have indirect and infrequent genetic exchange with each other. The degree of interchange within the metapopulation is often crucial to the survival of some endangered species.

You will also come across the term **population density**. This is used where we have no need or have made no attempt to count every individual in a population (for most plants and animals this is impossible). Instead, we count their numbers in a specified area or in some circumstances, a known volume.

us to calculate the **net reproductive rate** over that interval:

Net reproductive rate $(R_N) = N_t/N_0$

The number of individuals alive (N_t) at the end of the time interval (t) is divided by those counted at the start (N_0). If the number alive at the end (survivors and offspring) is greater than at the start, the population has increased $(R_N > 1)$. The population declines when the number of deaths is larger than any additions $(R_N < 1)$.

R_N is called the net reproductive rate because it includes both deaths and births over the sampling interval. Again, it is the rate of change per individual averaged over the entire population and if it remained fixed, we could readily predict future population sizes.

Can you see when R_0 and R_N will equal one another? They are the same when the census interval is equal

to the generation time. Both then measure the average rate of change per individual over one generation.

We can follow the change in the population size on a graph. If we set R_N to a fixed value, and then plot the size of the population with time, we get a characteristic curve (Figure 3.1). In this example, we start with 100 individuals and set R_N to 1.2. After one breeding season there will be 120 fish in the pond. Note that some may have died in the meantime, but the net increment is 20 individuals. Next time around, there are 24 additional fish, and in the third generation 29. You may like to derive a second set of data to check the shape of curve produced with a different value of R_N.

This example illustrates an important point about the pattern of population growth. Despite R_N staying constant, the increase in population size is not the same from one interval to the next. Instead, the size

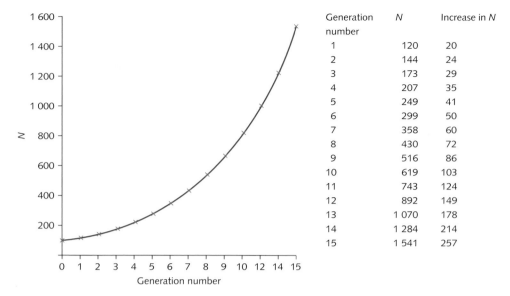

Generation number	N	Increase in N
1	120	20
2	144	24
3	173	29
4	207	35
5	249	41
6	299	50
7	358	60
8	430	72
9	516	86
10	619	103
11	743	124
12	892	149
13	1 070	178
14	1 284	214
15	1 541	257

Figure 3.1 A model of population growth in an organism with discrete generations, starting with a population of 100 and a constant reproductive rate of 1.2. The results used to plot the graph have been rounded up or down to give whole numbers.

of the increment itself increases because population growth is accelerating—there are more individuals breeding at each successive interval, and collectively they produce more offspring.

This gives the curve its characteristic shape, termed exponential growth for populations where births and deaths occur at any time. This shape will be achieved by any population with an R_N greater than 1. Even with R_N values only slightly above 1 the curve is the same, though its rise is shallow. Given long enough it will still grow to a very large population. Similarly, an accelerating decrease follows when R_N falls below 1, and the population tips towards extinction.

We have represented this as a simple deterministic model, calculating the population growth as a series of discrete steps, using a fixed value of R_N. This is the easiest way to calculate the change from one time interval to another.

Populations with continuous change

The picture becomes more complicated when we model populations where births or deaths occur continuously. Many species of fish grow and reproduce for several years. But death can occur at any time and reproduction may also be a more or less continual process, not confined to a particular season. Then, both the birth rate (or natality rate) and the death rate (or mortality rate) vary continuously. The difference between them is expressed as r, the **rate of change per individual** for a particular point in time:

$$r = b - m$$

When the natality rate per individual (b) is greater than the mortality rate per individual (m) the population is increasing. The model of population growth becomes:

$$\frac{dN}{dt} = rN$$

The change in the population size (dN) after the time interval (dt) equals the average rate of change per individual (r) multiplied by the number of individuals (N). It differs from our previous example only in deriving the change over a very small interval of time (dt). These are described as instantaneous rates and r is more properly described as the instantaneous rate of change per individual (or the intrinsic rate of increase).

If we fix r at a particular value, the population grows exactly as described before. Again this model produces exponential growth as long as the instantaneous birth rate is greater than the instantaneous mortality rate. When they balance ($b - m = 0$) the population does not grow ($r = 0$). In real populations r will change as conditions change, as b and m fluctuate.

Factors affecting the rate of population growth

The maximum value of r (r_{max}) describes how rapidly a population would grow under ideal conditions, when b is maximized and m is minimized. This differs between species according to the details of their life history. Species that take a long time to reach reproductive age, or that produce few offspring on each occasion have slower rates of population growth.

Elephants and other large-bodied animals typically have low r_{max} values because they produce few offspring over a long time. The gestation period in the African elephant is relatively long (22 months), followed by a juvenile stage lasting 10–12 years, before the individual becomes sexually mature. They may live for an average of 30–40 years (though females can still be fertile at 60 years old) and produce just one offspring at a time. On average the African elephant has an r_{max} value of 0.06 per year. At the other extreme, mice live short lives, quickly becoming sexually mature, and have multiple births following a brief (21-day) gestation period. Many small mammals, including mice, have a high r_{max}, ranging from 0.3 to 8.0 per year, and their populations can increase very rapidly.

Why is r_{max} rarely achieved in the wild? Usually because conditions for growth and reproduction are far from perfect. Natality is reduced with poor nutrition or disease, or when partners simply fail to meet. Mortality rates depend on a wide variety of factors, and any rise in the death rate depletes the numbers reproducing. Nevertheless, the size of r_{max} is of great interest because very different r_{max} values imply very different life history patterns and very different biologies. Each species will invest its resources in growth or reproduction according to a schedule that is part of its adaptive strategy, and selected by its habitat. As we shall see later on, such strategies need not be fixed—some species change their reproductive strategy and biology to suit changeable habitats (Box 3.3).

Populations in limited environments

In a famous calculation, Charles Darwin estimated that a single pair of elephants would have 19 million descendents in just 750 years, if all reproduced at their maximum rate. Yet we are not overrun by elephants and common sense tells us that some factor in their habitat must limit their population growth.

Rapid population growth can be sustained only for as long as conditions allow. Resources may become scarce, space may become limited, or waste may accumulate in the habitat. These checks on growth, collectively known as 'environmental resistance', place an upper limit on the population size, termed the 'carrying capacity' (K)—simply the maximum number the prevailing environmental conditions can support.

What will happen when a population is close to its carrying capacity? Any shortage of resources will mean competition between individuals, with some getting less than they need. A lack of food, for example, might prevent or delay some becoming sexually mature, whilst others might die. Either impact reduces the size of the next generation. As a population approaches its carrying capacity, more individuals compete for fewer resources and the intensity of competition grows.

In many species competition between individuals of the same species (intraspecific competition) is the major factor limiting population size (Box 4.2). In these cases population growth is described as density-dependent because the intensity of environmental resistance changes with the population density. Population size increases rapidly as long as resources are abundant, but rising numbers increase competition and population growth will then slow.

We can include these checks in our model, by setting an upper limit on population size. For example, if our pond can accommodate 500 fish ($K = 500$) and 100 are already present (N), there is 'space' for just 400 more. Put another way, of the total capacity, one-fifth is occupied and four-fifths are available:

$$\frac{K - N}{K} = \frac{500 - 100}{500} = \frac{4}{5} = 0.8$$

So environmental resistance increases as the unused capacity gets smaller—when there are 300 individuals present, the unused capacity is reduced to 0.4. We can build this into our model very easily:

$$\frac{dN}{dt} = rN\left(\frac{K - N}{K}\right)$$

Figure 3.2a, derived from this model, shows how increasing environmental resistance slows growth by its impact on r. The model shows the population rising smoothly and resting perfectly at its carrying capacity. This s-shaped pattern (Figure 3.2a) is termed logistic growth, and the population comes to rest at K by the effects of intraspecific competition in a limited environment. A shortage of food, for example, may reduce egg production (b falls) and/or increase mortality (m rises).

We can see the effect more easily by plotting net incremental growth (dN/dt) against population size (Figure 3.2b). At first the number of added individuals rises as N increases. The greatest increase (23) is at half of the carrying capacity when N and the unused capacity for population growth are both still large. Thereafter rates of increase decline because competition

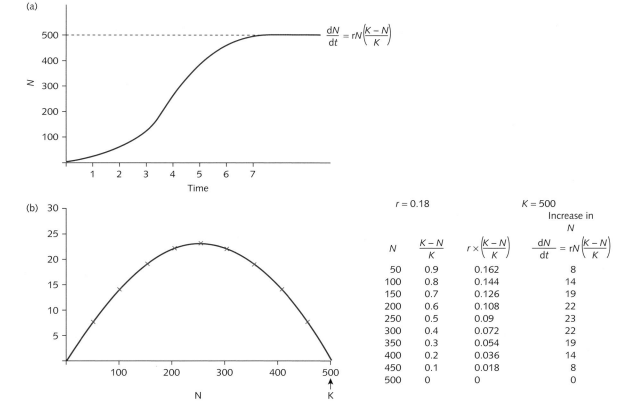

$r = 0.18$ $K = 500$

N	$\frac{K - N}{K}$	$r \times \left(\frac{K - N}{K}\right)$	$\frac{dN}{dt} = rN\left(\frac{K - N}{K}\right)$ Increase in N
50	0.9	0.162	8
100	0.8	0.144	14
150	0.7	0.126	19
200	0.6	0.108	22
250	0.5	0.09	23
300	0.4	0.072	22
350	0.3	0.054	19
400	0.2	0.036	14
450	0.1	0.018	8
500	0	0	0

Figure 3.2 A model of a population growing in a limited environment. (a) In this example we model overlapping generations, with the carrying capacity (K) set to 500. The rate of increase per individual in the population (r) is 0.18. Population growth is reduced by the increasing effect of environmental resistance as the carrying capacity is approached. (b) We can show the effect of environmental resistance by calculating the size of increase at particular population sizes. For each N, the environmental resistance is calculated as the proportion of capacity still available ((K − N)/K)—for 50 this is nine-tenths ((500 − 50)/500 = 0.90). Multiplying this by 0.18 (r) gives the actual rate of increase (0.18 × 0.90 = 0.162). Finally, multiplying the result by N gives the net increase at that population size (0.162 × 50 = 8). Notice that the largest increment is when N = 250, half of the carrying capacity. The increase in N declines as the habitat fills until no increase is possible at K.

lowers b and increases m. Increment size now falls as the population gets larger, until, at the carrying capacity, $b = m$ and no net additions occur.

This deterministic model helps us to analyse the elements which set the limits on growth and identify the nature of intraspecific competition for a real population. In reality, time-lags and a variety of other factors cause populations to overshoot their K and oscillate around it, but these complications are omitted here. We can test our ideas with experiments—consider what happens if the carrying capacity now changes. If we expand the space or the availability of some limiting resource, b should rise or m should fall, and the population grows to a new ceiling. Alternatively, a lowering of K leads to a reduction in the population as mortalities outpace births.

Not all plants and animals have their populations limited by intraspecific competition. For some, competition with different species (**interspecific competition**) is more important (Section 4.3): predation or parasitism, where one species is consumed by another, may limit the population growth of both consumer and consumed.

What determines the carrying capacity?

An abundance of key resources may mean that competition is rarely significant for some plants and animals. Many insects show little constancy in their population size. Their habitats change too rapidly for their numbers to settle at an equilibrium with their environment. Frosts, floods, fires, and other large-scale upheavals cause major changes in their habitats. These shifts in the carrying capacity bear no relationship to the size of the resident population. In other words, N is independent of density, so they are described as **density-independent** populations. Not surprisingly, species living under these conditions undergo major population fluctuations.

Even without dramatic changes in their environment, the abundance of some species varies wildly from year to year and some important pests occasionally undergo explosive population growth. Several species of locust have periodic population outbreaks, when massive swarms sweep across thousands of kilometres of Africa. For a long time we were unable to predict such outbreaks, but today we can model their swarming behaviour and predict outbreaks using satellite data on vegetation cover and rainfall in their breeding ranges.

The simple model we have described above is most relevant to species closely dependent on the long-term availability of resources in their environment—particularly the larger plants and animals that grow slowly and live longer. Their numbers do not fluctuate wildly and they are termed **density-dependent**.

What determines the maximum number of individuals that a habitat can support? For larger organisms with relatively stable populations we can make a reasonable prediction based on average body size. A large adult mass needs sufficient resources to grow and sustain itself. Meeting this demand requires space: a sufficient volume of soil from which to extract water or area over which to forage for food. The amount of space needed to support each individual plant or animal thus determines an ecosystem's carrying capacity for that species. For this reason, the average density of many large species at K shows some correspondence with adult body size. As we see later on, this has important consequences for the larger plants and animals we seek to protect from extinction.

3.3 Harvesting a population

Back at the pond, we will assume that growth of our fish population is density-dependent, limited only by intraspecific competition. This pond is a very stable environment and the population is close to its carrying capacity. We want to maintain its potential to provide catches from one year to the next, but we also want to take the maximum number of fish every season. The **maximum sustainable yield (MSY)** is the largest number we can harvest year after year, that which can be replaced by reproduction in each season.

We already know this number. The graph of population increase in a limited environment (Figure 3.2b) showed that the greatest increment was at half the carrying capacity. At this *N*, the population produces the largest number of offspring, before *r* starts to be reduced by intraspecific competition. This is said to be its **optimal yield**—the highest sustainable rate of population increase under a given set of environmental conditions—sustainable because we can harvest this number of fish year on year and because the population would rapidly grow up to its carrying capacity were we to stop fishing.

Fishery models

Calculating a maximum sustainable yield is feasible if we know the carrying capacity of the environment. This may be easy in a pond, but is highly problematical in the open sea. Besides *K*, we need to know the size of the **catchable stock**, those fish large enough to be caught in our nets and which are, in effect, the population size (*N*) of interest to us. Many disputes about fishing quotas revolve around interpretations of such data and the methods used to estimate the MSY.

Of course, the fish are never actually counted. Commercial fisheries are interested in the yield in weight, or **biomass**, and fishery models measure this, rather than the numbers caught. Nevertheless, the principles and basic ideas of the models remain the same—we simply replace individuals with units of weight.

Assuming there is no net migration into the fish population, biomass is restored by the addition of new individuals and also by the growth of new tissues by existing fish in the stock:

$$F + M = G + R$$

Here the biomass caught by the fishery (*F*) and the loss due to natural mortality (*M*) are balanced by tissue growth (*G*) and by recruitment (*R*) of new individuals. Fish are only recruited to the stock when they are large enough to be held by our nets. Their numbers have then to be converted into the equivalent biomass.

In most fisheries we have no control over natural mortality. We can only regulate our fishing effort so that fishing mortality (*F*) does not exceed *G* + *R* − *M*, and the catchable stock (*N*) does not decline—that is, we adjust our effort according to the level of natural mortality. The MSY is achieved when the largest harvest is taken without the catchable stock falling from one season to the next. This is the basis of the **surplus yield model**, so-called because the yield is equivalent to the growth and recruitment surplus to that needed to maintain *N* at its most productive. In effect, the population remains close to a constant size by replacing its losses through growth and recruitment. By noting the effort (time) needed to catch a fish, and the size of the catch from one year to the next, the fisherman at the pond is using the same model . . . though perhaps only intuitively.

This yield is maximized when the population is half of *K*, when *G* + *R* are at their greatest rate of increase (Figure 3.2b). The model estimates whether the yield is optimal by looking at the relationship between yield and fishing effort: any additional effort that consistently produces no increase in catch means the optimum has been passed. The catchable stock is then being depleted and each additional unit of effort brings a diminishing return (Figure 3.3).

In open-sea fisheries, the size of the catchable stock used to be estimated from catches landed at the quayside. A decline in the catch per trip (or more accurately the catch per unit fishing effort) over a number of seasons implied that the stock was declining. Fishery managers knew a pattern of falling yields indicated overfishing. Catches then included a proportion of the population whose reproduction and growth were needed to replace losses to *F* and *M*.

In fact, such patterns are notoriously unreliable. Many commercial fisheries are based on highly variable populations, whose size fluctuates with changes in weather, the movements of ocean currents, and a range of other factors, not least the behaviour of the fish. One of the most important variables is the recruitment to the stock from one year to the next. Because we are not able to predict this consistently, setting fishing quotas for depleted stocks such as

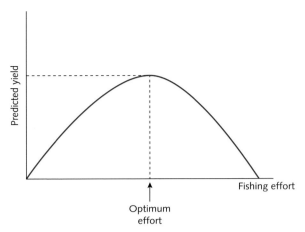

Figure 3.3 The estimate of the maximum sustainable yield from a fishery, based on the fishing effort. This is derived from Figure 3.2b. Fishing yields will decline above the optimum effort because we are now catching the stock needed to maintain itself. Below the optimum, the scope for higher yields is indicated by increased yields with more effort.

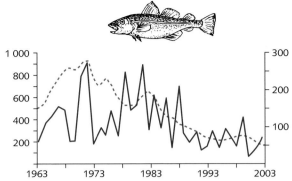

Figure 3.4 The variability in recruitment of North Sea Cod and the declining biomass of its spawning stock since 1963. Recruitment (red line) has been highly variable but in decline for the last 20 years. There has been a similar fall in the total biomass of its reproductive stock (blue line). Over the same period, F—the proportion of fish removed by our nets—has risen from 0.4 until, at the turn of the century, nearly all fish land on our plates ($F = 1.0$). As recently as 1980 the fishery yielded 300 000 tonnes but in 2002 this was just 41 000 tonnes. The Newfoundland fishery closed in 1992 when yields fell to 22 000 tonnes. Its population has yet to recover. The North Sea fishery remains open and fishing quotas are still being broken.

Atlantic and North Sea Cod (*Gadus morhua*) continues to be a source of international friction, even though everybody accepts this stock is severely overfished. Decision makers also know that weak and late regulation has led to the closure of cod fisheries elsewhere, which have yet to recover (Figure 3.4).

Measuring recruitment rates is crucial in intensively fished populations because these determine the size of catchable stock, more so than the carrying capacity. Most exploited populations are far below their K value, and any depressive effect of competition on r will be small. The critical problem then is to sustain recruitment, to maintain a viable reproductive population and the supply of fish growing up to this size.

Because of these uncertainties our simple model is only partially effective. Its analytical simplicity does not include the important details of tissue growth and recruitment rates, and the effect of body size on reproductive output (Case study 3). Since our prime concern is the weight caught, fish should be harvested only after they have had time to grow, not as soon as they have entered the catchable stock. Thus, a realistic estimation of maximum sustainable yield needs to incorporate rates of growth as well as recruitment.

A second group of models, **dynamic pool models**, do this. They use growth rates to calculate the biomass harvested from each age group. Many fish grow at a relatively constant rate throughout their life, so their size is a direct reflection of their age, and average weights can be estimated from the length of time an individual remains in the stock. We can then work out the optimal survival time for an individual to deliver a maximum yield. At the optimum fishing effort, residence times allow fish to reproduce and to grow to an optimum size. The model estimates the appropriate fishing intensity to give the maximum sustainable yield, based on recruitment and the survival time of the average fish in the catchable stock.

Just as a viable population must remain for growth in numbers, so individuals should be allowed time to grow new tissues. At high fishing intensities, few fish escape capture and few survive long in the catchable stock. Even if numbers are replenished, biomass is not, because fish have only a short time to grow (Figure 3.5).

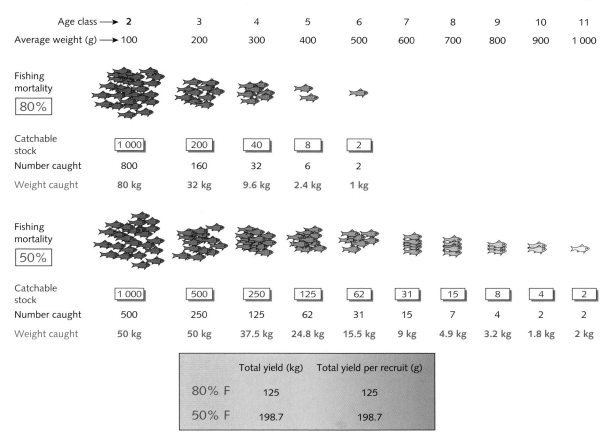

Figure 3.5 The effect of fishing intensity on yields to a fishery. Here we have two models, each starting with a cohort of 1 000 recruits, but with differing fishing intensities. In the first, 80 per cent are removed during each fishing season. The cohort is fished out in six years. At the lower intensity the cohort continues for 11 years. The weight of a fish depends upon its age. At 80 per cent *F*, most individuals are taken when they are still young, so in age class 2, for example, 800 fish are caught weighing an average 100 g, yielding 80 kg. At 50 per cent *F* the yield is only 50 kg, but more fish are left to grow on. The net effect is a larger total yield and yield per recruit at the lower *F*. At low fishing intensities high natural mortality will reduce the catchable stock and we could harvest more. A dynamic pool model provides plots of yield per recruit against fishing intensity, to find where the maximum yield is achieved—where the balance between growth and catches is optimum. These simplified models above assume no natural mortality and allow all uncaught individuals to grow to full size.

3.4 Growth rates, age, and recruitment

Dynamic pool models are a derivation of the surplus yield model, but they divide the population into age groups, and then work out the biomass each con-tributes to the yield. Such models require consider-able detail to make realistic predictions. We need data on the number of individuals being recruited, the rate of mortality in each age group, and the average weight of each age class. Ecologists summarize such data in **life tables**.

A life table estimates the chances of an individual of a particular age dying, or surviving for a length of time. You may be familiar with such statistics from comparisons of life expectancies for various age groups or peoples in different countries. Indeed, ecologists have borrowed many of these techniques from insurance actuaries who use life tables to calculate the risk of our dying while covered by their policies.

Life tables come in two basic forms: **cohort** and **static life tables**. In the first, a group of individuals born at about the same time (a cohort) is followed through its life and the numbers surviving in each age class are recorded. From this we calculate the survival probability and life expectancy for an individual of a certain age. A static life table derives the same data but from a census of the whole population, looking at the proportion in each age class, on one occasion.

Life tables tell us much about the dynamics of a population. With the data for several years, we can decide whether the mortality in a particular age group is significant for the overall population size. For example, larval mortality limits recruitment for many fish and it is often these numbers which determine the size of the catchable stock.

Life tables also reveal the **life history strategies** adopted by an organism. All species try to maximize their population growth but they differ in how they allocate resources between stages in their life cycle. Some produce large numbers of offspring, of which few survive to mate. At the other extreme are species that produce just one or two offspring each time, but which are highly likely to go on to reproduce themselves.

The structure of a population, the proportion of individuals in each age class, reflects these strategies. They also tell us much about the potential of a population for growth. Rapidly growing populations are typically 'bottom heavy', dominated by the new arrivals in the youngest age classes (Box 3.2;

BOX 3.2   **Age structures**

A population with overlapping generations will have different proportions of individuals in different age classes. In most populations this distribution flexes with changes in the environment, but two distinctive age structures are seen when growth is either close to a maximum or close to zero.

When there is large scope for population growth and the environment remains relatively constant, the population will increase at a constant rate per individual (r is fixed). It assumes a **stable age distribution** with birth rates and death rates constant in each age group (Figure 3.6). Such a population will be growing exponentially, with fixed proportions in each age group. The highest proportions of individuals are in the youngest age groups because of the high reproductive rate, and the distribution is characteristically 'bottom-heavy'. This pattern remains stable as long as there is no significant change in the environment and no change in migration rates in and out of the population.

A population may also have a stable age distribution when it is close to its carrying capacity, when it has little capacity for growth, and the birth rate is equal to the death rate ($r = 0$). Again, the environment must be largely unchanging,

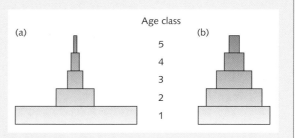

Figure 3.6 Age distributions. (a) A stable age distribution has a fixed shape—the proportion of individuals in each age group remains unchanged. This is because additions and losses to each age group are constant. Overall, the whole population has a fixed rate of births and deaths, and is growing exponentially. For this reason it is dominated by the younger age groups. (b) A stationary age distribution is also fixed, but now the birth rate and the death rate balance each other, and the population is close to K, its carrying capacity. This is largely a theoretical distribution (partly because the shape is bound to change as the environment changes), but now, with resources scarce, we see a more even distribution between the age classes.

but now the proportion of individuals is more evenly distributed between age classes—the birth rate is lower, and the younger groups (which tend to have the highest mortality) do not dominate the population. This is termed a **stationary age distribution** (Figure 3.6b).

Different human populations approximate to both these types—stationary age distributions are a feature of those countries whose population has remained relatively constant, or even declined slightly. Nations undergoing very rapid

population growth typically show a stable age distribution dominated by the younger age groups.

In animal populations, these distributions are not always indicative of the population growth rate. One obvious case is the fish we harvest—these typically have an age structure also dominated by younger individuals because our fishing techniques aim to catch only the older, larger fish. Unfortunately, despite the dominance of the youngsters, these populations are not growing rapidly.

Figure 3.6a). So are heavily harvested populations. Alternatively, a population close to its carrying capacity, when its birth rate is matched by its death rate, has a more even age distribution, with greater proportions in the older cohorts. Few are being added, since competition favours the old and the large (Figure 3.6b).

You can get a sense of this difference by walking through temperate woodlands of different ages. An undisturbed woodland, several centuries old, is typically dominated by a small number of large, old trees with very few saplings between them (Figure 3.7a). The young trees are waiting for gaps to appear, allowing them to grow on and compete for space

(a)

(b)

Figure 3.7 (a) Mature lowland forest—the Bialowieza National Park, Poland. This forest, on the Belorus border, is perhaps the only remaining fragment of pristine lowland forest in Europe. It is home to the remaining population of the European bison. (b) Regenerating woodland, with large numbers of small saplings.

and other resources. In contrast, a woodland re-establishing itself has large numbers of saplings, derived from the seed of a few nearby trees (Figure 3.7b). This is a population growing rapidly, when resources, primarily light, are plentiful.

Life table analyses can be crucial for maximizing yields in both fisheries and woodland management. Knowledge of their patterns of growth and the age of first reproduction allows us to estimate optimal yields. The dynamic pool model uses these kinds of data with sub-models to calculate rates of growth and chances of survival. Given this emphasis on detail and data from real world populations, it is more a simulation than an analytical model.

The same approach is used in forestry management, though here we benefit from a much more complete data set (Figure 3.8). Managers can measure competition directly, as the effect of tree density

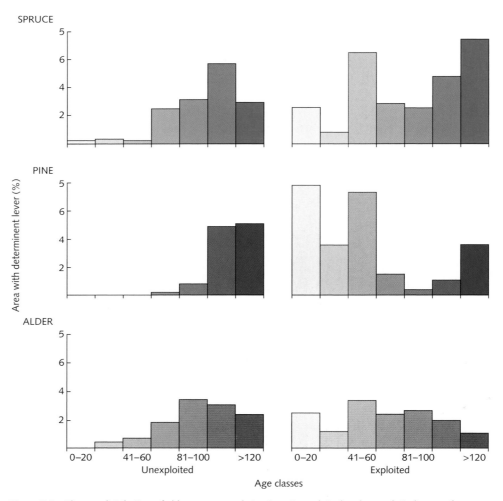

Figure 3.8   The age distribution of alder, spruce, and pine trees in exploited and unexploited areas of the Bialowieza national park in eastern Poland. The unexploited population is derived from a 'primeval' forest that has remained largely undisturbed since the fifteenth century, a community close to the natural forest that once covered most of lowland Europe. The exploited stands have been harvested since 1915. Notice that the age distribution includes a greater proportion of younger trees, established here largely through active replanting.

on individual growth. They also have more control over the population, including its recruitment (by their planting strategy), rates of growth (through their application of fertilizers), and competition (through their control of the density and species composition of the woodland).

3.5 Life history strategies

The stages in the life cycle of many organisms are very well-defined, marked by different body forms and ways of life. A frog begins as an egg, continues as an aquatic tadpole, and metamorphoses into a terrestrial adult. Seeds sprout wiry saplings adapted to life on the forest floor, but these will grow into massive trees if a gap opens in the canopy.

The life history strategy of a species will have evolved to ensure the survival of as many offspring as possible. During their life cycle some organisms radically change their habits and physiology to exploit different niches. For example, many amphibian tadpoles begin as herbivores, grazing the algae attached to stones in their pond, only becoming carnivores as they approach metamorphosis. The adult primarily feeds on terrestrial invertebrates, some distance from the pond. Similarly, adult dragonflies feeding on insects flying above the pond are not competing with their carnivorous nymphs feeding on other insects in the water below.

What are the benefits of partitioning the life cycle in this way? Most likely, making the best use of the resources available. Different stages feeding in different habitats or on different diets reduces the competition between age classes. In many species of insect, larval stages are often given over entirely to feeding and growth, whereas the adult, whose role is to disperse and reproduce, may not feed at all: adult mayflies emerge from the pond and fly only to mate and to lay eggs.

Other aspects of an organism's life history are adaptations to maximize survival. Large numbers of eggs anticipate large losses, but the massive wastage in most years is offset by the odd occasion when conditions allow many to grow into adults. The alternative strategy is for parents to invest time and energy in protecting and supporting the few offspring they have produced. We can think of this in terms of costs and benefits. Natural selection will favour a strategy if the benefits outweigh the costs: lots of eggs each costing little may mean high but acceptable losses—acceptable if enough survive to pass on the genetic instructions for this strategy. This code will then persist into the next generation and will come to dominate the genotype of inferior strategies. Producing only a few offspring is viable if the risks are low (or can be lowered).

These reproductive strategies are an indication of the scale of mortality rates in different age groups, and of the chances of going on to reproduce. Through natural selection, risks are weighed and the costs are balanced against the likely benefit. A change in the environment checks growth or reproduction and the balance shifts, and a new strategy is favoured.

A female cod can produce 4 million eggs at a single spawning, but devotes few resources to each egg and little effort to its aftercare. On average, perhaps three or four eggs will grow to sexual maturity. In contrast, an elephant produces a single calf, which it has fed in the womb, will suckle at the breast, and nurture for several years thereafter. During this time, the calf benefits not only from the resources its mother provides, but also from her protection and her knowledge of the environment. We should not think the elephant is more successful than the cod: indeed, there are certainly more cod than elephants in the world. They represent two different strategies—each adapted to the demands of their particular habitat.

The spectrum from *r*-selected to *K*-selected strategies

A life history strategy consists of several elements: the average length of a life, the proportion of time spent in each stage, the balance of the sexes, the age at first reproduction, the number of reproductive events, and the number born on each occasion. Together, these determine the speed of population growth and, consequently, have considerable adaptive significance.

Consistent patterns appear when we compare the strategies adopted by various organisms. In contrast to those mating only once, species that reproduce repeatedly take longer to reach sexual maturity and produce few offspring on each occasion. Their eggs are usually larger and may receive more parental care. Each egg, therefore, represents a much larger cost to the parent or parents.

Life history traits as adaptations to the permanence of the habitat are perhaps most obvious amongst the plants (Section 4.3). Where conditions are transient and opportunities for population growth are short-lived, mortality rates are high and natural selection favours species that reproduce rapidly. This implies a rapid completion of the life cycle and a short generation time, which in turn usually means a small body. Being small and producing many offspring are both aids to dispersal, necessary if a species is to colonize transient habitats. As resources become available elsewhere, such opportunist species can quickly move to exploit them. These are the characteristics of many pest or weed species, organisms adapted to habitats where there are few checks on their population growth. Natural selection here favours those with a short generation time and prolific reproduction, species which can quickly take advantage of any available resources. Because of their high r value these are termed 'r-selected' species. Often these will tolerate a wide range of conditions, due in part to their genetic adaptability (itself a product of their fast reproduction rate).

In a more predictable habitat there is little advantage in being adaptable. Nor is there any advantage in being small. Well-established and consistent habitats are dominated by large species with long generation times. Here the favoured strategy is to stand and fight, to crowd out competitors and invaders. The advantage lies with the highly specialized, those able to make efficient use of the scant resources, and most likely to produce offspring. These 'K-selected' species are adapted to compete in habitats close to their carrying capacity, species that can postpone reproduction until conditions are favourable but which can reproduce many times. In this case, mortality rates are typically highest among the younger age groups which can secure few resources from the older competition.

Compare the mouse, the elephant, the oak tree, and the cod. Many small mammals, insects, and the annual plants found in disturbed soils show r-selected characteristics. Large mammals and long-lived trees are typically K-selected, growing slowly as individuals and as a population.

This is a very broad schema, and many organisms are not readily classified as perfectly r- or K-selected. Although the predictability of its habitat must inform the life history strategy of most species, many plants and animals show characteristics which sit between the two extremes, whilst others, such as the dandelion, have races following different strategies in different habitats (Section 2.3). Cod, along with many other fish, show r-characteristics in their reproductive strategy, reflecting the uncertainty in recruitment from one season to the next. However, they will grow on as adults for many years and reproduce numerous times, which is more of a K-selected character.

Through natural selection, a species will match its reproductive strategy to the prevailing environmental conditions, if these remain predictable over a number of generations. Where conditions are not so reliable, having alternative strategies may be advantageous, and some species are able to hedge their bets, with some of their offspring adopting one strategy and the rest another. In this way at least some survive to reproduce (Box 3.3).

BOX 3.3   **A frog for all seasons**

Investing in reproduction or investing in growth are not always simple alternatives for an organism. Natural selection optimizes life history strategies to yield the most offspring, either favouring growth and a delayed reproduction or, alternatively, limited growth and early reproduction. This allocation of resources is crucial to any strategy and will depend on costs relative to benefits, risks and opportunities, and the ease with which critical resources are acquired.

In a changeable world (or under intense fishing pressure) opting to reproduce quickly may be the best option if delay could mean a lost opportunity, perhaps because of high mortality rates. But early reproduction often means a small adult size, a single reproductive event and few offspring surviving to reproductive age. The benefit is that some individuals carry the parental code into the next generation. Delaying reproduction may allow for a larger size and a more prolific reproductive effort but the cost may include, ultimately, the risk of not mating at all. This latter strategy will work if there are reliable and consistent patterns of change, for the species to adapt to the seasonal signals and to match its reproduction to the most favourable periods.

In some parts of the world, seasonal changes are highly unreliable, and adopting a single strategy may be disastrous. Then natural selection will favour short-term, phenotypic adaptability. Take the case of the West African reed frog *Hyperolius nitidulus* (Figure 3.9). Like most amphibians, these need freshwater to spawn and to support tadpoles prior to their metamorphosis into froglets. This requires a minimum period before they can leave the water and a further interval before they can reproduce. If the rainy season is too short or if they metamorphose too late in the season they will be unable to spawn until the following year.

Kathrin Lampert and Eduard Linsenmair have studied differences between *Hyperolius* which emerge early in the rainy season and those appearing much later. Late arrivals are unlikely to spawn in the year they emerge, and will have to survive the dry season if they are to reproduce. To do this they have to develop a physiology and anatomy to withstand the impending hot and dry conditions. Part of this conditioning involves, surprisingly, sunbathing. By gradually exposing themselves to the full sun for longer periods, *Hyperolius* froglets induce changes in their skin to make it more reflective. They can then aestivate (become inactive during the dry conditions of the summer) and survive without water for some time.

As ponds become available at the beginning of the next rainy season these froglets become active again and quickly grow to maturity and spawn. At this point, their tadpoles can adopt one of two strategies—grow quickly to reproduce before the rainy season ends, or emerge later, grow slowly, and condition themselves for the next dry season (Figure 3.10). The fast-growing, early-reproducing frogs make no attempt to adapt for the dry season—their skin has little reflective ability and they hide in the damp vegetation to avoid sunlight. When they mature they tend to be small, especially the males. Fully adult frogs will not survive the dry season and must lay their eggs before they die. However, their offspring must now complete their metamorphosis and condition themselves before the dry season arrives. If the rainy season and the ponds remain for some time this strategy will be successful. An early dry season will prevent fast growing froglets from fully conditioning themselves and only the slow-growing, well-conditioned froglets will survive.

The problem is the unpredictability of the rainy season and the duration of the ponds. For early emerging tadpoles, adopting a rapid reproductive strategy is the best option. Those emerging late can only prepare for aestivation. But those emerging in the middle produce offspring that may follow either route. The parents spread their risk, ensuring that some will carry their genes into the next generation

Figure 3.9   West African reed frog (*Hyperolius nitidulus*).

(continued overleaf)

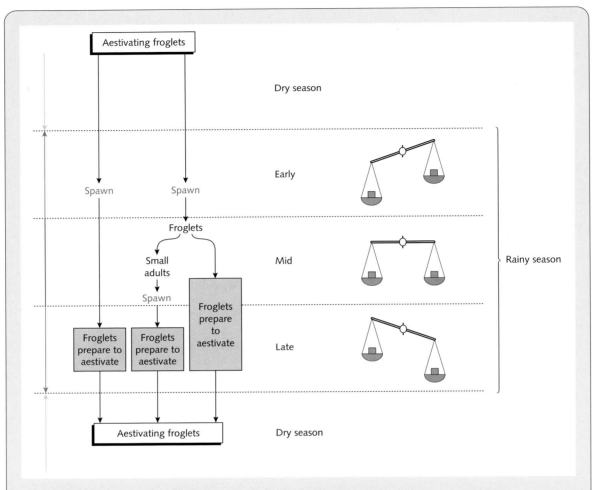

Figure 3.10 A long and reliable rainy season can allow the froglets of *Hyperolius* to grow quickly and produce a second generation within a year. However, the later they emerge in the rainy season the more risky this strategy becomes. In the middle of the rainy season, a proportion of the tadpoles grow slowly and prepare to aestivate, while others remain small and spawn. In this way a parent can hedge its bet—getting in a second generation if the rainy season is long, but ensuring that some offspring are prepared for an early dry season. Froglets which emerge later have no option and must condition themselves for the imminent dry season. The balance, favouring either reproduction (red) or growth (conditioning—green), depends on the probable time remaining of the rainy season and available for conditioning.

however long the rains and the ponds last. No surprises: this strategy means these *Hyperolius* avoid putting all their eggs into one basket.

Lampert and Linsenmair's work shows how one species can adopt different life history strategies according to the most likely environmental conditions. If this remains uncertain *Hyperolius* produce offspring to suit either eventuality, hedging their bets. Although this is most likely a phenotypic adaptation, theory suggests that over many generations a change in the genotype would be favoured if the seasons were consistently inconsistent. Evidence from a series of breeding experiments indicates that this may indeed be the case.

3.6 Genetics and population size

Having an adaptable life history strategy, one which can flex with small-scale change, is clearly an advantage when a habitat is changeable. However, for many species, the key elements of their life cycle are not so flexible. For these, a different life history requires genetic change, and this, ultimately, can only come from mutations.

Without mutation, individuals simply represent new combinations of existing alleles. For traits where there is little or no variation there will be few alternative phenotypes to distinguish between individuals. Since mutation rates are low, and many mutations are deleterious, species are limited to the particular environmental conditions for which their ancestry adapts them (Section 1.2). Environmental change can be rapid, but genetic change is usually slow and largely determined by the mutation rate for a particular locus.

A single population has only some fraction of the total variation available within a species and a small population has a small fraction. It can provide only a limited variety of genotypes in the next generation. Endangered species are reduced to small numbers, often with limited genetic variation to draw upon and consequently little capacity to adapt to new selective pressures.

This may be offset if there are frequent migrations between populations, introducing new alleles into a local population. Usually there will be a number of populations, a metapopulation, composed of relatively distinct populations with different degrees of isolation (Box 3.4; Case study 8). In some cases, populations rarely exchange genes with their neighbours and must rely on their own genetic resources to survive. More often there will be migrations between populations, and a gene flow between them, termed admixture. So important is this source of genetic novelty that many plants and animals have strategies to ensure that gametes are swapped between populations—male elephants move between groups of females and many flowering plants produce pollen well before their own ovules become receptive, to prevent self-fertilization. Unless gametes are swapped with genetically different individuals, the full benefits of sex are not being enjoyed.

Without this exchange, populations become genetically uniform and small populations lose genetic information very rapidly through **genetic drift** and **fixation** (Section 1.3). Remember that a new generation represents a sample of the genes of their parents' generation since only a proportion of the parental code will be represented in the offspring. Drift occurs when these sampling differences are preserved (and become more pronounced) down the generations and the population becomes genetically distinct. Such a population has evolved, but not by natural selection: drift is non-adaptive evolution and is simply a result of chance. The smaller a population becomes the smaller the total variation to sample from, and the more likely it is that some alleles will be lost entirely from the population. A population reduced to a single allele for a locus—that is, all individuals show the same trait—is described as fixed for that allele. Fixation happens faster in small populations, and rarer alleles are lost most readily. Smaller numbers also mean fewer possibilities for mutation and any fixed trait is then less likely to be reversed.

Genes that have major selective significance are less likely to change. Most mutations are disruptive and more often reduce the fitness of the phenotype. Even when the change is advantageous, such rare alleles have little chance of being sampled if population numbers are small. Yet variation, and the capacity to adapt to a changing world, is essential for the long-term survival of a population and a species. For this reason much of the effort in conserving endangered species is directed at preserving as much genetic variation as possible.

Occasionally, the lack of complete genotypic separation can be used to salvage an endangered species. In 1997 seven female Texas cougars (*Puma concolor stanleyana*) were introduced into Florida to provide mates for a dwindling population (around 30 individuals) of Florida panthers (*P. c. coryi*). The result was a series of successful matings and the birth of

BOX 3.4 Gene flow on the ice floes

Measuring populations has moved into the space age and, in the case of some large carnivores, demonstrates the extent to which populations within a metapopulation can be defined. Tracking polar bear movements with satellites can also help to explain patterns in the genetic code inside the cells of the different populations. Together with colleagues from research institutes in Norway, Russia, and Alaska, Mette Mauritzen tracked the position of 105 female bears roaming the ice sheet above Norway and Russia for three years (Figure 3.11).

Male bears lose their radio collars too readily because their neck is wider than their head and so data was only collected for females. This was used to map the range of each female and thereby assign her to a population. It also allowed the researchers to record when females wandered into the home ranges of neighbouring populations (Figure 3.12) and therefore measure the potential gene flow between populations. In this way, they were able to construct a picture of the dynamics of the whole metapopulation.

The Svalbard population was the largest, containing an average of 54 females. These bears were highly aggregated and had the smallest home ranges, well segregated from the other groups. This was the only near-shore population, and such bears characteristically return to the same patch of coastline each summer (Box 9.2). Exchanges between populations were highly seasonal, and not surprisingly, reflected the changes in the extent of the pack ice. Svalbard exchanged most individuals with its neighbour, a population (with 33 females) largely confined to the ice floes of the Barents Sea. The Barents population (described as 'pelagic' —literally open sea) migrated over large distances and spent little time on land. Interestingly, the Laptev females, consisting of just three collared bears, showed virtually no exchanges with its neighbours in the Kara Sea.

Some bears were clearly greater travellers than others and individuals were swapped most frequently between the pelagic populations. This is supported by previous genetic studies, which showed that the allele frequencies of populations in this part of the Arctic did not differ between populations. It seems that these exchanges prevent populations from becoming genetically distinct.

Figure 3.11 Placing a tracking collar on a sedated female polar bear on closed drift ice in the Barents Sea, April 1998. The adult female and her two 1-year-old cubs are attended by Andrei Boltunov, Andrew Derocher, and Øystein Wiig (left to right). The white collar with its satellite transmitter lies between the two cubs.

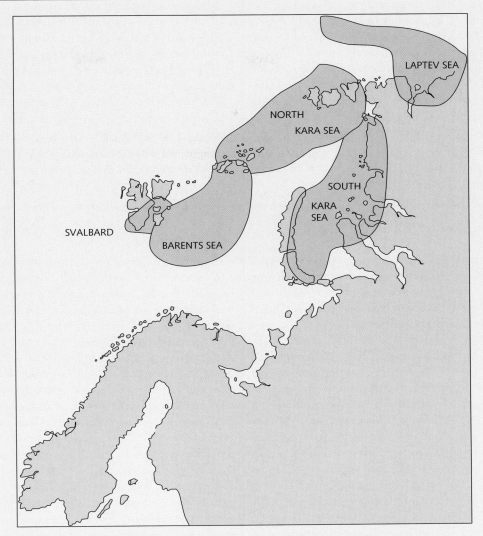

Figure 3.12 The five polar bear populations identified by satellite tracking in the Norwegian and Russian Arctic. These are for female bears only and say nothing about the mobility of the male bears that may range wider. Four of the populations rely on the extent of sea ice, and their ranges expand and contract with the ice. The Svalbard population is near-shore and will retire to the islands when the ice retreats. This is the period of fasting for the bears, when they are unable hunt from the ice floes.

In contrast, the North American populations do show genetic differences from each other. The greater prevalence of land in this part of the circumpolar region means that most of the populations are near-shore and likely to have smaller home ranges. It seems that either side of the Arctic, the distribution of land and ice defines the degree of isolation and genetic exchange between polar bear populations.

This distribution reflects the abundance of food resources and the access each population has to its prey. The behaviour of the bears and the size of their home ranges are largely defined by the geography of the landscape through which they move. Perhaps the genetics of their metapopulation is then best understood by looking in an atlas or peering down from space.

12 kittens from five cougar mothers. Fortunately the kittens retained panther characteristics and appear to have restored both genetic variation and a better gender balance to the small population. In this case the small genetic difference between the two cats has improved the prospects for the endangered ecotype. What is less clear, however, is the extent to which the distinctiveness of the Florida panther population will be compromised by using Texas cougars as surrogate females repeatedly.

More intriguing still are the attempts to resurrect the thylacine (*Thylacinus cynocephalus*—Section 2.3) using DNA collected from museum specimens. In this case, an entirely different species (but another marsupial) will be used as the mother, both to supply the egg cell and then to raise the embryo. By 2005 the DNA had been collected but the project came to a halt when its preservation was found to be poor. Shortly afterwards the project itself was resurrected by enthusiasts from several Australian universities.

3.7 Survival and extinction

An understanding of the reproductive and genetic characteristics of species we exploit or which face extinction is critical to our attempts to manage their populations. In the case of the North Sea cod (*Gadus morhua*) both apply. Commercial stocks of cod have already been lost from one of its most important habitats (the Grand Banks of Newfoundland), and European Union scientists believe that only a total ban in the North Sea would conserve the populations there. Although the EU Fisheries Commission pressed for a reduction in fishing mortality by 80 per cent (and the catch by two-thirds), the eventual agreement in 2003 was for a temporary reduction in fishing mortality of 65 per cent and the catch to be cut by half. The intensity of our fishing has been a potent selective force on the cod—the population of the North Sea today reaches sexual maturity 2–3 years earlier than it did 50 years ago, simply because of the short time they survive in the catchable stock. Such intense fishing pressure favours individuals that mature early and reproduce quickly, and it is their offspring we are catching today.

For both exploitation and conservation we have to maximize a species' reproductive rate and reduce its mortality rate. Adult populations of *r*-selected species fluctuate greatly as resources wax and wane, and they are often able to accommodate change or move to other locations. For *K*-selected species it is juvenile abundance which varies most. This is because births exceed deaths only when there is spare capacity in the ecosystem. It is the young which suffer at other times and their numbers decline if the carrying capacity is reduced. The poor dispersal of some *K*-selected species, and their close adaptation to particular conditions, makes habitat loss a key reason why they are so prone to extinction. Their size is another, implying a low rate of population growth and a low population density. Thus much of our conservation effort is directed toward large species whose reproductive rate is low and whose recovery from small numbers is slow.

Because of their size and activity, *K*-selected species often play a critical role in shaping their community. Browsing by elephant and giraffe is essential to keep the savanna grasslands free of thorn bushes and scrub, allowing other animals to graze (Figure 3.13)—the loss of elephant can lead to changes in the plant community so that many species, including some of the acacia trees upon which they feed, would disappear with them. Elsewhere, high elephant densities in forests and woodlands can lead to significant habitat change. Organisms with such critical roles are called keystone species (Section 5.2), species whose presence or absence determines the nature of the habitat.

Conserving a species in the wild means conserving its habitat. Life in a zoological or botanical garden may be the only option for a species whose natural habitat has disappeared, but we are then, in one sense, only conserving its genetic code (Box 3.5). For them to have a long-term future, a habitat large enough to support a genetically viable population is needed.

(a)

(b)

Figure 3.13   (a) Elephants grazing on *Acacia* trees, Masai Mara, Kenya and (b) the scale of their damage.

BOX 3.5   Test-tube conservation

Conservation aims to protect rare and endangered species within their own habitats, surrounded by the community and conditions under which they arose. However, this is not always possible, and *in-situ* conservation then has to be augmented by *ex-situ* conservation where individuals are removed from their original habitat and held in protective custody until they or their offspring can be released back into the wild. In the process, we can assist their reproductive efforts, and choices, to maintain or strengthen their genetic fitness. Such captive breeding programmes use stud books or pedigree charts to record partners and to minimize inbreeding.

Not so long ago the private lives of rare species frequently made the headlines, with precious individuals moving between zoos for carefully arranged liaisons. Today sperm, ova or embryos are more likely make the journey. *In-vitro fertilization* techniques avoid many of the problems that can limit success and even side-step some reproductive barriers to using closely related species. In this way, embryos of the Bengal tiger (*Panthera tigris tigris*) have been raised in Siberian tigers (*Panthera tigris altaica*).

Plants are particularly amenable to reproductive technologies because most of their cells are **totipotent**, capable of regenerating a complete clone of the parent. This is one form of **vegetative propagation**, used in horticulture and agriculture to produce cuttings and grafts to obtain several plants from a single specimen. The capacity to regenerate vegetatively varies from species to species and

(a)

(b)

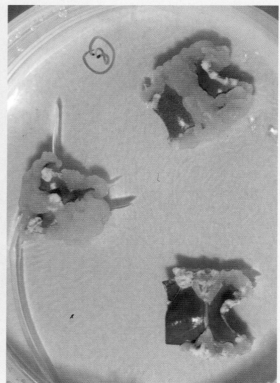

Figure 3.14   (a) Café Marron (*Ramosmania rodriguesii*). (b) Undifferentiated cells (callus) grown for the research programme seeking to reverse the genetic defect which prevents successful pollination.

from tissue to tissue. Some need to be coaxed using **tissue culture**, encouraging a small group of cells to grow on a nutrient gel. The gel is laced with plant growth hormones to promote controlled cell division that leads to the development of the different cell types that form the adult plant. These techniques have allowed us to propagate rare species of cacti and orchid threatened by collection from the wild. Now the market for these rarities can be flooded with millions of commercially grown plants, reducing the price and making the illegal trade in collected specimens uneconomic.

The Café Marron plant (*Ramosmania rodriguesii*) is a relative of cultivated coffee and endemic to the tropical island of Rodrigues. Like many other island endemics, Café Marron has suffered from the impact of introduced species. Rabbits and goats almost grazed the plant out of existence, and young plants were unable to compete with another European invader, privet (*Ligustrum vulgare*). By the 1940s there was only one remaining specimen of Café Marron, and it was later presumed extinct. However, in 1980, a boy presented a sprig of the 'extinct' bush for the school nature table, setting off a hunt to find and protect Café Marron.

Botanists eventually found a goat-ravaged bush and carefully fenced it off. It was only then that they discovered that the only specimen left on Earth was unable to reproduce due to a genetic defect which rendered it infertile. Although the pollen landed on the stigma it never grew a tube down the style to fertilize the flower. In 1986 three cuttings were taken from the plant and a team at the Royal Botanic Gardens, Kew attempted to propagate it by both conventional and tissue culture techniques (Figure 3.14). Fifteen years later, in November 2001, 11 young plants were returned to Rodrigues, in a reintroduction programme.

This was something of a hollow victory, however, because the reintroduced plants were all clones carrying the same genetic defect. Viswamabharan Sarasan considered how the research team might remove the barrier between pollen and the ovule. He and Carlos Magdalena dissected flowers and placed the pollen directly on the ovule, providing the pollen with a short cut. After many attempts they eventually had two successful 'pollinations' which formed fruit in August 2003. Seeds from the fruit have been germinated (once again using tissue culture techniques to maximize success) and these genetically unique individuals may at last give Café Marron a long-term future.

Fragmentation, extinction, and colonization

Extinction is a necessary part of the natural process of change. Many people are aware of the abrupt periods in the past when massive numbers of extinctions occurred on Earth (Section 9.2). Fewer realize that we are going through such a period now. Although we are unsure about the causes of past extinction events, today we can be certain that the loss of natural habitats is the prime factor.

As environments are degraded or destroyed the original habitat is broken up into fragments or patches, each with a relatively small carrying capacity. In the process, the distance between patches increases, making migration more difficult (Sections 8.1, 9.1). Small, isolated habitats are much more vulnerable to change, and less likely to maintain the conditions required by a large or specialized species. A once continuous population is now divided into a fragmented metapopulation whose overall size has been reduced.

Small and discrete populations can be severely threatened by relatively minor environmental changes. A small habitat patch provides little protection when conditions deteriorate. For example, a flood will have a minimal impact if resident animals can escape to higher ground within the patch. But these refuges are scarce in small patches and populations are more easily lost. Even without complete destruction, the carrying capacity of the patch has been reduced. Populations of all sizes may suffer in variable environments, but the implications are far more serious when numbers are low.

Individuals in small populations face other difficulties, not least in finding a suitable mate. Not every adult they encounter will be fertile or free from disease and some will never produce viable offspring. In many species a single male can inseminate many females, so a population's reproductive capacity depends on the number of females, and in future years, on the number of females born. Together, the size, age, and sex structure of a population govern its potential for growth. Consequently, the size of the breeding population (referred to as the **effective population size**) is always smaller than the actual population, and often substantially so.

Breeding from a small number of individuals means their genetic differences become less marked over generations. As time passes, the likelihood of mating with a partner sharing much of the same genetic code increases and, with very small numbers, mating may only be possible with close relatives. **Inbreeding** occurs when two individuals with similar genotypes mate—the more ancestors two parents have in common the more code they will share. Then there is a significant risk of each parent contributing the same recessive gene to their offspring: in this homozygous combination, the coded trait will be expressed in the offspring's phenotype. Since many recessive traits are deleterious, the result can be small, weak, and infertile offspring that ultimately cause a depression of population growth. **Inbreeding depression** is a common problem in domesticated plants and animals bred from a limited stock, when individuals do not grow vigorously and compete poorly with out-bred offspring. We see the same effect in human populations where mating has been confined to small groups, isolated by geography, custom, or religion. Even in larger populations, some people assume a status that separates them from the majority and choose only to breed with a small group of limited partners.

Whilst the size of the metapopulation may be large, the effective breeding population within any single patch can be very small. As habitats disappear, so the metapopulation becomes more fragmented and the gene flow between populations declines. As well as inbreeding depression, genetic isolation may also produce locally adapted races—characters selected within a population adapt them to local conditions and distinguish them from their neighbours. The result can be **outbreeding depression**, where matings between two populations are relatively unsuccessful—any offspring are not well adapted to either habitat and cannot compete with locally bred offspring. This is sometimes a problem with *K*-selected species, particularly various trees that have

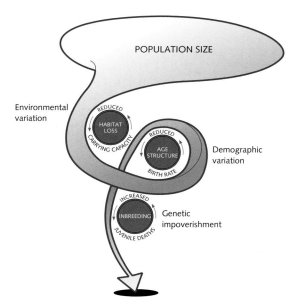

Figure 3.15 The extinction vortex. Habitat loss reduces the carrying capacity for the population, demographic factors reduce their birth rate and genetic factors may increase juvenile mortality. Because these all reduce population size, and because smaller populations exacerbate the demographic and genetic effects, an endangered population will accelerate towards oblivion.

evolved local races. Such local differences are the first stages in speciation (Box 2.4), with the result that reunited populations may not breed with each other or, as we see below, perhaps should not be allowed to.

This combination of difficulties—a susceptibility to environmental variation, the problem of sustaining population growth with a small number of individuals, and the genetic consequences of a reduced breeding stock—can lead to a downward spiral in numbers, where one factor feeds on another, accelerating the species fall toward extinction. This has been called the extinction vortex (Figure 3.15).

The elephant and the rhinoceros

The creation of the Royal Chitwan National Park in Nepal has saved the greater one-horned rhino (*Rhinoceros unicornis*) from near-certain extinction (Figure 3.16). This impressive species was found along the margins of lowland rivers in the Himalaya,

Figure 3.16 Indian rhino (*Rhinoceros unicornis*) in Chitwan National Park, Nepal.

and until recently it had a population that ran into the tens of thousands. Hunting and habitat destruction during the last 100 years had reduced it to just two groups of about 100 individuals by the early 1960s. The loss of their habitat, as well as the pressure of hunting, both for sport and to prevent damage to tea plantations, were the main reasons for its collapse.

In Chitwan the rhino was reduced to an effective breeding population of just 21–28 individuals. Today, with substantial protection, the total population is above 500, whilst the Kaziranga National Park in India has 1 300 animals. Much of the credit must go to the rhinos themselves: their spectacular recovery after the ending of poaching and hunting has been possible because they have retained a very high genetic variation.

We would normally expect genetic variation to be low in a population reduced to such a small size. Ironically, the Indian rhino seems to have held on to much of its variation because its decline was so rapid. Two demographic factors contributed to this—the rhino have a long generation time (11–12 years) and females produce only one calf every four years. Thus the extent of inbreeding was limited to a short period during their recent decline. Kaziranga has around half of the global population and currently its population is growing at 5 per cent per year. According to the Worldwide Fund for Nature (WWF) the reserve is close to its carrying capacity (Table 3.1), but poaching is still estimated to take 5 per cent of its numbers each year.

TABLE 3.1 The global populations of five species of rhinoceros in the world in 2005, based on data from the International Rhino Foundation. The total population size (N) is for wild populations only and an asterisk denotes that a species has a poor record of breeding in captivity. The Sumatran rhino has the most fragmented population and there is some evidence that it suffers from inbreeding depression.

Species	Total N	Number of populations	Number in captivity
African			
Black	3 610	75	250
White			
(Northern sub-species)	10	1	10*
(Southern sub-species)	11 300	248	750
Asian			
Javan	60	2	0
Sumatran	>300	12	7*
Indian	2 400	7	150

In Asia, the security of the rhino populations and the loss of their habitats are closely interlinked, as forest clearance and agriculture have fragmented habitats. Just 60 Javan rhinos (*Rhinoceros sondaicus*) remain in two populations—in Lam Dong province in Vietnam and in Ujung Kulon reserve in western Java, with the latter (containing 50 individuals) thought to be the only viable population. There have been fierce arguments as to whether to leave them all in the wild, possibly combining the two populations on a single reserve, or whether to transfer some to zoos for captive breeding. Combining the populations to improve their genetic variation is perhaps even more of an issue for the smallest rhino, the Sumatran (*Dicerorhinus sumatrensis*). Its meta-population is divided into populations on the islands of Sumatra and Borneo and, like the Javan rhino, may now represent two sub-species.

African rhino are found in grassland, woodland, and thorn scrub habitats, much of which remain. Here, it is poaching for their horn which has devastated the two species (Table 3.1). Despite legislation and a policy in some countries to shoot poachers on sight, the high price of the horn ensures that the slaughter continues. Because of their masculine associations, the horns are prized as handles for daggers (or jambiyas) and, when ground to a powder, as an aphrodisiac. While the horn probably makes the dagger unusable, the powder has no likely physiological effect on the libido of the human male. As much as half of all traded rhino horn (1.75 tonnes) was going to Yemen prior to 1990 but since 1997 this country has imposed penalties for using horn in daggers and the trade has fallen dramatically.

Perhaps the conservation of the rhino would be better served by a more radical approach. South Africa argues that the material from de-horned rhino (cut, under anaesthetic, before poachers get to the owner) should be traded. This might lower global prices and make poaching less lucrative. Additionally, there would be a requirement to mark and register official stocks and only these would be legally traded. As it is, the high price fetched by the horn (up to 60 000 US dollars per kilogram in the Far East) is a major temptation to local peoples with meagre incomes, who probably see the rhino as a local resource

Figure 3.17   Black rhinoceros (*Diceros bicornis*).

to be harvested, in the same way that other peoples catch fish.

Of the hundreds of thousands of black rhino (*Diceros bicornis*) that once ranged over sub-Saharan Africa, 65 000 were alive in 1970 (Figures 3.17 and 3.18). Twenty years later there were just 3 800 distributed among 75 populations but only 10 of these populations had more than 50 individuals. Because of this, animals are moved between reserves in East Africa, to improve, or at least maintain, the genetic constitution of each population. Based on their horn shape and body size, seven sub-species or local races have been identified. One is now extinct. Present evidence suggests these have become genetically differentiated only recently, and that breeding between them could be a viable conservation strategy.

The picture is less encouraging for one race of the white rhino. Although the southern race (*Ceratotherium simum simum*) has been through a major reduction in numbers, it has recovered to become one of the most successful conservation initiatives. This race was once thought to be extinct until a small number were found in a remote area of Natal, South Africa at the beginning of the twentieth century. They were protected and have today grown to around 11 300, with many translocated to various reserves in the region. In contrast, the northern race (*C. s. cottoni*), historically the most abundant sub-species, has suffered a much more dramatic fall and is on the brink of extinction. In 1960 there were 2 000 left in the wild but this fell to 20–30 in the 1990s.

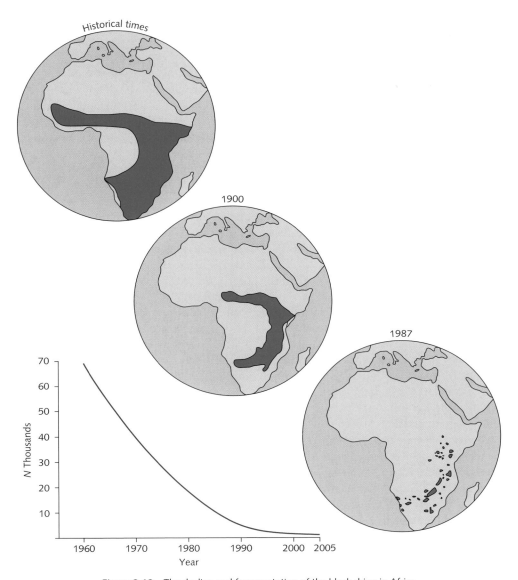

Figure 3.18   The decline and fragmentation of the black rhino in Africa.

Although numbers rose slightly in the early part of this century, a resumption of poaching alongside continuing political unrest saw numbers collapse to just 10 in the Garamba reserve of the Congo (Table 3.1). Analysis of their mitochondrial DNA shows the two races of white rhino are genetically distinct, probably because their geographical ranges have not overlapped in recent times. Conservation biologists have decided not to interbreed them, fearing the offspring would show outbreeding depression.

For the different species and the different races of rhinoceros, the details of their decline differ. Conservation efforts must be informed by this detail, by their ecology and genetic history. But we also have to pay due regard to their status in the environment, especially the value local peoples place upon them, either as a resource or as part of their culture. A conservation strategy that ignored all of these complicating factors would be unlikely to give the rhinos a long future.

As for the horn, so for the tusk. The international trade in ivory has led to the slaughter of large numbers of Asian (*Elephas maximus*) and African (*Loxodonta africana*) elephants. Laos and Cambodia once had 40 000 elephants, but now have fewer than 4 000 and Vietnam fewer than 100. Again, much of the decline in the Asian elephant is associated with habitat loss, whereas poaching has been the prime factor in Africa. Early in 2006, the 13 Asian nations with elephant populations agreed to develop strategies promoting elephant conservation that will cross national boundaries.

Unlike their Asian counterparts, both male and female African elephants bear tusks, and so both are slaughtered. Large tusks attract a larger price, so poachers favour the larger adults, especially the males. An elephant might live between 40 and 60 years, though today most adults are under 30 years old. Unfortunately, males do not begin breeding until they are around 30. Changing the age structure of these populations not only affects their future growth, it also has implications for the social hierarchy of the herds. These are normally led by elderly females, the matriarchs, whose knowledge of waterholes and foraging routes is passed on to the younger members of the group. Today most herds in Africa are led by females in their early twenties and teens.

The loss of the elephant also has important implications for the African landscape. On the savanna, their browsing on shrubs and trees is crucial to maintaining open grassland and prevent the invasion of thorn and scrub. The grasses that come to dominate provide food for grazers such as gazelle, wildebeest, and zebra. Less obvious are the many smaller vertebrates and invertebrates which, directly or indirectly, rely on open bush. Elephants and giraffes are keystone species, maintaining the grassland community of the dry savanna and open scrub in the wetter parts of Africa.

However, there are complications when the elephant population is close to its carrying capacity. Elephants will push over trees to browse and their trampling damages grassland regeneration, particularly around waterholes (Figure 3.13). Richard Hoare describes how the focus of elephant conservation in eastern and southern Africa has changed, from controlling poaching in the 1980s to the current conflicts between elephants and humans. Following the international ban on ivory trading and the reduction in poaching, elephants are increasingly seen as agricultural pests in these areas.

The problem stems from elephants exceeding the carrying capacity of their reserves, spilling over into unprotected ranges where they meet an expanding human population. The density of elephants in some of these reserves has caused a loss of plant biodiversity and serious habitat damage. On farmland outside the reserves, a variety of control measures have been used to keep elephants away from crops—from organizing hunts to repellent sprays and fencing. Part of the answer may be to designate newer and larger reserves, but coexistence will need the cooperation of local peoples, and perhaps an expansion of the schemes that exploit elephants for tourism and for hunting.

In Zimbabwe, the government has designated areas where native agriculture runs side by side with a game management in an effort to control poaching. Government marksmen carry out organized culls, taking selected elephants. This helps to maintain the integrity of the elephant herd by managing its age structure, while meat from the kill is distributed among villagers, and the skin and tusks are sold on. The proceeds are being used to fund conservation and to provide compensation to farmers in the form of additional food. Local people are thus being encouraged to tolerate and, indeed, value their wildlife resources.

Richard Hoare describes how organized sports hunting of elephant has also directly benefited local peoples. In the Addo Elephant National Park in South Africa, each elephant is estimated to generate 300 000 dollars per year in the economic activity associated with tourism. However, culling is not allowed here and Amanda Lombard and her co-workers have measured the impact of the exceptionally high population density (2.2 elephants per square kilometre) on the rare endemic plants within the reserve. This ecosystem is part of the Eastern Cape succulent thicket habitat (Case study 5), which includes plants found nowhere else in the world. Many of these have declined dramatically in the last 30 years as the elephant population has grown. Now the reserve has a population of close to 400 elephants (from a founding population of just 11 when the park was set up in 1931) and these represent 78 per cent of the total herbivore biomass in Addo (a figure presumably based only on large herbivorous mammals). This is an imbalance that cannot be sustained in the long term since this keystone species has become a destroyer of this habitat. Lombard suggests expanding the reserve and creating botanical reserves, as well as accepting more general conservation priorities for its management. This could include allowing black rhinos access to the botanical reserves.

The balance has shifted. Elephant populations in various African states have been downgraded as conservation priorities and, in 2002, the UN CITES conference in Chile agreed that South Africa, Namibia, and Botswana could sell 60 tonnes of their ivory stocks in 2004. The delay was to ensure that adequate monitoring procedures would be in place to prevent poached ivory finding its way onto the market. One promising scheme is to create a databank on the genetic identity of the different herds, to allow the origin of each tusk to be traced back to a legitimate source.

The elephants and the rhino illustrate our ambivalent relationship with such talismanic species. For some they embody the ethos of nature conservation, for some they are symbols of power and masculinity, and for others they are still a resource to be exploited. Different publics make different demands upon them and it may be that the cultural differences between nations will determine these species' future —in the same way, perhaps, as other peoples use their cultural traditions to assert their fishing and whaling rights.

● SUMMARY

Simple analytical population models help us to understand the key features of population growth but lack sufficient detail for us to make useful predictions about real world populations. Optimal yield is about half the population's carrying capacity. In a relatively stable environment this also represents the maximum sustainable yield a fishery could expect to take, when the harvest is matched by recruitment or growth of new tissues. To find this balance, we need data to match our fishing effort with rates of replacement of tissues and individuals.

For both fisheries and single species conservation, we have to consider the details of age and sex structure within a population, and the life history and reproductive strategy of a species. Fragmented and small populations within a metapopulation may suffer genetic inbreeding if there is little migration between populations. Conservation efforts are often concentrated on the large, slow-growing, and competitive K-selected species, closely adapted to their habitat. These are most likely to suffer when there is significant habitat loss. On the other hand, r-selected species are adaptable and more able to move between habitat fragments.

The various species of elephant and rhino in Africa and Asia illustrate some of the conflicting issues in single species conservation. In Asia habitat loss has been the principal cause of decline in both animals, and the recovery of rhino species has been particularly effective inside protected reserves. Control of poaching in Africa has also allowed recovery but there are now problems of over-population of elephant on some reserves.

● FURTHER READING

Alstad, D. 2001. *Basic Populus Models of Ecology*. Prentice Hall, New Jersey. A detailed but accessible introduction to the range of basic population models.

Kaiser, M. J., Attrill, M. J., Jennings, S., Thomas, D. N., Barnes, D. K. A., Brierley, A. S. Polunin, N. V. C., Raffaelli, D. G., and Williams, P. J. 2005. *Marine Ecology*. Oxford University Press, Oxford. An excellent introduction to marine ecology providing detail on fisheries and fishing techniques.

Milner-Gulland, E. J. and Mace, R. 1998. *Conservation of Biological Resources.* Blackwell, Oxford. A wide-ranging collection of essays on the conservation of endangered species and habitats, including some of the social and economic aspects of over-fishing.

Scalet, C. G., Flake, L. D., and Willis, D. W. 1996. *Introduction to Wildlife and Fisheries*. W H Freeman, New York. A systematic textbook that provides a foundation in both species conservation and resource management.

● WEB PAGES

For details of the current status of large endangered species:
http://www.panda.org/about_wwf/what_we_do/species/

International Rhino Foundation:
http://www.rhino-irf.org

World Conservation Union—home of the red data book:
http://www.iucn.org

Convention on International Trade in Endangered Species:
http://www.cites.org

International Council for the Exploration of the Sea—data on a range of marine resources from an ecological perspective, especially fisheries:
http://www.ices.dk

visit the Thylacine Museum and see the movie:
http://www.naturalworlds.org/thylacine/

CASE STUDY 3

Muddying the waters

The fisheries models we have described here are perhaps too simple—the clarity they bring to the principles is provided at the expense of the detail needed to manage real world stocks. These principles are useful if the various 'stakeholders' with an interest in the fishery—the managers, fishermen, and consumers—are to understand why no fishery stock is limitless. But if stocks are to be exploited sustainably, yields and fishing quotas have to be based on the detail of the fish's ecology, requiring simulation models to make realistic predictions.

At their most basic, fishery models pay little attention to the effects of fish size and population structure on the re-productive capacity of the stock, even though this information can be critical for their proper management. Traditional models using only the biomass of the spawning stock miss important differences between reproducers—the small or the large, the healthy and the less-than-healthy adults. Consequently, important differences between populations are not considered and the predictive power of the models is reduced. Cod, for example, increase their spawning as they get larger so that older fish produce more eggs, spawn over a longer period, and produce several batches during the season. As in many other open-water species, the size and age distribution of female cod in a population's reproductive stock determines its potential for recruitment.

Adding this realism can make for very complex models. Using data from the Icelandic fishery, Beth Scott and her colleagues have incorporated the important variables for

Atlantic Cod into a computer model to predict the reproductive output from fish within one of three age/size classes (Scott *et al.* 2006). An individual's reproductive effort depends upon its size, its condition, and its history of spawning. Using data on the average size and condition of fish in each cohort from catches, the model predicts both the size and viability of an egg batch and when it is likely to be laid (Figure 3.19). Crucially, it does this for different times during the breeding season incorporating the changes in the condition of the fish through the season. Rather like the dynamic pool model, cohorts are then summed for all age-classes to give the **stock reproductive potential** (SRP), the overall capacity of a population to produce viable eggs and larvae. In simple terms, a reproductive stock will need to have many young, small adults if it is to match the output of a few large and healthy fish.

Scott and her team used their model to examine the response of different population structures under different fishing intensities. They ran the simulation with six population structures that were realistic for six levels of fishing mortality (*F*), each under two different scenarios: the first had a population size and age structure that generated a recruitment to match the annual fishing harvest (a perfect surplus yield). This was termed the constant biomass scenario (*CB*) because the biomass of this stock never declined. For an intensively fished stock, this would mean recruiting many juveniles, and an increase in the survival of these smaller fish. The second scenario had recruitment entirely independent of the size of the spawning stock. Assuming a fixed natural mortality (*M*), this means any increase in *F* causes the size of the spawning biomass to decline—the declining biomass scenario (*DB*). *CB* and *DB* are treated as two extremes, a range that is likely to encompass the real population.

Because fish condition and the size and quality of eggs changes with time the model calculates the reproductive output for each day of the spawning season. It does not predict the absolute numbers of potential recruits, but the relative performance over time to show how recruitment potential and the length of the spawning season change under different fishing intensities for each population structure in each scenario (Figure 3.20).

Overall, the predictions from the model are stark, mostly intuitive but occasionally surprising. Generally 'older, larger and better conditioned fish produce more batches of eggs and more eggs per batch . . . [eggs] which are larger and more viable . . . ' (Scott *et al.* 2006). They also spawn for longer. A population that is intensively fished has an SRP between one-half and one-quarter of an un-fished population, depending on egg viability. However, high *F* also makes

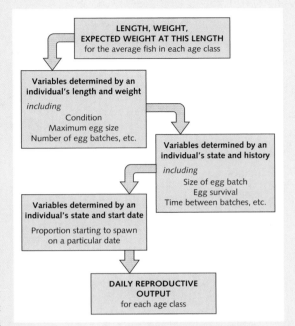

Figure 3.19   The model developed by Scott *et al.* (2006) calculates the daily reproductive output for each age class of cod, using data from the North Atlantic fishery. Four modules in the program use this input to calculate a series of variables that predict the number of potential recruits (number of eggs multiplied by their survival rate). For example, a fish in good condition has a higher-than-average weight for its length and devotes more resources to producing eggs. Size and condition also determine when a fish will spawn. The model predicts the output for each age class on each day of the spawning season. Adding these together for all age classes gives the total SRP.

the role of smaller fish critical for recruitment (Figure 3.20). With an age structure dominated by smaller individuals, the spawning season may be reduced by up to four weeks. The smaller fish now become the most important group for reproductive output. Any benefit of improved fish condition on the SRP also depends on this population structure.

Detailed as it is, the Scott model does not include all the potential factors that might determine longer-term recruitment: for example, water temperature can alter these relations—every 2 °C rise in temperature reduces time to maturity by one year so that recruitment to cod stocks increases in the northern and colder waters when average temperatures are raised (Drinkwater 2005). Further south, however, persistent warming of the Irish and North Seas

(continued overleaf)

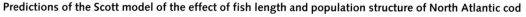

Predictions of the Scott model of the effect of fish length and population structure of North Atlantic cod

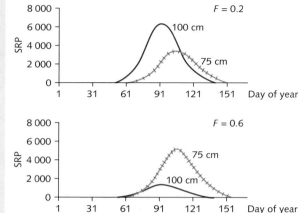

At an $F = 0.2$ (fishing mortality of 20 per cent) larger cod (100 cm against 75 cm) reproduce earlier, for longer and produce more offspring.

With the uneven population structure created by a higher fishing intensity ($F = 0.6$), smaller fish become more important. Both sizes start to reproduce at the same time but small cod spawn for longer and are responsible for more recruits—simply because there are many more fish in this size class.

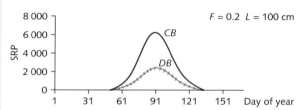

Effect of stock biomass

The reproductive output of 100 cm cod are predicted to be three times as large when there is a constant spawning stock compared to a declining stock, even though spawning begins and ends at the same times.

Figure 3.20   How fish condition, age, and population structure can affect recruitment potential (outcomes from the Scott model for North Atlantic Cod).

by 3 °C could cause these stocks to disappear entirely, as the waters become too warm for cod (Section 2.5). There is evidence for such effects already: most of the spawning activity has moved north as the North Sea has warmed, probably in response to available larval food. It is to these areas that, in turn, juveniles and mature fish return to spawn (Rindorf and Lewy 2006).

If nothing else, the Scott model demonstrates that fishing quotas which ignore the age structure of the population will miss important effects. The size of a spawning stock is easy to understand, but its potential for reproduction is not, confounded as it is by the interactions between fish size, condition, egg viability, time of spawning, and population structure. Whatever their size, a spawning stock in very poor condition, even if our nets bulge with their biomass, could suffer almost complete recruitment failure in difficult years.

Convincing the fishermen and the consumer that the Atlantic and North Sea Cod are plunging towards the same depths as the Newfoundland fishery was always going to

be difficult. The total collapse there and the speed of the decline here suggest that our failure then to protect stocks will not prevent us now from making the same mistake.

References

Drinkwater, K. 2005. Cod stocks: winners and losers in the climate change sweepstakes. *ICES/CIEM Newsletter 45*, 6–8.

Rindorf, A. and Lewy, P. 2006. Warm, windy winters drive cod north and homing of spawners keeps them there. *Journal of Applied Ecology 43*, 445–453.

Scott, B. E., Marteinsdottir, G., Begg, G. A., Wright, P. J., and Kjesbu. O. S. 2006. Effects of population size/age structure, condition and temporal dynamics of spawning on reproductive output in Atlantic cod (*Gadus morhua*). *Ecological Modelling 191*, 383–415.

● EXERCISES

1. Why will intraspecific competition decrease the further a population is from its carrying capacity?

2. Why should the average size of a fish beginning reproductive activity become smaller in heavily fished populations? Why is this likely to be a genotypic rather than a phenotypic effect?

3. Use the equation in Figure 3.2 and the following data to derive the maximum rate of increase for a starting population of 50 when:
    (a) $K = 1000, r = 0.36$
    (b) $K = 1000, r = 0.18$
    (c) $K = 500, r = 0.36$

    You will probably find it useful to create a table like that in Figure 3.2(b) for each calculation.

    What do you notice about the effect of r and K across these examples? What is a quick way to calculate the maximum rate of increase in these simple models?

4. Study Figure 3.7 and answer the following questions:
    (a) Why is the balance weighted in favour of reproduction at the start of the rainy season?
    (b) Why is the balance weighted in favour of growth (conditioning) at the end of the rainy season?
    (c) What are the costs and benefits (risks and returns) for froglets attempting to produce a second generation in the middle of the rainy season?

5. Populations of the woodlouse *Porcellio scaber* were sampled from two different habitats on the same day. The number of individuals in different size classes (which approximate to age classes) were as follows:

		Population	
		A	B
Size (age) class	1	297	203
	2	91	112
	3	32	87
	4	10	41
	5	1	19

    (a) Would such data enable us to construct a cohort or a static life table?
    (b) Draw an age pyramid for each population. Which of the two populations most closely approximates to a stable age distribution? Give your reasons.
    (c) One of the habitats sampled was a frequently-disturbed compost heap and the other a long-established rough pasture. Which population most likely came from each habitat? Give your reasons for your choices.

Tutorial or seminar questions

6. Consider how the virtual extinction of cod and the depletion of many other fish stocks might affect the polar bears of Svalbard. What are likely to be the consequences for other species that interact with cod?

7. In parts of Southern Africa, the pressure of elephant numbers can lead to significant habitat change, with grave implications for other species. Review the economic, social, and political implications of some nations being allowed to harvest ivory and sell this to earn foreign currency.

8. Recent evidence suggests that polar bears today have significantly shorter seasons to roam the ice floes in some parts of the Arctic. What are the likely implications for the genetic health of the various populations that make up the metapopulation east of Svalbard?

Interactions

'They made us many promises, more than I can remember, but they never kept but one; they promised to take our land and they took it.'

Chief Red Cloud of the Oglala Teton Sioux

CHAPTER OUTLINE

- Resources—how they are acquired and allocated.

- Forms of cooperation—mutualism and commensalism.

- Competition between individuals and between species to secure resources.

- Life history strategies—the natural selection of life cycles.

- Plant defence mechanisms against herbivores.

- Predator–prey interactions and population stability.

- Parasitism and pathogenicity as an extreme form of predation.

- Biological features of pests—how native and introduced species become pests.

- Conventional and biological pest control.

← Cheetah after a succesful kill.

Red Cloud speaks as a victim of intraspecific competition—when individuals of the same species compete for limited resources. We might call it war, because individuals had grouped themselves together and acted cooperatively to acquire or defend a resource. To Jacob Bronowski war was not a human instinct, but rather, a 'highly organized and cooperative form of theft'.

Humanity is not alone in cooperating in this way. Many animals, from ants to chimpanzees, form themselves into bands to improve their chances of winning a competitive battle. In a world of shortages, individuals and societies need strategies to secure their resources, for this generation and for generations to come.

Of course, it is too simplistic to describe war as a contest for scarce resources. It may be a prime cause, but conflicts also begin when societies seek to promote or preserve their structures and their traditions. Individually and collectively, people make moral and ethical judgements to fight and sacrifice resources to defend an idea. For both humans and animals, defending a society often means defending the system by which resources are partitioned amongst its members.

The problem for the individual is to make a living, to acquire the resources needed to survive and to thrive, and to ensure that its genes make it into the next generation. This means securing the energy and nutrients necessary to produce gametes and to nourish its offspring. Societies are often organizations that facilitate this process. An ant society is highly integrated, with most individuals sacrificing their own reproductive effort to promote that of a single queen.

Mammalian societies are much looser arrangements, but they too increase their efficiency by dividing the work and by partitioning resources amongst their members.

Seen in these stark terms, life appears to be very brutal. Yet the great variety of life on the planet, and the beauty of its plants and animals, has arisen from this struggle for resources. Cooperation can benefit both parties, among related individuals sharing part of their genetic code, or even between different species. Such cooperation is essential for the survival and reproductive success of a variety of organisms, and close associations form when two species become dependent on each other. In other cases the benefits are one-sided, and one organism uses another as a resource. All of these interactions evolve and develop through natural selection, and help to tie communities of species together.

This chapter looks at the range of these interactions, considering the costs and benefits to each partner. Each individual seeks to maximize the proportion of its genes in the next generation, to maximize its reproductive success, whether by preying on or cooperating with another species. In some cases, the trade-offs may not be obvious and it is only by unpicking the detail that we discover how a species benefits from its interactions. We begin by exploring some cooperative associations, and then look at the competitive battles between individuals and between species. In more direct conflicts, one species loses out because another exploits it—when a predator consumes its prey or a parasite thrives on its host—associations we sometimes use to control pests and disease.

4.1 Acquiring resources

There are some needs which are common to all living organisms. The two most fundamental are food and water: water, because it is the medium in which the chemical reactions of life take place; food, because it supplies both the materials for these reactions and the energy that powers them.

Food and water are often in short supply, either periodically or continually and an organism needs strategies to acquire and to conserve supplies of both. All species require space in which to secure their resources. Plants need soil and air from which to capture nutrients, water, and sunlight. Animals too need space—even sea anemones will jostle each other for room, pushing (albeit slowly) for a prime location. More obviously, dragonflies dart up and down their stretch of riverbank, defending a territory in which they feed and mate. Even a partner can be regarded as a resource, necessary to produce the offspring that will carry some of their parents' genetic information into the next generation. 'Resources' thus refers to all of those elements of an organism's habitat that it needs to survive and to procreate.

Many resources are so abundant that an organism can afford to be profligate, making no special effort to conserve supplies. Oxygen is a prime example. In contrast, energy is limited for most individuals and they have to allocate what they have acquired to best effect (Box 4.1), partitioning their supply between reproduction, growth, or further energy acquisition. Like the different reproductive strategies we have seen in various species (Section 3.5), these acquisition strategies help to define the ecological niche of a species.

BOX 4.1   Allocating resources

When a resource is in short supply, be it energy or some key nutrient, an organism needs a strategy to make efficient use of what it can acquire. Since its prime objective is to reproduce and ensure the survival of its genetic code, this means allocating its resources to maximize its reproductive success.

Often the dilemma is whether to play the short or long game. Resources could be used to produce offspring in the short term, with what is at hand. Alternatively, they might be devoted to survival now, perhaps allowing the organism to grow larger and seek further supplies that could support several reproductive events later on. Natural selection will favour different strategies in different environments, according to how predictable they are (Section 3.5, Box 3.3).

Any strategy has to balance the costs of acquisition against the potential return. A predator must evaluate the energy costs of capturing a prey against the benefit it will get from consuming it. Clearly, if the energy invested in searching for, handling, and consuming a prey is less than that assimilated from eating it, the predator will show a net loss and have less energy to devote to reproduction, or the next chase. An active predator, therefore, is continually reviewing how easily food is being found and will tend to stay where its prey is abundant. This is an unconscious cost–benefit analysis, with the consumer moving on when search costs start to rise too high.

All that we say of animals we could also say, in slightly different terms, about plants. Deciduous trees shed their leaves in winter because of the cost of respiration. Simply keeping the leaf alive during the short days outweighs the return from any photosynthesis (Section 6.2). Similarly, plants have

different strategies for investing in root growth (to acquire nutrients) or timing their shoot growth according to a variety of factors (including other species) in their environment.

In fact, any organism has to meet a series of costs that it cannot avoid, simply to stay alive (Figure 4.1). Only a fraction of the energy in its food will actually be assimilated (A)

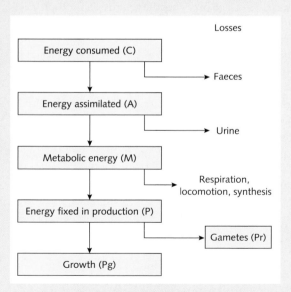

Figure 4.1   The allocation of energy to growth and reproduction in an animal. Note that the same diagram could be used to describe the allocation of any key nutrients, such as nitrogen or phosphorous.

(continued overleaf)

and then only a small fraction of this is actually fixed in the tissues (Box 6.2). Energy has to be used in metabolism, simply to maintain the tissues, in movement and in growing new tissues. This is the energy of respiration (R), eventually lost as heat. The difference between A and R can be used to produce gametes (P_r) or grow new tissues (P_g). The balance is thus

$$P_r + P_g = A - R$$

When food is in short supply and energy reserves are low, this allocation of resources becomes critical. Even if energy itself is not limiting, a lack of some key nutrient, such as water, may reduce its assimilation. Under these conditions an individual may only grow very slowly, perhaps in competition with others for the key nutrient. Long-lived species able to match their growth against resource availability will be favoured in these habitats. Their strategy is to delay reproduction until its chances of success are most favourable. Alternatively, in unpredictable habitats brief periods of plenty have to be exploited quickly, favouring species with short generation times that can produce gametes rapidly.

4.2 Cooperation

One strategy is to share some of the costs of securing a resource with the neighbours. A group of female lions combine their strength to overpower a wildebeest. On their own they have little chance of bringing down a healthy adult; but together they are successful and share the costs (and risks) of killing their prey. This works because the wildebeest represents a meal for the whole pride and far more meat than a single female could eat in one sitting. The costs and the benefits are shared. A range of animals cooperate in this way, from whales swimming in coordinated manoeuvres to confine a shoal of fish, to ants swarming over a large beetle.

This is cooperation between individuals of the same species, but we need to know how much this benefits each member of the team. One method is to measure an individual's reproductive success. We then judge the fitness of its resource acquisition and allocation strategy by the number of offspring it leaves to the next generation.

Very often, a group of animals consists of close relatives, parents and siblings (primarily sisters and daughters in a pride of lions), which, through their cooperation, improve the chances of their shared genetic code passing into the next generation. In other cases, genetic relatedness between the individuals is low but there are still benefits in cooperation.

For a wildebeest, living in a large group affords protection and reduces the costs of watching out for predators. Further, one individual in a herd of a thousand has less chance of being eaten than when it is the sole object of a predator's attention. For these animals, the benefit comes from dividing the workload, being able to feed when others work as lookouts, and the reduced chance of being attacked by a predator.

Some of the more complex cooperative strategies are best understood by looking at the genetic relationships between members of a group and the trade-offs that each is making. Working with a brother or sister can improve the prospects for the genes they share. Egg production in a colony of ants is delegated to one individual, the queen. No other females are reproductively active (Figure 4.2) and each worker is a sterile female sharing a large proportion of her genotype with her sisters and her mother, the queen. The workers support the queen who will pass on their genes.

The role of each worker is controlled by what they are fed. Particular chemicals, termed pheromones, govern the development of larvae and their subsequent behaviour. The prime source of these pheromones is the queen and these signals move through the nest via contact between workers and

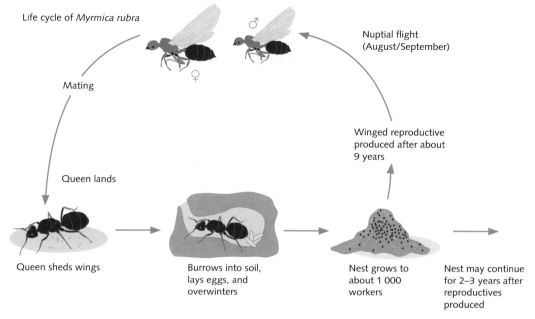

Life cycle of *Myrmica rubra*

Nuptial flight
(August/September)

Mating

Winged reproductive
produced after about
9 years

Queen lands

Queen sheds wings

Burrows into soil,
lays eggs, and
overwinters

Nest grows to
about 1 000
workers

Nest may continue
for 2–3 years after
reproductives
produced

Figure 4.2 Life cycle of a temperate ant *Myrmica rubra*, a species common in woodland and grassland areas, where it can be found in loose soil or rotting logs.

with the food. In this way, both the numbers of each caste and their activity is regulated. If the queen is lost, each female larva stands some chance of becoming queen, if fed the correct diet.

The queen controls the sex of her offspring by allowing the egg to be fertilized by sperm stored from her nuptial flight. Males are actually produced from an unfertilized egg and so, ironically perhaps, only carry the chromosomes derived from the female queen. The males become adult when it is time to form a new nest and are winged to disperse and find a female.

Although the queen has the power to decide the gender it is wrong to assume that she controls the nest. In fact, the overall coordination derives from the combination of the chemical communication system and a large number of individuals doing a small range of predictable tasks. What may appear as chaotic behaviour from viewing a small number of workers can make sense when the colony is seen as a whole: order emerges from the interaction of a large number of repetitive elements.

Together, the collective effort of the female workers ensures their shared genetic code has a greater chance of survival, better than if they wasted energy and resources competing with each other. Interestingly, all the sperm produced by a male are identical, so each worker has at least half of her genes in common with her sisters, the genes inherited from their father.

An individual worker may be dwarfed by her monarch, but she is not sacrificing herself or her genes for her queen. In fact, the genetic cost–benefit scales are tipped in favour of the code carried by the workers, not the queen. Their effort is for their brothers and sisters who will eventually leave the nest and carry the shared code into the next generation. Similar strategies are found in other insects, including bees, wasps, and termites, and they have produced some of the most elaborate and highly organized of all animal societies.

As you might expect, insect societies are also some of the largest. For example, one ant society from the Jura, in France consists of 300 million individuals

Figure 4.3 The Argentine ant (*Linepithema humile*), an alien species with a level of intraspecific cooperation that can disrupt Old World ecosystems.

house for masters whose own workers are mostly soldiers. The cost of having specialist fighters is offset by not having to produce eggs to make workers. Other species seize chemical control within a nest when the original queen is replaced because her own offspring are induced to kill her. A parasitic queen is then introduced and protected and fed by these workers, who raise her offspring. As Richard Dawkins points out, each parasitic queen incurs small costs in subverting the chemical control of the host nest, yet gains immensely as the duped workers now support the spread of their genetic code. It is an interesting question whether we should call this caste war, given that two different species are fighting over the workers.

Cooperating with different species

Cooperation between individuals of the same species may occur because both require the same resources. With two different species, however, requirements differ and cooperative associations will only form if they each derive a different benefit from the partnership, or at least the benefit derived by one causes no cost to the other.

A **mutualistic** association confers benefits on both species. Many insects have formed close mutualisms with plants, feeding on their nectar and conveying their pollen. Flowers are structures evolved to attract these go-betweens and nectar is the insect's reward. The transfer of its male gametes, the pollen, is the return for the plant. The anthers of the flower dust the insect with the pollen as its feeds, which carries this to the female parts (the style and stigma) of the next flower (Figure 4.4). Very often, the flower is simply a flag that food is available, and a variety of insects (and others) may visit to feed. Large amounts of pollen may be needed to ensure that the gametes reach a receptive flower of another plant. Besides the flower itself, the plant incurs the cost of producing the nectar, and many also produce a dummy pollen to reduce losses to the insects that will consume pollen grains.

Pollination is less haphazard if particular insects can be repeatedly attracted to the flowers of that species. A plant which only admits insects of a certain

divided between 1 200 ant-hills, covering around 70 hectares. The whole arrangement is connected by 100 km of tracks. Information, food, and larvae move down these tracks, with the whole colony divided into sectors. There are around 15–20 main nests, and a number of smaller ones used to rear larvae in the summer months. Where food is short, resources, in the form of food and labour, are moved to give a more equitable distribution.

Another, even larger colony of Argentine ant (*Linepithema humile*) has been discovered along the northern coast of the Mediterranean (Figure 4.3). Stretching over 6 000 km, from Italy to the Atlantic coast of Spain, it comprises millions of nests and is the largest cooperative unit ever recorded. We are still learning how coordination on such an immense scale is achieved. Its success here, outside of its natural range, is due to behavioural changes in *Linepithem*, including a reduction in inter-nest aggression.

Large ant colonies are known throughout the world. Some species do not make permanent nests, but instead move *en masse*, continually foraging and setting up bivouacs for the night. Some raid the nests of other ant species and have castes adapted for fighting the battles. Slave-maker ants rifle the nest, removing larvae and pupae that they then rear in their own nest. The emerging workers become slaves, keeping

Figure 4.4   A hoverfly pollinates a feverfew flower.

size, or which releases its pollen to those with a certain configuration of insect mouthparts, will pass more of its pollen to the correct flowers. By concentrating on specific messengers, plants can reduce their pollen costs and improve their chances of successful fertilization. Differences in the configuration of flower parts or the scent used to attract the insects help to ensure that only certain visitors receive the pollen or the nectar. The result is a mutualistic association, where both the plant and its pollinator evolve closely together, from which both benefit.

The yucca (*Yucca filamentosa*) from the deserts of western North America is an example of how close these associations can become. The plant can only be pollinated by the female yucca moth (*Tegeticula yuccasella*). The moth does not feed on the flower, but collects its pollen and forms it into a ball. She then flies to another flower and lays eggs in its seed buds. Thereafter, she climbs up the stigma and applies some pollen, before flying off to repeat the process in other flowers. By fertilizing the flower the moth ensures her larvae will have an abundant supply of

ripening seeds. Much of the seed will not be consumed by the larvae and so the plant benefits from the care taken by the moth. Notice also that the moth spreads her risks by not laying all her eggs in one flower. This mutualism is as complete as it is complex, for without each other, both the yucca and its moth would become extinct.

Ants make very bad pollinators. Their subterranean lifestyle exposes them to microbial pathogens and they also secrete a range of protective substances, including antibiotics, which can reduce the viability of pollen. However, an ant colony can make a very effective defence system, not least because of its high degree of cooperative behaviour. The bull's horn acacia (*Acacia cornigera*) of Mexico cultivates an association with the ant *Pseudomyrmex ferruginea* by producing protein-rich nodules at the tips of its leaves. The ants collect these to supplement the protein content of their diet. Additionally, the plant provides carbohydrate from special nectaries located at the base of its leaves and shelter in the form of large hollow thorns. Inside these the ants build nests within easy reach of food and from which they will defend their home and larder. Few herbivores will feed where busy and aggressive ants are swarming. Both ant and acacia depend on each other—they have an obligate association because each needs its partner to survive.

Plants also enter mutualistic partnerships with animals to disperse their seed. Fruit is produced to attract consumers, but only when the seed is ripe and ready to be moved. By this stage, the fruit has produced the sugars to reward the carrier, and advertises this by a change of colour. Until then, the fruit remains inconspicuous and bitter.

The close associations between insect and plant can be easily upset and this has happened in the South African fynbos (Section 5.1). Here one-third of its flora relies on ants to disperse their seed, especially those within the protea family. Proteas recruit the ants by providing seeds with little food parcels (elaiosomes) attached to them. Once home with their 'take-away' meal the ants eat the elaiosome and either leave the seed in the nest or discard it along with the colony's waste. In this way the seeds are dispersed and inadvertently sown by the ants. Within the nests, the seeds are protected from the periodic fires that surge through the low, dry vegetation and provided with nutrients in an otherwise nutrient-poor environment.

For one particular species of protea, *Mimetes cucullatus*, this dispersal strategy is under threat from the invasive Argentine ant (*L. humile*). This species has rapidly spread through the South African shrubland following its accidental introduction a century ago. It favours large-seeded proteas like *Mimetes* and simply bites off the elaiosome, dropping the seed on the ground, where it may be lost to fire or seed-eaters. Additionally, the ants nest beneath large rocks, so any seed which might be taken back to their nest have little chance of becoming established. Consequently there has been a 94 per cent decrease in these plants where *L. humile* has become dominant (Figure 4.5).

Caroline Christian has described the changes that Argentine ants are causing to the species composition of these communities, with the small-seeded proteas (which are dispersed by generalist ant species and are unaffected by *Linepithema*) proliferating. Another consequence is the displacement of native ants, espe-

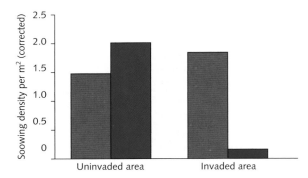

Figure 4.5   The effect of the Argentine ant (*L. humile*) on the regeneration of large- and small-seeded proteas in the fynbos of South Africa. A decrease in seedling density of large-seeded proteas (blue bars) is seen where the invasive ant is present. The small-seeded proteas (red bars) remain unaffected as *L. humile* does not compete with the ants that disperse them.

cially *Anoplolepis custodiens* and *Pheidole capensis*, which were previously associated with the large-seeded proteas.

Many mutualistic associations are not obligate (i.e. not a requirement for each of the partners) though one or both of the partners may fare less well without the interaction. One example is the association between many higher plants and soil fungi, called **mycorrhizae** (Section 7.1), in which the strands (hyphae) of particular soil fungi enter the root system of the plant. The plant benefits from the extension to its root system, improving its drought resistance and its capacity to absorb nutrients (particularly phosphates). The fungi benefit from the readily available source of energy provided by the plant.

The association between photosynthetic organisms and fungus is most highly developed in the lichens (Figure 4.6). Around 20 000 types of lichen are known, but this is another association which is not obligatory for its partners—both fungi and algae will grow in the absence of each other. Nevertheless, the association is so close that the lichen produces asexual reproductive structures (soredia) that contain both algal and fungal cells. The fungus provides a superstructure (the fungal mycelium formed of hyphae) that supports the photosynthetic partner, either an alga or a species of cyanobacteria. The

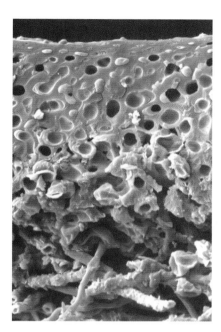

Figure 4.6 Electron micrograph of a cross-section through a lichen *Lobaria pulmonaria*. The lichen consists of a super structure of fungal mycelium (consisting of thread-like hyphal strands) with the symbiotic algal cells embedded within it.

fungus also produces complex organic acids, which dissuades consumption by herbivores, helps to protect against desiccation, and may also release minerals from its substrate the algae can absorb. In turn the fungal partner benefits from the sugars produced by the algal photosynthesis. In some cases,

the fungi may 'cull' older algal cells to absorb their nutrients. Some biologists have therefore argued that this mutualism comes close to a form of controlled exploitation of the algae by the fungi.

Not all associations benefit both parties. We saw earlier how the co-evolution of a number of species of orchid with bees and wasps has led to the flower mimicking the female insect (Section 2.3). While female bees show no interest in the flower, its sight and smell are irresistible to males who will try to mate with the flower for some time—long enough for pollen to be dumped on his back. Ecologists would regard this as an example of **commensalism**, an association in which only one partner, in this case the plant, benefits. For the bee, the benefits are more apparent than real, but whilst a commensal relationship may actually incur little or no costs on the unwitting partner, they may be essential for the other. For example, the brown-headed cowbird (*Molothrus ater*) follows grazing livestock, feeding on the invertebrates they displace. The birds benefit from the activity of the grazers, but the livestock incur no costs from the presence of the cowbird. Similarly, scavengers such as vultures benefit from the killing power of large carnivores, feeding at no cost to the predator that has since left the remains of the kill.

Commensalism is often a loose association: for the cowbirds the species of grazer it follows is immaterial. Similarly, many dung beetles are not fussy about whose dung they lay their eggs in . . . although others are very discerning indeed.

4.3 Competition

Competition is central to the theory of natural selection since the acquisition of resources (food, shelter, partners, and so on) determines who gets to reproduce. The fittest survive and secure the means to produce offspring which themselves mature into adults.

Competition between individuals of the same species is termed **intraspecific competition** (Section 3.2, Box 4.2). **Interspecific competition** is between individuals of two different species. When resources

are limited, the best strategy may simply be to take as large a share as possible, though it can be more subtle and may take different forms. One species may simply dominate a resource, say a water supply, either by depriving others or by making more efficient use of what it acquires. Interference competition occurs when an individual prevents another from exploiting the resource by its activity or behaviour. In scramble competition all individuals have equal access to a shared resource and there is a free-for-all to gain a share.

These competitive battles can be classified according to the way they are fought (Table 4.1).

Intraspecific competition

With intraspecific competition, direct combat over a resource, be it a mate or food, is both costly and carries considerable risk. Only when the stakes are high—perhaps when a harem of females is to be won—will competitors actually lock horns. Death rates amongst adult males can then be significant. For this reason, many species have mechanisms to diffuse such confrontations, using size, coloration, or ornamentation to signal the status of the owner or their position in the hierarchy. Even so, the horns of many large mammals are not just elaborate ornaments but have evolved for defence and attack (Figure 4.7).

TABLE 4.1 **Types of competition**

Chemical competition

Production of a toxic or deterrent chemical to exclude competitors

Consumptive competition

Competitive use of a renewable resource (e.g. food)

Encounter competition

Physical defence of a resource, possibly by aggression

Overgrowth competition

Successful competitor overwhelms the opponent in size or number

Pre-emptive competition

Rapid colonization of space when it is available

Territorial competition

Defence of territory, breeding, and feeding areas

Figure 4.7 Musk oxen (*Ovibos moschatus*) in a head-to-head confrontation. Horns serve several purposes. Their size and shape are primarily determined by the battles between males for a female. They are also an indication of the status of the male in the hierarchy of the herd and are recognized by both males and females.

There are also striking examples of intraspecific competition among plants. Competition between seedlings and young plants for space, light, and nutrients is fierce. Many plants produce vast numbers of seed but only a small fraction will grow to maturity. The horseweed (*Erigeon canadensis*) sheds 100 000 seeds per square metre, to produce, perhaps, a thousand seedlings of which only a few will become mature plants. Sometimes the end result are regularly spaced adult plants. The creosote bush (*Larrea* spp.), from the deserts of North America, is so uniformly spaced that it appears to have been deliberately planted. This spacing is entirely natural. Each plant occupies a uniform area from which it collects its resources and which it may defend by secreting an inhibitor to suppress the growth of its neighbours. This type of interference competition using chemistry is termed **allelopathy**.

Chapter 3 explored how limited resources result in density-dependent population growth and the stabilization of a population's size around a carrying capacity (Section 3.2). We have also seen how differences within a population might, with some degree of isolation, lead to character displacement and the formation of a new species. Specialization on a particular part of a resource spectrum can then result in two new niches being defined which may lead to speciation (Section 2.4; Case study 2). Populations of the common garden snail (*Cantareus aspersus*) adapting to local conditions may be producing distinct ecotypes (Box 4.2). In doing so, some populations have resource allocation strategies, behaviour and growth patterns that make them more likely to win an intraspecific competition for limited food.

BOX 4.2   **Building a bigger and better house**

Competition between neighbours, intraspecific competition, can be subtle but it can also be very obvious. Often it is the larger individual that secures the resource and it is their genes that pass into the next generation. With so much at stake, competitors frequently evolve a range of devices and strategies to give themselves an advantage. For example, the common garden snail, *Cantareus aspersus*, suppresses the feeding of other individuals with the slime they secrete as they move around (Figure 4.8).

All snails need large amounts of calcium to build and reinforce their shells, their principal means of protection from predators and from water loss. Without calcium, the proteinaceous horn on the outside of the shell is thin, weak, and little more than camouflage. A properly reinforced shell will protect the soft tissues and serve as a reserve of calcium for a range of physiological functions. If calcium is not available, an individual has to balance the various demands on its limited supply—between strengthening the shell, courtship, producing eggs, and several key metabolic processes. If calcium is abundant, young snails will grow large, reproduce quickly, and have robust shells.

Interestingly, it seems the rate at which the shell is reinforced is genetically determined and different populations have different priorities for their calcium intake. Populations from sites where calcium is abundant tend to produce thick, strong shells even when fed a low-calcium diet. Perhaps there has been no selective pressure for these snails to be more frugal in their allocation. Others, typically from calcium-poor sites, are more parsimonious, producing relatively thin shells even when fed a calcium-rich diet.

How do these strategies compare when individuals from different populations compete against each other? In a series of trials, Larry Richmond and Alan Beeby have found one population that tends to beat all comers in a race when their calcium supply is limited. Individuals from the chalk downland of southern England grow larger and denser shells than those from calcium-poor sites, when these are competing directly for a limited supply of food. The chalkland snails may simply be quicker to the food, thereby reducing the calcium available for their neighbours. Or perhaps, they make better use what they consume.

(continued overleaf)

Figure 4.8 Where the swift and the strong survive—calcium is a valuable resource that determines the outcome of competition between individual common garden snails (*Cantareus aspersus*).

In most competitions there was no predictable outcome and no consistent winner. Invariably each population would have one or two rapidly growing individuals with the majority being quickly left behind. It seems that an early spurt of growth secured the later success of these faster growers. The chalkland population, however, adopt a different calcium allocation strategy, reinforcing their shell at the expense of their soft tissues. This may be a response to the hot and dry conditions that develop on the downs in the summer, possibly to allow them to aestivate or reproduce quickly before soil conditions become too dry for feeding or for their eggs to hatch.

Alternatively, it is their competitors that show the important adaptation. Several of these populations were collected from highly polluted habitats, where individuals may have been selected for their capacity to tolerate toxic metals. Calcium is used by a range of invertebrates to isolate and excrete excess metals, in particular zinc and cadmium. It may also be used to accelerate the loss of lead, even though this could reduce the calcium that can be added to the shell. In such habitats, slow shell growth may be a necessity.

Again, both explanations may apply. On a high lead diet, the chalkland juveniles accumulate higher concentrations in their soft tissues and have the biggest reduction in shell growth rate. By comparison, a population with a long history of exposure to lead, from North Wales, always grows shells slowly, and this changed little when the toxic metal was added to their diet.

What makes the race more interesting is the limits to shell growth across the two populations imposed by the amount of calcium available. Over 100 days (covering their fastest shell-building phase), 10 juveniles competing for a fixed amount of food collectively increased their shell height by a total of 18 mm. This total shell growth was the same, no matter which two populations were competing, and even if one population dominated. It seems that the weight of calcium available sets an upper limit on the total shell build. Which individuals grew the most shell probably depended on who got to the food first. The chalkland snails may just be faster than most.

Figure 4.9 Bracken (*Pteridium aquilinum*), an invasive fern found worldwide, seen here invading a colony of thrift (*Armeria maritima*). In this photograph *Armeria* still dominates.

Interspecific competition

In the best locations space is limited. Plants and sessile animals are immobile and compete by pre-emptive or overgrowth competition to swamp out potential and existing competitors. Bracken (*Pteridium aquilinum*) is a highly invasive weed that uses this form of inter-ference competition. Its dense foliage and deep litter force out existing plants and prevent others from be-coming established, forming a uniform carpet where few other plants are visible (Figure 4.9). Its success has led to bracken becoming a global pest.

In an extended competition between two species for limited resources, natural selection can produce either of two outcomes:

- Each species becomes more highly adapted to one part of the resource spectrum and shows the char-acter displacement of a specialist. This may lead to a progressive reduction in niche overlap, which may eventually result in complete niche separation (Section 2.5).

- One species extends the range over which it feeds, making more efficient use of that resource than its competitor. It may do this by interfering with the growth of its competitor or by simply producing more offspring per unit of resource. In this way it displaces its competitor.

The second outcome is a demonstration of the com-petitive exclusion principle (Section 2.5): where there is significant overlap between two species for a key or limiting resource, the less efficient species will be lost from the community. As long as no other factor defuses the fight, two species cannot occupy the same niche at the same time. A number of experiments using micro-organisms, plants and cereal beetles, have confirmed the competitive exclusion principle, at least in the simple ecosystems created in the laboratory (Figure 4.10).

Sometimes, these laboratory fights were resolved by the winner reproducing fastest (resource competi-tion) or by the winner eating the opposition, or at least their larvae and pupae (interference competition).

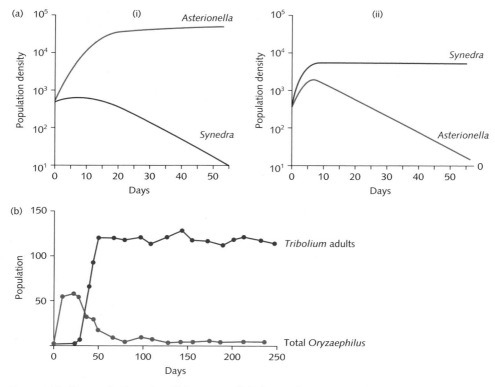

Figure 4.10 The results of a series of laboratory trials looking at the competitive battles between two species that occupy similar niches. (a) Between two diatom species (unicellular algae with a siliceous case), *Asterionella* and *Synedra*, in cultures where silica is limiting. (i) In most trials between the two species *Asterionella* won, whether *Synedra* had a higher initial population or not; (ii) however, when the temperature was raised to 24 °C *Synedra* won. (b) Between two cereal beetles, *Trilobium* and *Oryzaephilus*. These species predate each other's larvae and pupae when placed together in a simple flour ecosystem. *Trilobium* won most battles because it ate more of the opposition. Coexistence was possible if *Oryzaephilus* larvae had a refuge from adult *Trilobium*. In this example, both species start with the same population size, and the growth in the *Trilobium* population is matched by the decline in *Oryzaephilus*.

The result of each battle was not always guaranteed, since changing the conditions of the experiment could sometimes reverse the result. Clearly, the outcome of these struggles depended on a range of factors, some external to the competitors.

We also have evidence of competitive exclusion from the real world when individuals from two species with similar niches meet for the first time, often following an introduction by humans. This has happened with the American grey squirrel (*Sciurus carolinensis*) in competition with the native Eurasian red squirrel (*S. vulgaris*) in the United Kingdom and northern Italy (Case study 4).

Whilst introduced species sometimes become pests and out-compete the native species, more often the introduction fails to succeed, presumably because they are less well adapted to their new habitat. This is not true of the common house gecko (*Hemidactylus frenatus*), which has spread across several Pacific islands in the last 60 years. Geckos are lizards able to cling to vertical surfaces and are commonly found on the walls of houses in the tropics, where they feed on insects attracted to the lights. *H. frenatus* (Figure 4.11) is native to Asia but was introduced into Hawai'i in the 1940s. Thereafter three other gecko species, previously introduced from elsewhere

Figure 4.11   The common house gecko, *Hemidactylus frenatus*.

in Polynesia, all suffered major population reductions in suburban areas. However, another Polynesian species, the mourning gecko (*Lepidodactylus lugubris*), survives well in these drier habitats and seems to be holding on, even though its numbers have been reduced. The eventual outcome may depend on interference competition again, since *Hemidactylus* is known to eat *Lepidodactylus* juveniles, at least in laboratory experiments.

Hemidactylus appears to be winning most of the battles in other parts of the Pacific. Its numbers have risen on Suva (Fiji) at the expense of other geckos. It was unknown on Vanuatu until 1971 but has since become the most abundant species over much of the island. Ted Case and his colleagues recorded the arrival of *Hemidactylus* in Tahiti in 1989, noting it had spread beyond the docks in a 10-km radius in just two years. Case and Kenneth Petren have since shown that urbanization has promoted the success of *Hemidactylus*. *Lepidodactylus* prefers complex habitats in which it can stalk its prey. The relatively simple, open habitats inside modern buildings favour *Hemidactylus'* more pursuit-driven mode of hunting.

Although the confrontations between *Hemidactylus* and its rivals are far from resolved, its successes and occasional failures show that competitive exclusion is not the only possible outcome of interspecific competition. In a variable environment, when conditions might change or where there is scope for avoiding competition, coexistence might be possible. Unfortunately, the success of *Hemidactylus* over much of the Pacific also demonstrates how readily species

can be lost when two competitors meet for the first time. Our propensity for global travel has increased the frequency of such unplanned meetings between native and introduced species.

Competitive types

Highly invasive species, such as the rat, often have a particular set of characteristics that fit them to their way of life. They approximate to the *r*-selected type (Section 3.5)—a species capable of rapid reproduction and dispersal, able quickly to exploit a range of opportunities. Species of this type can accommodate a range of habitat conditions and will grow and reproduce quickly to exploit opportunities where there is an absence of competition. The true *r*-selected species allocates much of its resources to a single reproductive event, whereas *K*-selected species will reproduce repeatedly and so spread their costs and risks.

The *r*-selected life history is not confined to small organisms with short lifespans and there are several examples of weedy and invasive trees, including birches (*Betula*) in temperate regions, and balsa (*Ochoroma lagopus*) in the tropics. Balsa will rapidly colonize open areas and exploit gaps in the forest canopy. The young trees grow quickly, producing a light wood with few chemical defences, to exploit the available sunlight and soil nutrients. On maturity, they produce numerous tiny seeds, a fraction of the weight of those from the competitive trees that dominate mature tropical forests.

The *r*–*K* division is not a complete description of life history strategies but rather a *continuum* between which there are many intermediate forms. Some species adopt different strategies at different stages in their life cycle. The century plant (*Agave americana*) is an example. This plant has a lifespan of decades (a *K*-selected trait), but its reproductive strategy is decidedly *r*-selected, as it flowers only once, after which it dies. Whilst the *r*–*K* continuum is a useful concept to help us understand resource and reproductive strategies, it is not one to which all organisms are closely tied.

As Richard Southwood has pointed out, life history strategies are selected not only by the predictability of the environment and the availability of resources, but also according to the presence of stressors—such

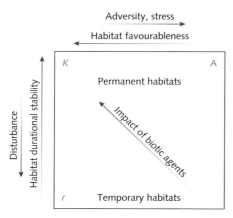

Figure 4.12 Southwood's habitat template. Life strategies are classified in terms of habitat stability and favourableness. The top left-hand corner represents habitats that are stable and favourable, the top right are stable but unfavourable conditions, and the lower left-hand corner unstable but favourable habitats. Habitats can be classified according to these three dimensions and species can also be placed according to the type of habitat they are commonly found in.

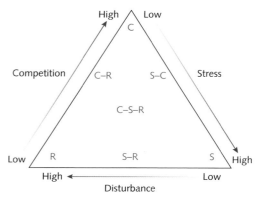

Figure 4.13 Grime's classification of resource acquisition strategies in plants. Ruderals (R) are roughly equivalent to *r*-selected weeds (e.g. *Stellaria media*). They are found in habitats that are regularly disturbed and are typically fast growing and short-lived. Competitors (C) are found in undisturbed habitats where resources are abundant. Their strategy is to try and crowd out other plants by rapid growth (e.g. stinging nettle, *Urtica dioica*). Stress tolerators (S) are slow-growing species that can tolerate shortage or other forms of stress in the soil; they prevent many other species from establishing themselves (e.g. sheep's fescue, *Festuca ovina*). Between these are various intermediate types.

as toxins and other factors which might limit growth. There are species that avoid having competition by occupying marginal habitats where these stressors are more prevalent, with adaptations allowing them to survive the hostile conditions. Southwood called this **adversity selection** (A), one of three dimensions in his 'habitat template' (Figure 4.12). *K*-selected species are the competitive species which dominate in favourable and stable environments, whereas *r*-selected species rapidly colonize unstable environments—provided there is an adequate supply of nutrients.

Philip Grime has developed a similar three-way classification based on plant-growth strategies. Again he notes that some plants are adapted to live in poor, highly stressed environments (Figure 4.13). Where resources are limiting, or there are other stresses (such as toxic metals or extreme soil pH), a species may only be able to grow slowly. Such species are termed **stress-tolerators** (S). Because few other plants can withstand these conditions they have few competitors and slow growth is not then a disadvantage. In contrast, **competitors** (C) outgrow or crowd out the opposition where resources are abundant. The final category is equivalent to the weedy-type, *r*-selected species, here classified as **ruderals** (R). These live in disturbed and unpredictable environments but where resources

may be occasionally abundant and available. They make poor competitors and so their strategy is to produce their seed before the competition arrives.

Again, there are intermediate forms between the three main types, indicating that there is a gradation between the extremes of each type. This scheme (known as the C–S–R model) emphasizes a plant's allocation strategy in relation to the available resources, the frequency of disturbance and adverse conditions, rather than its reproductive or dispersal strategy within a particular habitat.

These demands will change during the organism's life cycle, and the conditions under which a seedling flourishes may be very different from those demanded by a mature tree (Section 3.5). For many adult plants the worst of the competitive fight was fought long before it became established. In a garden lawn, for example, species of grasses and broad-leaved plants form a tight-knit community, a difficult place for young seedlings. Not only is there fierce competition for water and minerals, there are other constraints such as space and access to light. Even when gaps appear, seedlings have to be tolerant of trampling, soil compaction, and the regular 'grazing' of the lawn mower. Gaps with different conditions favour different species. Dry compacted gaps favour plantains (*Plantago* spp.), moderately damp areas daisies (*Bellis perennis*) and the wetter areas buttercups (*Ranunculus* spp.). Occasionally, in some lawns at least, the gaps will actually be colonized by grass, but these species also vary according to the nature of the gap. Fine-leaved grasses such as bents (*Agrostis* spp.) and fescues (*Festuca* spp.) fare better than annual meadowgrass (*Poa annua*) where water and nutrients are in short supply.

Yet the resources demanded by these different species are relatively similar and many are in direct competition with their neighbours for whatever is available. This presents us with something of a paradox: why should there be so many different plant species, even within the small area of a lawn, when they all require more or less the same of their environment? Furthermore, how do they manage to coexist without one or two species becoming dominant and excluding the rest?

David Tilman offers one possible explanation. He suggests that the amount of a resource needed by a species, relative to its availability in the environment, is critical: coexistence is possible when each species is prevented from dominating a resource completely, when the growth of each competitor is checked by a shortage of another resource. Let us say that the outcome of a competitive battle between individuals of two plant species depends on two key resources—for example, water and a nutrient. For each resource, either species would survive if its minimum requirement was met, but it cannot dominate the supply because its growth is checked by a shortage of the second factor. According to Tilman, coexistence is possible if each species is limited by a different factor, though each gets most of that which it finds most limiting. Both can then sustain population growth. Because different resources limit them, each can persist but will leave enough of a shared resource free for its competitor. This might help, in part, to explain the remarkable diversity of plants in the tropical forests, on soils that are relatively infertile (Section 9.5). Indeed, there is no reason why this principle should not also extend to animal communities.

Much of the diversity of a plant community derives from the co-evolution between the plants and their associated animals. The evolution of higher plants has been intimately linked with that of terrestrial animals, as pollinators, seed dispersers, and herbivores. Its dependence on animals may lead to a plant being only locally abundant or declining at particular times. Equally, the pressure of being grazed or browsed may well be important in preventing it becoming dominant.

Animals too can be checked by their consumers: the attentions of a predator or parasite may prevent competitive battles reaching a definitive conclusion. Overall, the abundance of any species will respond to a range of key factors, including other species, many of which are continually changing. This is one reason why communities are dynamic—variable conditions favour different species at different times. Some species are abundant because of their past successes, others are about to have their moment. We should see the forest or the lawn as an on-going competitive battle, with no foreseeable end and no winner.

Some species evolve extreme adaptations which enable them to survive in extreme environments, away from competition. Here, the costs of evolving physical structures and physiological processes are outweighed by the benefits of living in a competitor-free environment. The scuttle fly (*Megaselia yatesi*) is an example (Figure 4.14). Discovered in 2000 by Barry Yates, the fly is a mere 2 mm in length and lives in the shingle ridges at Rye, on the south coast of

Figure 4.14 *Megaselia yatesi*—the scuttle fly that triumphs in adversity.

England. Yates' scuttle fly is well-adapted to life underground. Its small wings allow it to navigate the gaps between the stones and the constant temperature, humidity, and lack of predators mean it can produce eggs with unusually thin walls, since there is little risk of these drying out or being eaten within the shingle. The resources saved can be channelled into growth and further egg production.

4.4 Consumerism

Herbivores consume plants and carnivores eat animals. Herbivory might be regarded as a form of predation (one organism eating another) but since it may not lead to the death of the plant we keep the distinction. We might also consider the action of parasites or even pathogens as a form of predation, but again the host may not be killed. As ever, we are faced with variations on a simple theme, where the demarcations between one category and another are sometimes arbitrarily drawn.

Herbivory

All plants have to maintain some sort of balance between their leaves and their roots: the leaves are net consumers of nutrients and net producers of energy-rich sugars, and the reverse is true of the roots. If this balance is not maintained then either respiration exceeds sugar production (when too many leaves are lost) or photosynthesis is inhibited by a lack of key nutrients (when roots are lost—Section 6.1). When they are eaten, plants alter their growth pattern to restore this balance, so leaves or roots are replaced as necessary.

One strategy is simply to grow quickly and produce seeds as rapidly as possible, before being consumed by herbivores. This is the strategy of ruderals, the weeds common in habitats where there is a relatively high frequency of disturbance. An example of this is chickweed (*Stellaria media*): rarely found in the closed community of the lawn, *Stellaria* is, however, common in the frequently disturbed borders around it.

An alternative strategy is to make a more substantial investment in leaves, roots, and shoots, perhaps as a biennial (living for two years) or a perennial (living for several years), possibly with more than one re-productive event or with some provision for asexual reproduction (such as a tuber or a runner). Another is to protect the tissues from herbivores or dissuade their attack. Plant defences are many and include both physical, chemical, and biological methods.

Spikes, thorns, and stings can be effective against larger herbivores and unsuspecting people, but are less effective against many invertebrates. Finer hairs, sometimes with sticky secretions, can prevent small insects from getting close to the leaf surface. A woody trunk protects against slug and fungal attacks at the soil surface. Tough silicate deposits on the leaves of some grasses limit the range of animals prepared to tackle them. Others simply toughen their leaves and deposit distasteful chemicals, such as tannins and phenolics, within them.

Plants contain some of the deadliest chemicals known (Table 4.2). These are known as secondary plant metabolites and have probably evolved from waste products deposited in leaves, to be lost when the leaf was shed. If herbivores found such compounds unpalatable then there would be some selective advantage in concentrating them in living leaves in need of defence.

Plants appear to allocate their chemical defences according to their growth pattern, and lower their production when no longer browsed, presumably to avoid the cost of production. Indeed, the presence of secondary plant metabolites can indicate if a plant is currently or was recently under attack. Thus, even if these chemicals started out as metabolic waste products, they must have since been refined by natural selection to be produced on demand.

This would explain the vast range of chemicals synthesized. A large variety of alkaloids, terpenes, phenolics, and others are responsible for a myriad of

TABLE 4.2   Secondary plant products

Class	Number of compounds	Occurrence	Physiological effect
Nitrogen compounds			
Alkaloids	10 000	Found in many flowering plants and roots, leaves, and fruits	Toxic and unpleasant tasting
Amines	100	Found in many flowering plants principally in flowers	Unpleasant smelling, some hallucinogenic
Amino-acids non-protein	400	Found in seeds of many flowering lants, particularly legumes	Toxic
Cyanogenic glycosides	30	Occasional (e.g. clover, Rosaceae, rose family)	Toxic (hydrogen cyanide)
Glucosinolates	75	Cruciferae (cabbage family) and other plant families	Bitter tasting, acrid smell
Terpenoids			
Monoterpenes	1 000	Found in many flowering plants, principally as essential oils	Strong smelling (not unpleasant)
Sesquiterpene lactones	5 000	Mainly in Asteracea (daisy family) and other groups of plants	Toxic, bitter, allergenic
Diterpenoids	2 000	Found in many flowering plants mainly in latex and resins	Toxic, sticky
Saponins	600	Widely found (in 70 plant families)	Haemolytic–damages blood cells
Limnoids	100	Mainly in the Rutaceae (*Citrus*), Meliaceae (mahogany), Simaroubaceae and Quassia families	Bitter-tasting
Cucurbitacins	50	Cucurbitaceae (cucumber) family	Bitter-tasting and toxic
Cardenolides	150	Mainly in Aponcynaceae (periwinkle), Asclepiadaceae (milkweed), and Scrophulariaceae (figwort) families	Bitter and toxic
Carotenoids	500	Widespread in fruits and flowers	Coloured pigments
Phenolics			
Simple phenols	200	Widespread in plant tissues	Antimicrobial
Flavenoids	4 000	Widespread in higher and lower plants	Coloured pigments
Quinones	500	Widespread (particularly in Rhamnaceae–buckthorn family)	Coloured pigments
Other			
Polyacetylenes	650	Mainly in Asteracea (daisy) and Umbelliferae (umbellifer) families	Some toxic

effects on herbivores. Not only does their taste seem to dissuade some herbivores, they also act as anti-feedants, nerve poisons, carcinogens, and, in human beings at least, hallucinogens. Some members of the clover and rose family release deadly hydrogen cyanide when chewed. Phenolic compounds and tannins bind to digestive enzymes, reducing protein assimilation. Animals grazing such vegetation lose condition and, once weakened, begin to lose their own competitive battles. Whilst this deters many herbivores, others have hijacked these defence mechanisms for their own purpose (Box 4.3).

 BOX 4.3   Objectionable behaviour

Caterpillars are soft-bodied, often fat, always slow-moving eating machines, and would seem to be an easy target for predators. But this vulnerability has led them to evolve a variety of ways to avoid being eaten. Some produce a mass of irritant hairs that defy all but the most persistent predator. Others swell their bodies or produce sticky strands to make them awkward to swallow. Some flash false eyes to appear part of something larger and able to defend itself. Others use cryptic coloration to camouflage themselves as leaves or twigs. Some use coloration to advertise the toxic threat they pose to potential predators (Table 4.3).

 TABLE 4.3   Animal defence strategies

Aggression

Threat behaviour, intimidation displays, and overt aggression are used to frighten off potential predators or competitors.

Aposematic coloration

Warning colours signal danger. These tend to be stark colours and patterns that predators associate with stings, poisons, or being distasteful.

Armour

Being tough or difficult to handle deters attackers and protects against potential injury.

Chemical defences

This can include passive defences such as the secretion or accumulation of poisonous and distasteful compounds within or on the body, or active defences in the case of venomous or unpleasant bites, stings, and sprays.

Crypsis (camouflage)

Merging into the background to avoid the attention of predators. Catalepsis (frozen posture) adds to the effectiveness of the defence.

Flocking, herding, and shoaling

Individuals group together to reduce their individual chances of being attacked. A large group is also less likely to be attacked if it intimidates or distracts potential predators.

Masting

The production of numerous progeny overwhelms the functional response of a predator.

Mimicry

Animals mimic living and non-living things. *Batesian mimicry* involves innocuous prey impersonating either a dangerous or an unpalatable organism. *Mullerian* mimicry reinforces aposematic coloration. Dangerous and distasteful organisms independently evolve similar forms of patterns and coloration.

Polymorphism

Groups of populations within a prey species avoid being eaten by looking sufficiently different from most of their species to go unrecognized by predators.

Bright and stark coloration signals that the owner is distasteful, can sting, or will poison. Once it is experienced, a predator will readily associate those markings with an animal to be avoided. Both vertebrates and invertebrates frequently use yellow and black stripes as a warning. The stripes of the cinnabar moth caterpillar (*Tyria jacobaeae*) signal the poisonous alkaloids it has accumulated from its diet of the ragwort (*Senecio jacobaea*), and the yellow-black-white of the monarch caterpillar (*Danaus plexippus*) flag the alkaloid it has concentrated from milkweeds. This poison causes the adult monarch butterfly to be readily disgorged by any bird that tries to eat it. The stark orange and black pattern of its wings is shared with another butterfly, the viceroy (*Limenitis archippus*), a species also distasteful to birds. Together the monarch and viceroy represent an example of **Mullerian mimicry**, where the common pattern means that predators more quickly learn to avoid any butterfly with these markings.

In contrast, some species bear the warning coloration but avoid the costs of producing or concentrating any poison in its tissues. This is termed **Batesian mimicry**. Hoverflies that sport a black and yellow livery are often mistaken for wasps even though they have no sting. Edible species copying distasteful species may save resources in the short term, but they also risk lessening the impact of the signal on a predator. If their numbers are large, and a predator has few or no encounters with the original distasteful species, an aversion to the warning coloration may not develop. Given long enough, natural selection will then favour predators not fooled by Batesian mimics.

When stripes and false eyes fail, the defensive behaviour of caterpillars can be equally colourful. Some will vomit on their attacker. As if being pelted with the half-digested contents of its gut is not bad enough, what was earlier on the caterpillar's menu can make matters even worse. For example, one species feeding on the coca plant (*Erythroxylum coca*) will shower their assailants with a cocaine-laced vomit.

Stephen Peterson studied similar behaviour in Eastern tent caterpillars (*Malacosoma americanum*) (Figure 4.15). These seek out young leaves of the black cherry (*Prunus serotina*) and then carefully mark out pheromone trails leading to the leaves with the richest supply of a compound called prunasin that generates cyanide. When attacked by ants, the caterpil-

Figure 4.15   Chemical attack! The Eastern tent caterpillar (*Malacosoma americanum*) bombards its assailants with cyanide-laced vomit.

lars will regurgitate the part-digested leaves, releasing the cyanide.

This strategy can have a much wider impact. In the spring of 2001, a spate of sudden and unexplained miscarriages in horses on central Kentucky stud farms was eventually attributed to a population explosion of *Malacosoma*. Terrence Fitzgerald analysed the potential concentrations of cyanide consumed, regurgitated, and excreted by the caterpillars. He found they actively chose young leaves with the most hydrogen cyanide, and between 63 and 85 parts per million (ppm) of cyanide could be found within their dried faeces. Whilst this may not be toxic on its own, Fitzgerald suggests that combined in quantity with leaves and young seedlings of *Prunus serotina* it could well prove harmful to vulnerable grazers and browsers.

The co-evolution of defences between plants and their herbivores is as impressive as that between plants and their pollinators. For example, ecdysone, the hormone that initiates moulting in insects, is produced by bracken, upsetting the moult cycle and maturation of most insects that feed upon it. Other plants produce precursors of hormones able to disrupt the reproductive cycles of the mammals that feed on them. We have used this for our purposes: the first contraceptive pill was based on naturally produced progesterone from Mexican yams. We derive several thousands of pharmacologically active compounds from plants, and traditional medicines based on such extracts are still the primary form of treatment for many peoples.

Plants producing secondary metabolites tend to be long-lived with many coming from relatively undisturbed habitats. They are typically competitive or stress-tolerant species, such as those of the Mediterranean maquis (Section 5.2) adapted for a long summer drought. Typically low bushes, most retain their leaves over several years. Species such as thyme, rosemary, and lavender produce a range of secondary metabolites which repel herbivores yet are attractive to us as flavourings or fragrances.

Some plants are far from passive in their interactions with animals. Carnivorous plants, more properly the insectivorous plants, are typical of environments where nitrogen is in short supply, often very damp and swampy places. Nitrogen is abundant in the insects that visit their flowers to feed and, using various mechanisms, several species have found ingenious ways of exploiting this nutrient source (Section 7.1).

Predation

The most direct effect of a predator on its prey is to reduce its numbers. But if this prey species is the only food source for the predator then falling prey numbers will in turn cause its population to fall. Who then is in control—the predator or the prey?

A series of mathematical models, based on the logistic equation (Section 3.2), have been used to examine predator–prey relationships (Figure 4.16). In their simplest form, these models assume that the predator population can only grow if there is

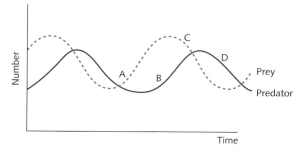

Figure 4.16 Predator–prey relationships. Changes in the population of a predator and its prey based on mathematical models where the predator exploits one prey species only. At point A the prey can increase because of low predator numbers. Predator numbers begin to rise at (B) as their prey become more abundant. Eventually predation checks the population growth of the prey (C). The decline in the prey is followed by a decline in the predators (D) and the interaction begins to cycle again.

enough prey to support its reproductive effort. When prey numbers are high, the predator has abundant resources from which to produce offspring. But any increase in the predator population must mean that more prey are consumed. If the reproduction of the prey species cannot match the losses to the predator, the prey population must decline.

In turn, this means that food eventually becomes limiting for the predator. Then its population growth is checked, and as the prey population declines further, predator numbers too start to fall. Once again, a point is reached when predation no longer limits the prey and the prey population can start to grow. Their increased numbers eventually support growth in the predator population and so the cycle repeats itself as the two populations stay locked in their deadly waltz.

In effect we are running two population models together, where the capacity for growth in one is determined by the population size of the other, and vice versa. In these simple terms, we can see the answer to our question is that prey and predator control each other. Because these populations begin out of step with each other, they continue to oscillate around each other, never settling down to constant population levels. If we do impose other checks (such as harvesting) on the population growth of either species, then the oscillations begin to disappear.

In fact, it is hard to show such cycles exist in nature and even more difficult to create them in the laboratory. Simple bench-scale ecosystems invariably lead to the predator eating all the prey and starving itself to death. Adding refuges for the prey, where some can escape predation, allows for oscillations to become established, but they rarely last more than a few generations before one or other of the populations goes extinct.

So, what is it about real ecosystems that allow predator and prey to coexist? Complexity is part of the answer—adding detail to the models helps to defuse the simple relationship between predator and prey numbers.

- First, we should not assume that each predator or each prey is equivalent. Some predators are better killers than others and some prey are more likely to get eaten. We know that predators often take the weak and the infirm, including older prey that may have already ceased reproducing.

- Second, many predators do not confine themselves to one prey species. Predators are often opportunists and will take what is easily available to reduce their costs in finding and acquiring their food. Then a predator population can maintain itself when one prey species is in short supply. By the same token, a predator will compete with several other species to utilize a prey resource. Together, this complex of interactions can mean the abundance of either the predator or the prey is not closely coupled to the other but reflect changes in the larger community.

- Third, a predator does not only respond numerically to an increase in prey numbers. Each individual may consume more (termed the **functional response**) and will only later produce more offspring (the **numerical response**). If the predator shows only the functional response, then prey numbers will decline without any change in predator abundance. Most often, the predator will respond both functionally and numerically, but each type of response will have delays associated with it. It may take time for the predator to increase its feeding rate (and perhaps switch its attention from an alternative prey). The significance of such time lags is often seen in major insect outbreaks in temperate forests, where considerable damage may be done before birds begin to feed exclusively on the pest.

- How quickly a glut of food translates into new offspring depends on the generation time of the predator, or whether there are young alive whose survival depends on the food supply. An increase in insect abundance may not translate into an increase in the adult population of the forest birds for several months or even a year.

- Finally, real ecosystems are not uniform in space but have patches where prey may accumulate or where they may escape predation. To reduce its search costs, a predator will stay where prey is abundant or of higher nutritional quality (Box 4.4). Thus some prey goes undiscovered or undevoured. Escapees are crucial if both the prey and the predator populations are to persist.

Taken together, these different factors mean that predator and prey numbers may not be closely linked, and explain why the simple oscillations of these models are rare in nature. Nevertheless, there are plenty of real world examples where predator abundance is the best explanation of variations in prey numbers and conversely, some where predator abundance follows that of one of its prey (Box 4.4, Figure 4.17).

Although predator and prey numbers are linked, we should also recognize that prey rarely sits around waiting to be picked off by a predator. A feeding aphid may seem oblivious to the ladybird consuming its sister a centimetre away, but neither it nor its genes have given up the battle. As for the ants, their individual fate is almost immaterial because their genetic code will be passed on by any of their sisters, mothers, or daughters that avoid being consumed.

In self-defence

The stakes are not the same for the predator and the prey. Whilst a predator may give up on a chase and only be short of breath, the prey has much more to lose. Richard Dawkins and John Krebs call this the 'life–dinner principle': for one it is a matter of life or death, for the other it is just a matter of dinner or no dinner. Although a predator cannot keep failing indefinitely, it need not win every chase. The prey,

BOX 4.4   **Seals wax and wane with the bears**

The picture of a predator population chasing prey numbers up and down their peaks of abundance suggests a simple relationship between consumer and consumed. The reality, as you might expect, is far more complex, even when the assumption made in the simplest models—of a single predator feeding on a single prey species—is largely met. Take, for example, polar bears (*Ursus maritimus*) feeding on ringed seals (*Phoca hispida*).

Estimates of both populations in the Beaufort Sea of the Canadian Arctic in the mid-1970s and 1980s provide the close correlation we might expect from the traditional predator–prey models. Indeed, the numbers of one seem to be a reliable predictor of the other. Ian Stirling and Nils Are Øritsland used aerial surveys of seals resting on the ice and counts of bears to estimate their population sizes, and suggested that the close match indicates some sort of equilibrium between the two. However, their work also showed these simple counts hide important details.

When the ice breaks up in the summer the bears cannot feed. Consequently, predation rates vary according to season: a bear will kill a seal on average every three days in April and May, but this falls to once every five days in July and probably to none at all at the height of the summer in August and September. Overall, the average bear is estimated to kill 43 seals each year in the Beaufort Sea region.

But not all seals are the same. Newborn pups have little fat and represent much less energy (perhaps 40 000 kJ) than a fully grown adult (627 000 kJ). Adults represent more calories but they are also more wary and harder to catch. In Barrow Strait, Hammill and Smith found that polar bears are responsible for virtually all predation attempts on ringed seals, and that 90–100 per cent of these are directed at the ice lairs used by pups. Indeed, on occasion, the bears ignored holes used by adult male seals—presumably the costs of the chase were too high, especially when easier pickings were available elsewhere.

When the lairs become exposed or easily found, successful predation attempts are much higher: for one season on

Baffin Island they were as high as 33 per cent, when a warm spring meant little snow cover. More usually the figure is 11 per cent. Because of this, the thickness of the snow and the extent to which the pup's lair is hidden is one factor governing loss rates to the bears.

Occasionally, bears will feed opportunistically on other species, but most of their predation attempts are directed at older and fatter pups, those that are relatively large though still naïve and easily caught. In fact, the bear's biology follows this closely—their annual cycle of fat accumulation and depletion matches the calendar of pup births and growth before weaning.

Stirling and Øritsland estimated how many seals were required to maintain a population of polar bears, with a roughly equal sex ratio of bears, given that adult male bears require around 1.5 times the energy intake of adult females. If the bears fed solely on the youngest seal pups, many more individuals would have to be killed to supply the population's energy needs—around 15 times more than if they were feeding entirely on adults. Given their preference for fattened pups, a more realistic estimate is that 4.6 times as many seals are taken when compared to a diet consisting entirely of adults. In fact, estimates suggest that predation rates vary according to the region of the Canadian Arctic—in some areas, almost half the unweaned pups are taken each year, but more normally the range is 14–27 per cent.

As it is mostly the young that are taken, these high levels of predation are sustainable because the reproductive potential of the seal population is not reduced by significant adult losses (Section 3.3). In some populations as much as 65 per cent of pups could be taken each year with no overall change in adult seal numbers. Perhaps the high degree of correlation between the two populations is, as Stirling and Øritsland suggest, because the polar bear populations are close to their carrying capacity, matching the maximum sustainable yield of the seal population. In this way, each population fluctuates with the fortunes of the other.

however, cannot afford to lose a single chase. Prey that escape may get to pass on their genes, and so will those predators that succeed most often. This selective pressure is so direct and so intense that a wide

variety of defence and attack strategies have evolved (Box 4.3, Table 4.3).

The pressure for improvement is likely to be greatest for predators and prey whose populations are

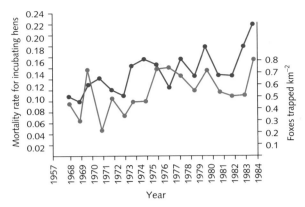

Figure 4.17 Foxes and hens: changes in the number of foxes trapped in five farms in Sussex and its correspondence with the number of grey partridge hens lost to predation by foxes. The level of predation is closely linked to the number of foxes, but this study showed that the size of the prey population was primarily regulated by two predators—foxes and the crows that take their eggs. However, when the partridge population is growing, abundance also depends on their food supply, primarily insects, and its effect on chick mortality. Blue circles show the mortality rate for incubating hens, primarily due to fox predation; red circles indicate the number of foxes trapped within the study area.

closely coupled, where change in one species has a direct effect on the reproductive success of another. Predators and prey then help to shape one another. The most extreme cases of this are parasites that have to be closely adapted to the environment represented by their hosts.

Parasites and pathogens

Most parasites have strategies that avoid some of the risks of wiping out the species on which they depend. If they are to complete their life cycle, their host must remain alive, and their presence should have little impact on its performance.

The range of parasite strategies and the intricacies of their life cycles are remarkable, from highly specific associations—where a parasite is locked into a single host—to generalist species that exploit a range of hosts (Box 4.5). Some of the most specialized parasites are the **parasitoids**—most especially several groups from the Hymenoptera (bees, ants, and wasps) and the Diptera (the true flies). Parasitoids practise a form of parasitism bordering on predation, laying a single egg inside the host insect, often a pupa or larva. As the larva grows, so does its parasite and all the effort of the caterpillar will ultimately benefit its passenger, in a form of controlled predation. The parasitoid then pupates inside the husk of its host, using this as protection until it can emerge as an adult. Since the caterpillar will never become adult, this parasitoid strategy is only viable with a low level of infestation in the host population.

The effects of most other parasites are far less severe, as long as their numbers do not rise too high. In a particular form of intraspecific competition, an existing parasite may prevent other individuals becoming established in its host. The host is then protected from too severe an infection and the parasite safeguards the resource it was the first to colonize.

Unlike predators that devour their prey and quickly move on to the next victim, parasites tend to establish long-term associations, whether they live on the outside (ectoparasites) or inside of their host (endoparasites). During this time they will reproduce frequently and produce eggs or larvae that attempt to migrate to another host (Box 4.5). Many parasites exploit the host's associations with other species to move between hosts (Box 4.5).

Some protozoan parasites blur the line between parasite and pathogen. A **pathogen** is any organism that causes a disease and, at high levels of infection, this would include many multicelluar parasites. There is, of course, a massive range of bacterial and fungal pathogens that infect humans, besides the viruses that hijack the gene replication machinery of our cells. Pathogenic disease may lead to the death of the host, and a parasite can only afford to adopt this strategy if it can transfer readily from one host to another. This is possible in many unicellular parasites because of their short generation times and rapid population growth. Malaria is caused by a series of sporozoan parasites that attack a wide range of animals. Together they represent the biggest killers of humanity throughout our history (Box 4.5).

At the other extreme, some parasites have sought to reduce the costs of reproduction by usurping the

BOX 4.5   The human ecosystem

Any quick review of the organisms that live within each of us makes it plain that we represent a valuable resource to other species. Over 100 animals are parasitic on human beings. In addition, a wide range of fungi and bacteria also make their home in or on us, with our bodies offering a variety of niches to be exploited.

Some of these associations are benign, with little overall effect on our well-being. Others are positively beneficial— many of the micro-organisms inhabiting our gut and other tracts are important for maintaining the internal environment. Indeed, they provide some protection against invasive pathogenic species. Such associations are mutualistic, benefiting both the resident flora and ourselves. Other organisms are true parasites, exacting a cost to the host. Occasionally, an organism may change from one to the other, from a benign form to a life-threatening infection. Amoebic dysentery, for example, occurs when *Entamoeba histolytica*, a normally well-behaved resident, begins to attack the wall of the large intestine, in a transformation we do not fully understand.

The variety of organisms that live within us is truly impressive, even if we confine the list to parasites alone (Table 4.4). Different species attack different parts of our bodies in different parts of the world. Which and where depends not only on the local climate, but also on the possible routes for infection. Local habits and hygiene determine our parasite burden, as do our associations with domesticated animals, soil and water.

Improvements in our understanding of these sources have enabled us to prevent infection, interrupting the parasite's life cycle, in some cases leading to their eradication. A number of viral diseases have been controlled in this way, most notably smallpox, a pathogen that now only exists in laboratory cultures. Poliomyelitis will soon follow, but other diseases have not succumbed so easily. Malaria continues to be the greatest killer of humanity, claiming 2.7 million people each year and affecting a further 500 million. The disease persists despite the chemical war of insecticides waged against its mosquito vectors or the drugs for the parasite itself (*Plasmodium* spp.—Figure 4.18). Both mosquito and parasite have continued to outwit humanity by evolving resistance to the sequence chemicals used against them. However, the gene sequence of *Plasmodium falciparum* has now been described and there is also a draft sequence for one mosquito vector, *Anopheles gambiae*, which together offer new lines of attack.

Our capacity to break the cycle of reproduction and transmission is the first step in control of any parasite. At some stage in their life cycle a parasite has to move from one host to another, with a particular stage adapted to make the journey. Most animal parasites have a large capacity for asexual reproduction, enabling them to generate large numbers of the infective stages, to improve their chances of moving between hosts. Most animal parasites are from ancient groups with relatively simple body plans so they can switch to asexual reproduction relatively easily.

As an ecosystem we are relatively short-lived and have a limited carrying capacity. Besides our behavioural or technological defences, our biology also has a highly adaptive immune system which the parasite has to overcome to establish itself. Many have sophisticated mechanisms for 'fooling' or suppressing the immune system. Another strategy is to be less selective and infect a number of different species, reproducing in a variety of hosts with similar physiologies. For example, trypanosomes (sleeping sickness, Chaga's disease) and *Leishmania* (Leishmaniasis) will attack most mammals. The final or **definitive host** (where the adult develops) depends on the **vector** (in this case, a species of blood-sucking insect) that carries the parasite from one host to another. More often, the vector transfers a larval form that will only mature in the definitive host. Humans are the sole definitive host for the tapeworms *Taenia solium* and *Taeniarhynchus saginatus*, derived from undercooked pork and beef respectively.

The beef tapeworm may produce 600 million eggs during its life, necessary because this is the resistant external stage that has to survive the external world and which has to be consumed by cattle. Life cycles dependent on such transfers could only evolve where there are close associations between hosts. Indeed, many human parasites use our domesticated animals as vectors (Table 4.4), and this implies a rapid evolution on the part of the parasite, within the 2 million years of *Homo* or the 10 000 years of *Homo sapiens*, the pastoralist.

One intriguing example is for the gastric bacterium *Helicobacter pylori*. Found in over half the human race, *H. pylori* occasionally becomes pathogenic, causing gastritis and eventually ulcers. Mark Achtman and his colleagues have traced the spread of humanity from its East African home using changes in the nucleotide sequence of the bacteria. Variation in the bacterial genome decreased with

(continued, p. 133)

TABLE 4.4   Some parasites of humans[a]

Species/group	Other hosts	Vectors	Principal sites in host	Distribution
Protozoa				
Trypanosoma (sleeping sickness)	Mammals	Flies, bugs	Blood	Tropical Africa, S. America
Leishmania (leishmaniasis)	Mammals	Sandflies	Blood	Tropics
Entamoeba (amoebic dysentery)	—	Housefly helps to transmit cysts	Large intestine, various organs	Tropical and warm temperate areas
Toxoplasma	Mammals	Domestic animals	All organs, especially brain	Global
Plasmodium (malaria)	Mammals	Mosquitoes	Blood/liver	Tropics, warm temperate
Flatworms and tapeworms				
Fasciolopsis (large intestinal fluke)	—	Resistant external stages. Aquatic snails	Intestine	S.E. Asia
Clonorchis	Dogs, cats	Resistant external stages. Aquatic snails, fish	Liver	S.E. Asia
Paragonimus (lung fluke)	Mammals	Aquatic snails Crabs, crayfish	Lung	Asia Americas
Schistosoma (bilharzias)	—	Resistant external stages. Aquatic snails, fish	Blood vessels of liver	Africa
Dyphyllobothrium	Dogs, cats	Resistant external stages. Aquatic snails, fish	Small intestine	Most of northern hemisphere
Hymenolepsis	Dogs, rodents	Resistant external stages. Fleas	Small intestine	Global, warm areas
Taenia	Pig	Cow, pig	Small intestine	Global
Roundworms				
Trichinella (trichinosis) *Ancylostoma* *Necator* (hookworms)	Mammals	Other mammals used as food	Small intestine and migratory. Migrates through different organs during development. Adult in small intestine	Gobal
Ascaris (intestinal roundworm)	—	Resistant external stages	Small intestine. Migrates through organs during development	Global
Wucheria, Loa (elephantiasis)	—	Various biting flies	Connective tissue, blood; migrates during life cycles	Tropics
Onchocerca	—	Black flies (gnats)	Connective tissue beneath skin	Africa
Dracunculus (Guinea worm)	—	Aquatic crustacea	Deep connective tissue	Topical Africa, Middle East

[a] This table gives only those animals whose association with humans is semi-permanent and for which humans are a definitive host. It excludes those who may feed on humans, but for whom the association may be intermittent. It thus omits external parasites, such as the human headlouse, bedbugs, ticks, and mites, as well as fleas, leeches, and vampire bats, that show less and less dependence upon humans as their main host. You may notice that this latter list gradually strays into the grey area between parasite and predator.

(continued overleaf)

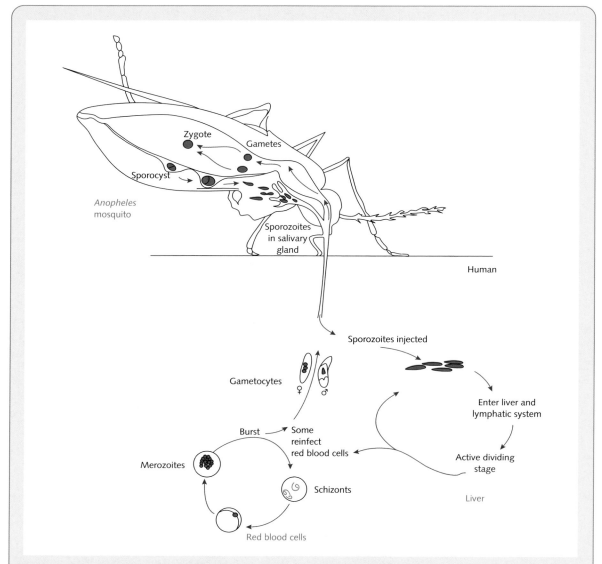

Figure 4.18   The life cycle of *Plasmodium*, the genus of protozoan parasite that causes malaria in humans and other mammals. The timing of its life cycle depends upon the species, but the stages are the same. Sporozoites, injected when the mosquito takes a blood meal, enter the mammalian liver and undergo rapid division. After they burst from the host's liver cells, some re-enter liver cells but others infect red blood cells. Here they become schizonts and undergo further asexual multiplication. When the merozoites are released from these cells, some reinfect other red blood cells, and some form sexual cells. If these gametocytes are then taken up by a mosquito they form gametes within its gut and fertilization produces a zygote. This encysts in the insect gut wall, eventually releasing the sporozoites that invade the insect's salivary glands, to be injected at the next feed. The insect itself is acting as an ectoparasite by taking a blood meal (or at least the female is, since only she feeds on blood). *Plasmodium* is parasitic on both the mammal and the insect: all its sexual stages take place in the mosquito and it is entirely asexual in the mammal, so we represent the vectors, with the mosquito as the definitive host.

distance away from East Africa, matching the same pattern in the genetic diversity of their human population hosts. This confirms their mutual spread out of Africa 58 000 years ago. Intriguingly, this partnership may have evolved out of a beneficial symbiosis, which would explain why *H. pylori* infections can be beneficial for some of their human hosts.

Parasites may play a fundamental role in the evolution of many species. The prevalence of sexual reproduction (Section 1.3), rather than the less costly, less risky asexual reproduction, may be a response to the threat posed by parasites. Within a population, the argument goes, individuals with the most common genotype are the ones most likely to be attacked because this is the genotype to which most of the parasites will be adapted. Variation may mean a parasite does not recognize some individuals. Novelty then has a selective advantage and sex is the prime means of generating it.

parental efforts of other species. Cuckoos (*Cuculus canorum*) are **brood parasites**, manipulating the behaviour patterns of other bird species to improve their own reproductive potential. Not only are its eggs laid in another's nest and reared by foster parents, the young cuckoo also practises interspecific competition by easing the host's chicks out of the nest. The adult cuckoo, meanwhile, freed from the demands of rearing her young, can go on to lay further eggs in other nests.

Several species of duck have a form of intraspecific brood parasitism, where eggs are laid in the nests of neighbours. All the birds benefit from the protection of the flock, but set against this is the cost of raising the offspring of a neighbour. Since many within the group will be relatively closely related, the genes shared with close neighbours offset this cost, at least to some extent. Some insects also practice brood-parasitism—several species of wasp have parasitic strains of queens who will lay their eggs in the nests of others. Taking this further, ants have learnt the value of regicide. The queen of *Lasius reginae* will kill the queen of another species (*Lasius alienus*) so that its workers care for her own larvae. Eventually the *L. alienus* workers die, but by that time the *L. reginae* workers have taken control of the nest.

Plants too can be parasitic. **Holoparasites**, such as broomrapes (*Orobanche*), have abandoned photosynthesis and acquire all their resources from their host plant (Figure 4.19). **Hemiparasites** hedge their bets and photosynthesize, both to supplement their energy needs or to sustain themselves in the

Figure 4.19 *Cistanche phelypaea*, a close relative of the broomrape, has no chlorophyll of its own and is totally dependent on its host plant—species of the plant family Chenopodiaceae.

Figure 4.20 Yellow rattle (*Rhinanthus minor*), a hemiparasite of grasses.

absence of a suitable host. Some, such as yellow rattle (*Rhinanthus minor*), exploit a range of species (mainly grasses), slowing down their host's growth (Figure 4.20). This helps to prevent vigorous grasses from out-competing other plants in a grassland community, and for this reason yellow rattle is sometimes included in 'wildflower' seed mixes used to produce low-maintenance grassland.

Mistletoe (*Viscum album*) is undoubtedly the most famous parasitic plant. A hemiparasite, it taps into the living vascular tissue of trees, even though it is able to photosynthesize itself. Bad infestations can reduce the growth of a tree by almost 20 per cent. Despite this, *V. album* is not a major economic pest, though there are more troublesome members of the mistletoe family (Loranthaceae). *Scurrula cordifolia* has spread from the Himalaya into northern India, where it attacks commercial orchard and forest trees. Control is difficult, because herbicides can harm the host tree as much as the mistletoe itself. However, *Scurrula* has an enemy of its own: another mistletoe, *Viscum loranthi*, which is unable to live directly on trees, but parasitizes the parasite, weakening and sometimes killing *Scurrula*. Now introduced into plantations and orchards, *V. loranthi* has proved to be a safe and efficient means of control because it has the most desirable property of the ideal biological control agent: it only attacks the pest.

4.5 Controlling pests

Understanding the close association between species and the details of their interactions enables us to manipulate these to our own advantage. Sometimes, this means correcting our past mistakes, if we have introduced a species that has become a pest. At other times, it means controlling native species that have become pests, such as weeds, *r*-selected opportunists which grow rapidly when resources are available and there is little competition.

A species is usually designated a pest because of its effect on human health, wealth or well-being. The term pest has no biological meaning—it is a label

we apply to species causing us to incur costs. This includes expenditure to safeguard threatened species, though sometimes our attempts to reverse our mistakes have made matters worse. The Indian mongoose, released onto several Caribbean islands to control the brown rat, has had a disastrous effect on their native bird, reptile, and amphibian fauna.

There are numerous examples of introduced plant species flourishing in their new homes. Sycamore (*Acer pseudoplatanus*) and *Rhododendron ponticum* are non-native species which have slowly invaded large areas of native vegetation in Britain and their competitive power has made them difficult to control. Similarly, Old World species have established themselves in New World habitats: cheatgrass (*Bromus tectorum*) arrived with European settlers and was first reported in Pennsylvania in 1790. Within a century it had reached the west coast of North America, covering perhaps 200 000 km² and excluding many native grasses in the process.

However, not all introduced species become pests. Many fail to establish in their new habitat, either because they are poorly adapted or because they cannot out-compete the native flora or fauna. Mark Williamson and Alistair Fitter suggest that, on average, 10 per cent of introduced species will become established, and of these successful colonists 10 per cent will become a pest. This is known as the 'tens rule' and although it cannot be used to predict the outcome of each introduction, it does indicate a regularity that seems to underlie invasions. Charles Boudouresque found that out of the 85 naturalized plant species within the Mediterranean Sea, nine had since become pests (Box 4.6).

BOX 4.6 Algal attack

Some visitors overstay their welcome in the Mediterranean. The exotic alga *Caulerpa taxifolia* (Figure 4.21) is one example, thought to have been accidentally introduced from an exhibit in the Oceanographic Institute of Monaco in the early 1980s. In 1984 a small patch was found growing just off the coast by the Institute. In less than 10 years it has spread to cover more than 16 000 ha of the sea bed, from southern Spain to Croatia (Figure 4.22).

It has now invaded coasts around California and Southern Australia, where it has spread equally rapidly, and because it contains a potent toxin (caulerpanyne) fish will not eat it. As *Caulerpa* becomes established it swamps other seaweeds and will quickly displace the inshore community. Alexandre Meinesz and his team are trying to find potential biological control agents for the weed.

In its native environment, *C. taxifolia* seldom exceeds 25 cm in height and will die when the sea temperature drops below 20 °C. Yet in the Mediterranean the plants grow two or three times as large and survive at temperatures as low as 10 °C. Meinesz suggests that the Mediterranean population might be an ecotype that arose in the aquarium environment (possibly even as the result of a mutation induced by its ultra-violet lighting). This ecotype shows no evidence of sexual reproduction and it seems to be an exclusively male form, so Meinesz believes the infestation is of a single clonal population, possibly a new species (tentatively named *Caulerpa xenogigantea*, literally 'large stranger').

Initially, control consisted of volunteers physically removing the plants, but there has also been a search for a suitable biological control agent. In 1992, Kerry Clark of the Florida

Figure 4.21 *Caulerpa taxifolia*—spreading throughout the Mediterranean Sea.

(continued overleaf)

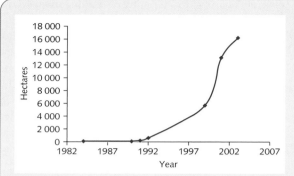

Figure 4.22   The dramatic rise in the area covered by *Caulerpa* since its accidental release in 1984.

Institute of Technology reported a number of marine molluscs that feed exclusively on *Caulerpa*. Meinesz's laboratory screened a number of these from *Caulerpa*'s home range, and identified two sea slugs, *Oxynoe azurepunctata* and *Elysia subornata*, as potential control agents. Both are voracious feeders, though *Elysia* will graze any species of the genus *Caulerpa* (and thus poses a risk to the native species of the genus). *Elysia* is also unlikely to maintain a viable population in the Mediterranean, where conditions for most of the year do not favour its feeding, growth, and reproduction.

A more practical solution might involve enhancing existing native populations of sea slugs that feed on *Caulerpa*, in particular *Oxynoe olivacea* and *Lobiger serradifalci*. These are currently found at low densities but will consume all species of *Caulerpa*. *Lobiger* may be less useful because it shreds the algae, leaving fragments that can regenerate into new plants. *O. olivacea*, however, is more thorough and more promising as a control agent. There is still a chance that boosting the population of a native mollusc might, in some way, alter the food web within its ecosystem. Using native species is less risky than using an exotic to catch an exotic.

Pest control requires difficult choices and hard work. Often the fastest (and cheapest) way to control a pest is to use chemical methods. Insecticides, herbicides, molluscicides, and so on can rapidly bring a pest outbreak under control, but many of the chemicals used in the past were general-purpose poisons that killed non-target, as well as target, species. Many have also been highly persistent—killing or impairing species long after they were applied (Box 6.4). This is one reason why several pests have developed resistance to some of these compounds. DDT resistance in malarial mosquitoes is one notable example and an impressive demonstration of natural selection producing new ecotypes within tens of years.

Nowadays, a more detailed study of the biology of pest species is undertaken to find out how a poison works and the best way of using it to control a pest. Increasingly, we are able to produce highly specific pesticides with low persistence. However, modern pesticides are expensive to develop and add considerably to the farmer's costs. Inevitably, there are also some ecological costs: even if the evolution of resistance can be avoided (and the experience from many case histories is that it cannot), there are a host of other impacts from chemical control methods— from the risks to workers to subtle shifts in species interactions.

Biological control offers one alternative approach, where another species is used to control the pest. Very often this amounts to simply reuniting a pest with its natural enemy (Box 4.6). This may take considerable research, requiring us to identify a pest's home range, the possible natural enemies there, and their capacity to control the pest's numbers.

The search for a new biological control agent begins with a detailed study of the ecology of the potential natural enemy and its chances of establishing itself in the new habitat. In particular, we need to know whether it will compete with the native fauna and if it can maintain a population without itself becoming a pest. Ideally, it will only consume the pest species, and that usually implies a specialist predator or parasitoid. To avoid the need for reintroduction, the natural enemy should not kill all the pests: rather, it should allow a low (and economically unimportant) pest population to survive. In that way both species remain at low numbers, with the natural enemy present as some safeguard against future

TABLE 4.5	Ten desirable attributes of a predator or parasitoid as an effective biological control agent

1. Rapid functional and numerical response to a rise in pest abundance
2. Able to find pest quickly (high searching efficiency)
3. Able to maintain a low pest population and sustain itself over the long term
4. Able to survive competition with native predators and parasitoids
5. High prey specificity with minimum impact on non-target species
6. Its activity and its life cycle show a close match with that of the pest
7. Easy to culture in the laboratory
8. Easy to release in the field
9. Cheap to use with rapid results to inspire confidence
10. No social nuisance

outbreaks (Box 4.6). Thereafter, our prime concern is that the natural enemy should not affect any other species in the community nor become a nuisance or create other problems. Ideally, it will be cheap and easy to rear and to release (Table 4.5).

We are not restricted to exploiting only predator–prey interactions in a biological control programme —competitive interactions can also be used. An ingenious application of intraspecific competition is used in the control of a highly destructive parasite, the screwworm (*Cochliomyia hominivorax*). The fly, native to central America, lays its eggs inside the open wounds of mammals, including humans. The larvae hatch and feed on the flesh, enlarging the wound and preventing it from healing. The stench from the wound attracts other adults, who add their eggs to the seething mass. In cases where there is a very high level of infestation the host may die.

The biological control technique used here attacks a weak point in the life cycle of the screwworm—the female only mates once. This pest is bred in vast numbers in a factory in Mexico and at the pupal

stage the males are separated. As their testes are forming they are irradiated and thereby sterilized. After growing on, sterile pupae are shipped to outbreak areas and dispersed, ready to emerge as properly formed adults. The female cannot distinguish sterile from wild males, so swamping an area with millions of factory-bred flies means that most females copulate to no purpose. The reproductive rate plummets and the population crashes. The sterile insect programme has eradicated screwworm from the southern United States and most of Central America, with its most recent success in Nicaragua in 1999.

The same strategy is now being used in Africa to control the tsetse fly (*Glossina* spp.), the vector of *Trypanosoma*, the protozoan parasite that causes sleeping sickness (trypanosomiasis) (Box 4.5). Trypanosomiasis occurs in 36 countries in sub-Saharan Africa where it affects over 500 000 people and 50 million cattle. There have been many attempts to eradicate it, mostly with limited success, but one successful project in Zanzibar finally achieved eradication in 1997, three years after the weekly release of around 70 000 sterile males flies began.

Sterile insect techniques might also prove useful in the control of West Nile Fever, a form of encephalitis caused by an RNA virus. An Israeli strain of the disease recently crossed the Atlantic, emerging in New York in 1999, and has since spread through North America. The disease is carried by birds which in turn infect mosquitoes (principally *Culex pipiens pipiens*) and these can transmit it to mammalian secondary hosts, including humans.

The growth of agricultural pest populations can be checked by changing cultivation techniques, such as the timing of our crops. Some cotton farmers grow an early 'trap crop' to catch over-wintering boll weevils (*Anthonomus grandis*). These infected plants are destroyed so that the main crop has a reduced chance of attack by the beetles. Another method is to grow strips of alfalfa to lure the weevils and other pests away from the cotton. The sequence of crops grown on an area of land can also help to avoid pest outbreaks, especially for those closely tied to one food plant. Removing their food source for one or more years, or indeed part of a year, makes it difficult for the pest to maintain a resident population.

Hedgerows and patches of wild vegetation are important reservoirs of generalist predators that can check pest growth before it takes off in the crop. In Britain, trials of specially planted grass banks in large fields have proved to be a useful control measure, increasing the patchiness of the habitat and reducing the distance between predator and pest at important stages in their life cycle (Section 8.1). Similar methods are used to attract specialist predators or parasitoids. In the vineyards of California, an introduced pest, the grape leafhopper *Erythroneura elegantula*, is controlled by a native parasitoid, *Anagrus epos*. This lays its eggs inside the leafhopper egg and overwinters as an adult, feeding on a native leafhopper that lives on bramble. Bramble patches around the edges of the vineyards allow *Anagrus* to sustain itself over winter, ready to attack the pest in the new growing season. The surrounding scrub vegetation increases the parasitoid population by as much as 5 per cent and this can be further enhanced by introducing a network of corridors within and between vineyards. These improve the spread of the parasitoid, but also attract the pest leafhoppers away from the vines and onto the more diverse vegetation beside it (Section 8.1).

The use of biological, cultural, and chemical methods together in a control programme is termed **integrated pest management** (IPM). This uses the principal advantages of each method to reduce costs and to give effective long-term pest control. Biological and cultural methods require a full understanding of the ecology of the pest and the natural enemy, and will, ideally, give sustained, low-cost control. Inevitably there are delays before an introduced natural enemy checks pest numbers and we

may then need to use chemical methods for a short period to prevent an outbreak. Chemicals may also be needed to allow a natural enemy to survive, perhaps by reducing predator or competitor numbers, or even to attract other control agents to a pest outbreak. Some of these techniques have been improved by developing pesticides that are highly specific to a target species. As a result, IPM can be used as an economic and viable alternative to the simple application of broad-spectrum poisons.

As our knowledge of predator–prey interactions increases so does the possibility of developing a new generation of agrochemicals designed to control rather than kill. Pheromones are already used in agriculture. For example, traps baited with the scent of the female codling moth are routinely used in orchards to lure males away from the trees. The recent discovery and synthesis of the female screwworm pheromone makes this a possibility for its control. We might also use pheromones to confuse: Jeremy Thomas and his colleagues recently isolated a panic-inducing pheromone which ichneumon wasps produce to confuse ants when they raid their nests. If synthesized, such a compound might prove a safe and effective ant-deterrent.

The interactions between individuals and between species maintain the structure of natural communities. Each individual is attempting to secure its own reproductive success and natural selection has explored a range of strategies in different species. The subtlety and intricacy of some of these should not surprise us—competition and predation are powerful selective forces and are likely to favour the novel. Understanding their detail enables us to control pests and to protect communities.

● **SUMMARY**

Organisms interact in various ways to secure resources essential for their survival and reproductive success. In mutualism, individuals of two species interact to the benefit of both partners. In commensalism, an association benefits one species but with no advantage or detriment to the other.

Intraspecific competition is the struggle for resources between individuals of the same species. Interspecific competition describes the interactions between individuals of different species competing for a limited resource or which inhibit each other. In simple laboratory experiments, competition for the same resource often leads to one species excluding the other, but interactions in the real world are far more complex, and change with a range of factors. Predation, herbivory, and parasitism are interactions where one species exploits another as a resource, and in which the performance of a host is impaired or the prey is killed. In simple models, the cycles of abundance of predators and their prey indicate that neither species controls the other. Again, the detail of real predator–prey relations shows how a range of factors serve to control numbers. Predation, like parasitism, is not confined to animals, and is found also in plants, fungi, and protists.

Whilst some introduced species may achieve pest status in their new environment, more frequently native species or the larger community successfully exclude the invader. Species become pests because they have escaped the checks imposed by the interactions within their natural community. Biological control uses a range of techniques to re-establish these interactions and limit their impact.

● FURTHER READING

Hill, D. S. 1997. *The Economic Importance of Insects*. Chapman & Hall, London. A comprehensive review of insect and other selected invertebrates which also includes medical, veterinary, and household pests.

Krebs, C. J. 2001. *Ecology: Experimental Analysis of Distribution and Abundance*, 5th edn. Benjamin Cummings, London. A comprehensive text which provides a useful insight into species interactions with each other and the environment.

Ruberson, J. R. 1999. *Handbook of Pest Management*. Marcel Dekker, New York. An overview of pests from arthropods through to vertebrates and weeds, with a strong emphasis on deterrence and integrated pest management.

Van Driesche, R. J. and Bellows, T. S. 1996. *Biological Control*. Chapman & Hall, London. This book reviews ecological aspects of controlling animal and plant pests by means of biological control.

● WEB PAGES

The World Health Organization provides up to date information on human pathogens and links to other health-related web resources:
http://www.who.int/en

Cornell University College of Agriculture and Life Sciences operates this useful site on pests and biological control:
http://www.nysaes.cornell.edu/ent/biocontrol/

For current information regarding the invasion and spread of *Caulerpa*:
http://www.caulerpa.org

As part of their Global Invasive Species, the Commonwealth Agricultural Bureau International (CABI) provides useful links concerning invasive species worldwide:
http://www.cabi-bioscience.ch/wwwgisp/gtc2b1.htm

CASE STUDY 4   Squirrel tales

Since its introduction in 1876, the American grey squirrel (*Sciurus carolinensis*) has replaced the native Eurasian red squirrel (*S. vulgaris*) over much of its range in the British Isles (Figures 4.23, 4.24). The grey's spread has been so dramatic that the UK red squirrel population is now down to just 161 000, compared to an estimated 2 520 000 greys. Red squirrels are protected by law and their recovery is part of the United Kingdom's Biodiversity Action Plan.

Grey squirrels are formidable competitors, able to exploit a wider range of food sources than reds. Greys readily eat acorns and do well on them, whereas reds are less able to detoxify the secondary plant products which inhibit their digestion. Reds fare better on conifer seeds and are now confined to the north and west of the British Isles where coniferous forests predominate (Figure 4.24), yet less than a century ago they were a common sight in mixed woodland across Britain.

John Gurnell, Luc Wauters, and their colleagues have studied populations of both species in the UK and northern Italy (where greys have been introduced with similar consequences) (Gurnell *et al.* 2004; Wauters *et al.* 2001). Using a combination of direct observation and radio-tracking, they discovered a considerable overlap in tree use by the reds and greys, and a 70 per cent niche overlap of food resources (Wauters *et al.* 2002). Bryce *et al.* (2002) found an equivalent level of overlap, though the Wauters study established that greys occupied larger ranges than reds. When the two species occurred together neither changed their behaviour, either to partition food or space, and so were competing directly (Wauters *et al.* 2001, 2002).

(a)

(b)

Figure 4.23   Interspecific competitors: (a) Eurasian red squirrel (*Sciurus vulgaris*) and (b) American grey squirrel (*S. carolinesis*).

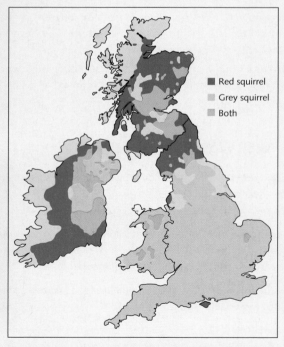

Figure 4.24   The distribution of red and grey squirrels in the British Isles. Greys outnumber reds by 66:1.

'Grey *vs* red' mythology has long held that greys are bigger and more able to maintain themselves through the winter. But a range of other factors could determine their relative fitness. One possibility was disease, and most especially the parapox virus, known to kill red squirrels and a disease unknown in the UK before the introduction of the grey (Tompkins *et al.* 2002). Earlier studies had found that many grey squirrels had been exposed to the virus and, crucially, were carriers of parapox in areas where reds were almost extinct. Conversely, greys coming from red squirrel strongholds showed no exposure to the virus (Bryce 1997; O'Teangana *et al.* 2000; Sainsbury *et al.* 2000).

The disease causes skin lesions which, at the height of infection, weep and ulcerate. Sick animals become weak and quickly pick up secondary infections. In the wild, parapox means certain death for red squirrels, whereas infected greys suffer no ill effects. Tompkins and his colleagues set up a series of infection experiments (in captivity the squirrels can readily be nursed back to health with the use of food supplements and antibiotics to control the secondary infections). One group was given the virus and the other a 'sham-infection' of saline (Figure 4.25a). The progress of the disease was then measured using a clinical scoring system which ranged from 0 (no effect) to 8 (severe symptoms). Any individual with repeatedly high scores was removed from the experiment and given immediate treatment.

All the infected reds scored 8 or more within four weeks of infection. After two weeks, both primary and secondary lesions were apparent, followed by lethargy, loss of appetite, and weight loss. Meanwhile, none of the control group scored more than 2 (there was a slight loss of condition in one of the control animals from handling stress). In an equivalent experiment with grey squirrels, no detectable symptoms were observed—proof that they could contract and carry the virus with no observable ill effects (Figure 4.25b). It thus seems likely that the parapox virus is a critical factor in some of the competitive battles between the two species, and a weapon carried unwittingly by the greys themselves.

The reasons for the decline of the red squirrel are becoming less of a grey area. Rather like Tilman's work, the details show how competitive interactions can be mediated by a number of contributory factors. Future conservation strategies might involve the control of grey squirrels and the use of a vaccine to protect the beleaguered red squirrel from parapox virus.

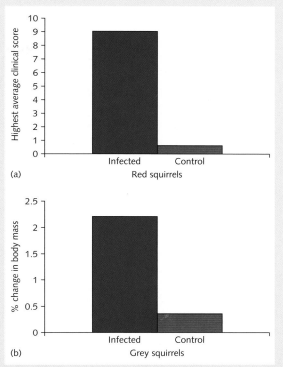

Figure 4.25   (a) Infection experiments carried out on eight red squirrels over a 60-day post-infection period (four infected with parapox virus and four given a control injection of saline). The progress of their disease was measured using an 8-point clinical score. The graph shows the highest average for treated and control groups. (The control group has a score of 0.5 as one of the squirrels suffered handling stress during the experiment.) (b) Greys do not exhibit symptoms of the disease so the percentage change in body mass (measured in grams) was measured. Twelve grey squirrels (six infected individuals and six controls) were weighed over a 43-day post-injection period. The growth of the greys did not differ between the two treatments.

References

Bryce, J. 1997. Changes in the distribution of red and grey squirrels in Scotland. *Mammal Review 27*, 171–176.

Bryce, J., Johnson, P. L., and McDonald, D. W. 2002. Can niche use in red and grey squirrels offer clues for their apparent coexistence? *Journal of Applied Ecology 39*, 875–887.

(continued overleaf)

Gurnell, J., Wauters, L. A., Lurz, P. W. W., and Tosi, G. 2004. Alien species and interspecific competition: effects of introduced eastern grey squirrels on red squirrel population dynamics. *Journal of Animal Ecology 73*, 26–35.

O'Teangana, D., Reilly, S., Montgomery, W. I., and Rochford, J. 2000. Distribution and status of the red squirrel (*Sciurus vulgaris*) in Ireland. *Mammal Review 30*, 45–56.

Sainsbury, A. W., Nettleton, P., Gilray, J., and Gurnell, J. 2000. Grey squirrels have high seroprevalence to a parapoxvirus associated with deaths in red squirrels. *Animal Conservation 3*, 229–233.

Tompkins, D. M., Sainsbury, A. W., Nettleton, P., Buxton, D., and Gurnell, J. 2002. Parapoxvirus causes a deleterious disease in red squirrels associated with UK population declines. *Proceedings of the Royal Society of London Series B 269*, 529–533.

Wauters, L. A., Gurnell, J., Martinoli, A., and Tosi, G. 2001. Does interspecific competition with introduced grey squirrels affect foraging and food choice of Eurasian red squirrels? *Animal Behaviour 61*, 1079–1091.

Wauters, L. A., Gurnell, J., Martinoli, A., and Tosi, G. 2002. Interspecific competition between native Eurasian red squirrels and alien grey squirrels: does resource partitioning occur? *Behavioral Ecology and Sociobiology 52*, 332–341.

EXERCISES

1. Construct a table in which the following 'predators'—cow, cheetah, parasitoid wasp, and tapeworm—are mapped against (a) the number of 'prey' they attack, (b) the number of 'prey' they kill. (Score number in terms of 'many', 'some', 'few', or 'very few'.)

2. Explain what is meant by the competitive exclusion principle.

3. Classify each of the following interactions as any one of: brood parasitism, commensalism, interspecific competition, intraspecific competition, mutualism, parasitism, parasitoid, or predation:
   (a) A barnacle on the shell of a live whelk.
   (b) An otter catching and eating a fish.
   (c) A bracket fungus growing on a birch tree.
   (d) A young stag attempting to take over an existing harem of female deer.
   (e) A wasp lays its eggs in a fly—these hatch and eat it alive.
   (f) A cuckoo chick within the nest of a reed warbler
   (g) A fig wasp laying eggs inside a developing fig.
   (h) Herring and black-blacked gulls fighting over scraps of fish.

4. Match the codes (a)–(f) shown on the Southwood habitat template with the following series of habitats, according to their degree of permanence, adversity stress, and availability of resources:
   (i) desert
   (ii) grassland
   (iii) heathland/moorland
   (iv) temperate woodland
   (v) tundra
   (vi) weedy field margin

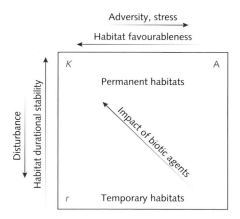

5. Some species are adapted to life in poor, highly stressed environments, where conditions are extreme (e.g. high or low temperatures, soils that are very acid or alkaline, etc.). Why should any species adapt to such demanding environmental conditions?

6. Match the following forms of competition—chemical competition, consumptive competition, encounter competition, overgrowth competition, pre-emptive competition, territorial competition—with the examples (a)–(f) that best describe them:
    (a) A humming bird defending its patch of flowers.
    (b) A hyena and vulture fighting over a carcass.
    (c) Creosote bushes suppressing the germination of seedlings around them.
    (d) Sycamore shading out oak samplings in a woodland.
    (e) Barnacles colonizing the surface of a rock.
    (f) Goats grazing vegetation used by Arabian oryx.

Tutorial/seminar questions

7. Historically, native Americans are likely to have had a very different community of animal parasites living within them than native Africans. Why?

8. Should we use DDT to control malaria in developing nations?

9. Review the possible reasons why red squirrels should not have evolved a greater tolerance to a diet rich in acorns or an immunity to the parapox virus before the arrival of the grey squirrel in Europe.

Communities

'You are left . . . with something rather like the skeleton of a body wasted with disease; the rich soft soil has all run away leaving the land skin and bone.'

Plato: *The Critias*

CHAPTER OUTLINE

- Are ecological communities a loose collection of species living in the same habitat and adapted to the same abiotic conditions? Or are they highly integrated and organized networks of interacting species?

- The convergence of mediterranean-type communities towards similar configurations of species.

- The role of species interactions in such communities.

- The predictability of species assemblages and 'rules of assembly'.

- The development of communities over time and space.

- The succession of species and the rules for assembling species into communities.

- Plant associations and their use in community classification.

- Disturbance and community development; human impact on Mediterranean communities.

← Remnant holm oak (*Quercus ilex*) amidst the garrigue that now dominates the limestone of the Corbières, in the Languedoc of southern France.

Emerging from an aircraft after flying any distance, we always compare the temperature of our destination with that we left behind. For those of us from the colder and wetter latitudes, this is nearly always a pleasant change.

On leaving the airport, we soon realize that the weather is not the only difference. Beyond the cultural, social, and architectural contrasts, those with an eye for the plant life around them quickly spot species unlike those back home. Some growing in the wild here may only be found in greenhouses elsewhere. Others we may never have seen before.

Of course, this should come as no surprise to us. We expect this wonderful climate to support a different plant life. If we gave it rather more thought, we would also expect the plants to show adaptations to the local heat and dryness. Indeed, we expect to see certain plants in certain climates—the almond blossom of the Mediterranean or the heather of the Scottish moors. Some plants are indicative, even evocative of their habitat, signalling the presence of whole communities of plants and animals.

This expectation underlies a fundamental question in ecology—to what extent are these collections of plants and animals predictable? Are species assemblages inevitable for a particular climate or a particular soil? And what links them together—is it the climate and soil, or their interactions with each other—the insects that feed on the plants and the birds that feed on the insects? An ecological community may be no more than a loose collection of species, found together because they are adapted to the same environment. But perhaps it might be something more organized and integrated, a complex network of interactions that only allows certain combinations of species to coexist. In a community structured principally by its interactions, the fortunes of one species will be closely tied to those of its neighbours.

We have seen that different species can converge on similar solutions to selective pressures—we compared the marsupial thylacine with the placental dogs in Section 2.3—but can whole communities also reach the same endpoint? If they did we should expect the same sort of community to form under identical conditions or to re-form following some major upheaval. Such a self-organizing community, repeating itself in time and space, would be strong evidence that only certain configurations of species, or at least, the niches they represent, can persist. This idea of an ecological community, ready to reassert itself, is, perhaps, what many people assume to be part of the 'balance of nature'.

In this chapter, we look at one experiment the Earth has been running over the last million years that allows us to test these ideas. Across the globe, where the tropics give way to temperate regions (Figure 5.1), five discrete areas have the climatic conditions typical of the Mediterranean Basin. Following the ice ages, each region had a different pool of species from which to draw colonists. We should not expect to find the same species in each community but there may be different species with equivalent functional roles. Here, we describe mediterranean-type communities, exploring species adaptations and interactions within each assembly, to try to decide whether community structures are repeated in each region. This allows us to study one community type in some detail, and to begin to understand how ecological communities are organized. We go on to look at how communities develop through time, how they construct and configure themselves, and whether their final assemblage of species is predictable.

For many people, the mediterranean climate is the closest to the ideal. The Mediterranean Basin was the cradle of Western civilization and the birthplace of the European package holiday. Under its benevolent influence some of the most influential cultures have grown up, from the Nile to Jericho, Athens, and Rome. Going further back, early remains of modern humans are found in the Near and Middle East, predating the present climate. Human hands shaped the ecology of these lands and the fate of several major cultures is written in its soils—from the annual rise and fall of the Nile to the doomed irrigation schemes around Babylon. Plato observed the wasting of the hills of Greece 2 500 years ago, though by this time agriculture was already failing to feed the people and the Greeks had taken to the sea, leaving Arcadia to trade and to conquer.

Today, all mediterranean-type communities throughout the world, and especially those in the

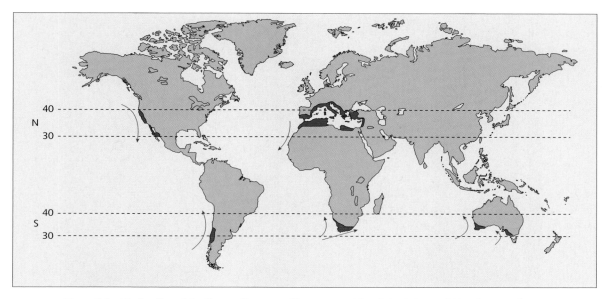

Figure 5.1 The location of the five mediterranean biomes around the globe, between 30° and 40° north and south of the Equator. The arrows indicate the cold surface currents which help to generate the summer drought, a key feature of their climate.

Mediterranean Basin itself, are under threat, whether from tourism, agriculture, or industrialization. Having changed with human cultures for thousands of years, the current changes in land use may yet be the most drastic, just as the climate of paradise itself is forecast to become less benign.

5.1 Mediterranean communities

The five regions we designate as 'mediterranean-type', defined by their climate, began to develop around 3 million years ago. However, these conditions have been interrupted by the colder intervals of each glacial advance and these communities have reconstructed themselves more than once, especially in the northern hemisphere. Indeed, the plant communities of today's Mediterranean Basin reflect a history of human activity, and are at most 7 500 years old.

Mediterranean climates represent transition zones between the moist temperate areas and the semi-arid regions of the sub-tropics. They are the most restricted of all climatic zones, confined to narrow bands around 30° to 40° either side of the Equator (Figure 5.1), on the western edges of continents.

Here, cold currents induce coastal fogs and mists so that moist air never progresses very far inland—the coastal fringe may be green, but beyond this there is often a desert. So one prime climatic feature defining the community is an extended drought during a hot summer. Plants then have to conserve water, many cease to grow, and others will shed their leaves. Winters are cool and wet, though rainfall is highly variable from one year to the next. Plant growth is largely confined to the spring and autumn, when it is warm and water is available.

Not only is the climate very similar in the five regions, three of them also share similar topographies. California, Chile, and the Mediterranean Basin have rugged landscapes with steep-sided valleys close to

Figure 5.2   The leaves of six typical sclerophyllous plants. Clockwise from top left: holm oak (*Quercus ilex*), kermes oak (*Q. coccifera*), olive (*Olea europaea*), California lilac (*Ceanothus*), rosemary (*Rosmarinus officinalis*), and juniper (*Juniperus phoenicea*).

the coast. This creates gradients of moisture and exposure, and valleys with different soils and contrasting plant communities. The result is a finely divided mosaic of habitat patches. In South Africa and southwestern Australia the landscape is more ancient, lowered and rounded by a long history of erosion. Here there are fewer contrasts and although all regions have relatively infertile soils, these soils are particularly low in phosphate (Section 7.1). The Mediterranean Basin is dominated by limestone, but this is much less common in the other regions and is absent from California and Chile.

The vegetation of the five areas share a common physiognomy, that is the dominant plants have similar appearances, and a similar physiology. Many are **xerophytic**, that is they are adapted to minimize water loss through transpiration (Figure 5.2) to survive the summer drought. Their leaves are characteristically small, tough, and leathery (termed **sclerophyllous**) and are retained throughout the year. Typically, the plant community forms open woodland dominated by short, evergreen trees; local names include maquis, chaparral, and matorral (Figure 5.3, Table 5.1).

Sclerophylly could also be an adaptation to the low nutrient level of the soils. Evergreen shrubs tend to dominate in poor soils because it would be costly to regrow leaves each year and to secure the necessary nutrients; better instead to maintain existing leaves, even if these will use more energy than they produce at times of drought. This might explain the dominance of sclerophyllous plants in southwestern Australia and South Africa and also the prevalence of evergreen trees in montane and temperate regions with nutrient-poor soils (Section 8.2). However, deciduous trees become dominant in mediterranean ecosystems where water is more available, so the summer drought certainly determines the extent of sclerophyllous scrub. Another possibility is that tough leaves are a means of dissuading insect attack, since leaves that have to last a long time (up to seven years in the case of kermes oak) are worth protecting from herbivores.

(a)

(b)

Figure 5.3   (a) The Maquis in the Languedoc of southwestern France; (b) Garrigue.

Where conditions are drier still, the sclerophylly becomes even more extreme. Then a shorter plant community is found, consisting of tussocks or low shrubs, or even tight cushions of thorns. These are called **chamaephytes** (literally 'dwarf plants'—Figure 5.4). Their thorns prevent grazing, but also help to create a microclimate of still air within the cushion, helping to reduce water loss. Some of these plants shed their leaves in the summer. Other adaptations, particularly in the Californian coastal sagebrush and the Chilean Jaral, include the use of succulent leaves and tubers to store water against the summer drought.

Figure 5.4   Phrygana on Kalymnos, a Greek island off the coast of Turkey.

These shared forms are not due to the same species being found in each region. Each has its particular collection of plants (and animals), which have colonized since the climate first developed and which have re-established themselves after each ice age. South Africa and Australia were colonized from tropical communities to the north. Most of the plant species of the Mediterranean Basin came from the temperate north, though several are derived from tropical Africa and the Near East. Interestingly, several important species were derived from the Cape region of South Africa, including the most emblematic plant of all, the olive (*Olea europaea*).

Given their different sources, we should not expect the same species to fill the same niches in each case, but the different regions may share equivalent niches. If they do, we should ask whether these niches are defined by the prevailing abiotic conditions or by the interactions between the species that assemble in each location.

Following their colonization, each region enjoyed a rapid speciation as these new communities developed. Of the five, the highest diversity is found in the Cape Province of South Africa followed by south-western Australia (Figure 5.5). The fynbos is one of the most diverse plant communities in the world, with a vast range of endemic species. Both regions have open plant communities, sometimes described as 'heaths', where soil nutrients are low and exposure is high (Table 5.1). These regions also share a high frequency of fire and a more predictable annual rainfall. This association of low soil nutrients, high environmental predictability, and high plant diversity is a pattern we shall encounter again in Chapter 9, one which is indicative of considerable niche differentiation (Section 3.5). It also points towards intense competition for the limited soil nutrients.

In the other regions, a large altitudinal range allows for several community types within a short distance of each other, from the valley floor to the top of a hill (Figure 5.6). Their relative isolation in steep-sided valleys has promoted local speciation and a high level of endemism. Today, all mediterranean regions are recognized as 'biodiversity hotspots'— ecosystems with a disproportionately high number of species; for example, the Mediterranean Basin has 10 per cent of known vascular plants in less than 2 per cent of the land area of the planet (Figure 5.7).

TABLE 5.1 **The main forms of mediterranean-type vegetation communities**

Location	Name
General type. Typically, low woodland (trees 2–5 m), evergreens with sclerophyllous leaves, beneath which is an understorey of annual and herbaceous perennials	
Mediterranean	Maquis (France)
	Macchia (Italy)
California	Chaparral
Chile	Mattoral
South Africa	Renosterveld
Australia	Mallee
More arid or disturbed types. Low and open communities (trees 0.5–2.0 m or low tussock bushes), often with drought-deciduous species, thorn bushes, and aromatic species	
Mediterranean	Garrigue (France)
	Phrygana (Greece)
	Batha (Israel)
California	Coastal sagebrush
Chile	Jaral
Low-nutrient soils supporting a heathland-type community. Low and open communities (between 0.2 and 1.5 m high), frequently showing a high degree of diversity and species endemism. Dominated by species of *Protea* in Africa and *Banksia* in Australia	
South Africa	Fynbos
Western Australia	Mallee heathland

The role of fire

This high diversity follows when no single species or group of plants becomes dominant, when any competitive advantage is kept in check. Along with the summer drought, fire is a key feature of many of these habitats, creating gaps which colonizing species can occupy.

Short-lived summer fires are commonplace in these communities, though rarely devastating. Because they shed their leaves infrequently, small amounts of litter accumulate beneath sclerophyllous shrubs and the fires are not sustained by a thick litter layer. Fire passes quickly through the canopy and is soon extinguished. Many trees, such as cork oak (*Quercus suber*), laurel (*Laurus*), and olive are unharmed by a fast burn whilst other plants sprout rapidly from crowns protected at or below the soil level.

Many also produce seeds that need scorching to induce germination, a useful trigger when space and nutrients are available. The Chilean matorral has plants well adapted to frequent fires, with seeds able to germinate within days of a fire passing, so that a self-replacing community develops. A high fire frequency in California invariably leads to a community dominated by chamise (*Adenostoma*) and the species

(a)

(b)

Figure 5.5   (a) The mallee heathland in southwestern Australia; (b) the fynbos of South Africa.

(a)

(b)

Figure 5.6 (a) The matorral of Chile; (b) the chaparral of California.

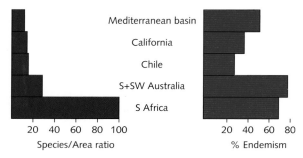

Figure 5.7 Plant species richness in the five mediterranean regions. All areas show a high species richness, but this is most marked for the Cape Province of South Africa and southwestern Australia.

composition also changes in the Mediterranean Basin: in the wetter northwestern corner oaks are replaced by more resistant pine species, such as the Aleppo pine (*Pinus halepensis*), whilst at the highest frequencies a low aromatic scrub community or *garrigue* becomes dominant (Table 5.1, Figures 5.3 and 5.8).

Fires have always occurred naturally in these regions, but human disturbance, including the deliberate setting of fires, is key to many of these plant communities. Much of the uplands bordering the Mediterranean were wooded until hominids began to use fire to clear the scrub and improve both their hunting and gathering (Table 5.2). The earliest indications of this strategy are found in northern Greece,

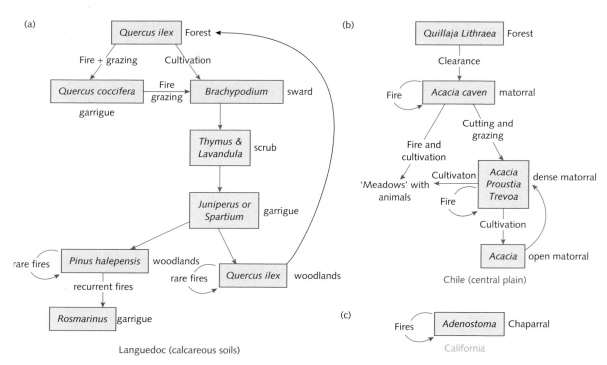

Figure 5.8 Changes in the composition of three mediterranean-type communities under different disturbances. In the Languedoc of southern France and the central plain of Chile forest gives way to a series of more open communities depending on the nature of the disturbance and its frequency. The matorral is now largely confined to steep slopes where there is little grazing pressure. The abandonment of grazing in the Languedoc has allowed the native holm oak (*Q. ilex*) forest to return, especially in upland areas. The chaparral of California is actually dominated by two species—either chamise (*Adenostoma fasciculatum*), as shown here, or scrub oak (*Q. dumosa*), and each will tend to sustain itself under the influence of fire alone.

dating back 1 million years, but fire was being used routinely in the eastern Mediterranean 10 000 years ago. The gaps encouraged rapid regrowth by a range of edible plants and these attracted game for the hunter. They also led to the domestication of grasses by early farmers (Box 5.1).

Frequent fires favour species adapted to its effects and these come to dominate the plant community, creating a uniform and largely continuous canopy.

At lower frequencies fire helps to maintain species diversity by creating gaps for invasive plants. Without fire competitive exclusion occurs (Section 4.3), and those forming the densest canopy come to dominate. In this way, one abiotic factor may be critical for the nature of the plant community in each region. However, other factors, including some form of human disturbance or grazing, can also create opportunities for less competitive plants (Box 5.3).

BOX 5.1   **Mankind and the Mediterranean**

Much of western culture has its origins in the Mediterranean Basin—Jewish, Christian, and Islamic traditions can all trace their origins back to the lands of the middle earth—and the ecology and geography of the region helps to explain much of its history. Indeed, Francesco Di Castri suggests that Mediterranean culture has evolved in step with the ecology of the Basin. Certainly the plant and animal communities of the Mediterranean Basin have been selected, directly and indirectly, by human activities (Table 5.2).

The current climatic conditions appeared after the first glaciation of the Pleistocene, around 1.64 million years ago, as *Homo erectus* arrived in the area. A series of glaciations followed and *H. erectus* was lost soon after a new species of *Homo* appeared in the fossil record—the Neanderthals. Around 40 000 years ago these were joined in the Mediterranean by a close relative, modern humans. With the end of the last Ice Age, the coexistence of these sub-species comes to an end, and it is the older residents that lose out.

Thereafter, from about 12 000 years ago, the lands around the Mediterranean Sea began to change, as agriculture and new tool technologies spread. We know that human beings had some measure of control over what was growing in different regions long before they actively cultivated the land. Wild barley may have been collected from the Nile Valley perhaps 18 000 years ago, and humans used fire to clear natural vegetation and encourage useful grasses. Sheep, goats, and gazelle were being husbanded in the Near East at the beginning of the Neolithic period, around 10 000 years ago (Figure 5.9).

Cultivation proper began in the 'Fertile Crescent' of the Near East at this time and developed independently in China and Mexico at later dates. Primitive cereals were exploited for the first time in Greece and the Levant around 8 000 years ago. This demanded a more settled way of life (Box 2.5) and a division of labour that operated with the seasons of the agricultural year. It also led to the establishment of markets and towns, where produce and services could be traded.

The general trend in the Mediterranean Basin was for technologies to originate in the east and diffuse westwards, with the sea as the highway that allowed trade and cultural exchange between its different centres. The human population grew considerably but the consequent expansion of agriculture took its toll, especially in the east (Box 5.2). Evidence from Greece suggests that major soil erosion began about 1 000 years after the onset of significant land use, a process that continued until about 600 AD. These losses were already substantial when Plato described them 2 500 years ago.

The history of the Mediterranean is of cultures meeting, often leading to confrontation and attempts to grab land and resources. These conflicts continue to this day. Because of the different traditions and languages packed into this small area, the Mediterranean Basin is both blessed and cursed by its cultural heritage. Yet, the fruits of this treasury have been bequeathed to the rest of the world, from the grape and the olive to mathematics, philosophy, and pizza.

(continued overleaf)

TABLE 5.2 The impact of three species of *Homo* on the ecology of the Mediterranean Basin. The development of agriculture, as hunter-gathering gave way to a more settled way of life, had a profound impact on mediterranean ecosystems. Humans used fire to clear scrub and simple tools to expand food production. This eventually led to established settlements and trade, but the ecological disturbance also promoted widespread soil erosion. Growing selected grains provided conditions that favoured the evolution of the earliest cereals.

Timescale (thousand years BP)	Homonid development	Technological advances	Ecological impact on the Mediterranean Basin
500	*Homo erectus*	Lower Palaeolithic Hunter/gatherers using hand-axes	Relatively minor—creation of gaps—later widespread use of fire. Gaps important for the spread of some grasses, especially *Avena sterilis* and *Hordeum spontaneum*, the precursors of modern cereals. Gaps also attract game. Fire may have been used to hunt
100	*Homo sapiens neanderthalensis*	Middle Palaeolithic Hunter/gatherers; flaked tools, first torch for carrying fire found in S. France	
40	*Homo sapiens sapiens*	Upper Palaeolithic Bone/antler tools; leaf blades, fire can be kindled	Fire used extensively, allowing food plants to flourish
20		Animal husbandry sheep and goats	Forest clearance on a large scale for grazing and cultivation
12	Population growth starts	Neolithic Revolution Agriculture, pottery, weaving	Primitive cereals to increase rapidly in Greece and Near East
9		Agricultural cultures in Egypt, Greece, and Persia	
7		Grapes and olives cultivated	
6		Shepherding in S. France	
5		Sequence of cultures flourish in E. Mediterranean and Mesopotamia; later W. Mediterranean	Increased aridity as plant cover is lost. Widespread soil erosion begins
1.4	Population growth slows		Forest clearance stops
0.1	Population growth confined to undeveloped nations	Industry and, latterly, tourism	Desertification in nations dependent on agriculture. Woodlands used for fuel; cultivation

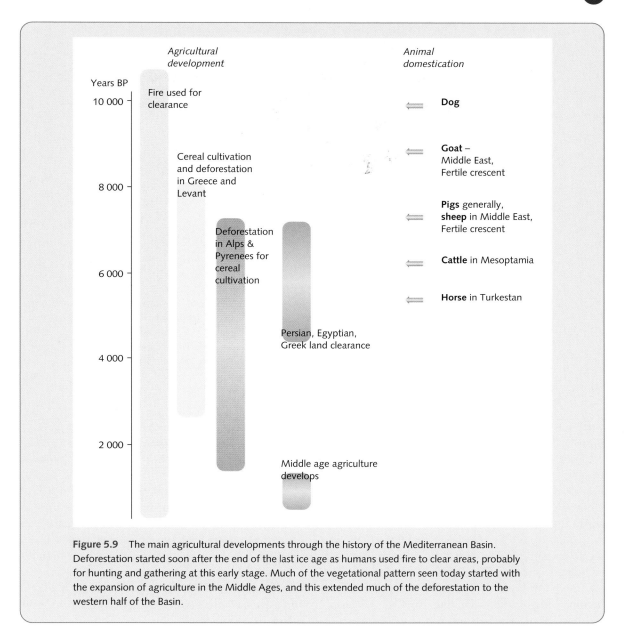

Figure 5.9 The main agricultural developments through the history of the Mediterranean Basin. Deforestation started soon after the end of the last ice age as humans used fire to clear areas, probably for hunting and gathering at this early stage. Much of the vegetational pattern seen today started with the expansion of agriculture in the Middle Ages, and this extended much of the deforestation to the western half of the Basin.

The scale of disturbance

These regions differ in their scale of human disturbance. Widespread human impact can be detected in the Mediterranean Basin from around 7 500 years ago, but our influence can be traced to each invasion by different species of *Homo* (Table 5.2). The typical garrigue-type community of upland coastal regions follows from changes in land use in the thirteenth century. Francesco Di Castri shows how these factors produce plant communities which are characteristic of each region (Figure 5.10).

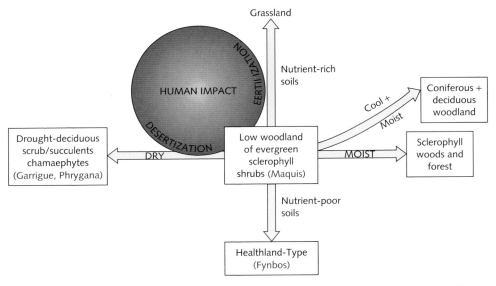

Figure 5.10 The main factors governing the plant composition of mediterranean-type plant communities. Assuming a drought-adapted maquis woodland as a basal community, different abiotic factors push the plant community into different configurations. Where it is moister and cooler, higher, more well-developed forest can grow. Increasing aridity promotes something closer to phrygana. Human impact, primarily the loss of cover (by fire and the grazing of animals), increases aridity and the fertilization of the soil—conditions which favour grasslands.

About half of the land in the Mediterranean Basin can be described as relatively undisturbed, confined primarily to steep or remote uplands. The cool and moister areas support cork oak (*Quercus suber*) and sweet chestnut forests (*Castanea sativa*). On the lower hills, much of the land had been used for grazing, resulting in soil erosion and more arid conditions. As this was abandoned, maquis scrub developed, dominated by evergreen species, including kermes (*Q. coccifera*) and holm oak (*Q. ilex*). At lower levels, other trees and shrubs become important including broom (*Cytisus*), *Genista*, olive, and strawberry tree (*Arbutus andrachne* and *A. unedo*), usually with an understorey of xerophytic grasses.

With significant disturbance, grazing, and trampling by sheep and goats, or the setting of fires, grasslands may develop, though more often a garrigue forms. This is a low open scrub of evergreen oaks, pistachio (*Pistacia lentiscus*, *P. terebinthus*), and juniper (*Juniperus oxycedrus*, *J. phoenicea*), but principally a ground cover of aromatic herbs, including thyme (*Thymus mastichina*), lavender (*Lavandula pedunculata*), and rosemary (*Rosmarinus officinalis*) (Figure 5.3).

An equivalent community, but with different species, has developed under similar pressures in Chile (matorral). This is not so in California where one or two species form large even-aged stands. Here, the open evergreen oak community has its counterpart as chaparral with scrub oaks (*Q. dumosa*) and aromatic shrubs (principally *Salvia mellifera* and *Artemisia californica*). However, the massive urban development in southern California, along with changes in agriculture and the prevention of natural fires have led to a major reduction of the natural sagebrush communities (Box 8.1). In South Africa and southwestern Australia human impact has not been so extensive and each region retains patches that are largely 'natural'.

Early agriculture had a significant impact on the vegetation of the Mediterranean Basin, removing the plant cover that otherwise protected the fragile soil from the searing summer heat and drying winds. The increased grazing and trampling from the stock led to further soil loss, though upland areas were later terraced to conserve soil and to create groves of olives, figs, almonds, and pomegranates. The grape, its cultivation and its fermentation, has since been exported to each of the other mediterranean climatic regions, and in each viticulture has become very important. Today, the area under the vine has expanded, livestock has been reduced, and the abandoned uplands have been left to regenerate a cover of maquis or garrigue (Box 5.2).

In effect, humans have exacerbated the summer aridity of the Mediterranean Basin, thereby favouring plants that can survive long droughts and simplifying the plant community in each location (Figure 5.8).

BOX 5.2   Drying the Basin

The Mediterranean Basin is far from a natural ecosystem, shaped as it is by the clearance of its forests and the agricultures practised around its rim. Most rural areas have assumed their present form since the Middle Ages, with local conditions of water availability, soil, and slope selecting a sustainable agriculture to feed a local community and perhaps to support trade. In some places the pressures of tourism, the demands for more agricultural land, or just new agricultural methods (Box 7.1) conspire to accelerate further clearance of scrub, woodland, and forest. Together with the threat of increasingly hot and drier summers, these changes may mean that the Basin will become less verdant and more desert-like (Figure 5.11).

But it did not start with our present civilization. As Jacques Blondel and James Aronson note, the forests have waxed and waned, and as major civilizations have waned so forests have managed to wax in some regions. Fragments and patches have disappeared and reappeared at different times over the last 10 000 years, largely following agricultural developments (Figure 5.10).

The forests of the Levant, including the extensive cedar groves for which Lebanon was famous, were exploited by the Pharaohs, by the Judean kings, and even by the Phoenicians themselves. The Phoenicians built a whole trading empire on their ships, but at the cost of their woodlands. Much of North Africa and the Middle East were still dense forest as late as 200 BC, and Julius Caesar used Tunisian forests to rebuild the Roman navy in 26 BC. The Iberian Peninsula lost large areas of forest as late as the fifteenth to seventeenth centuries, as Spain and Portugal in turn built their navies.

Figure 5.11 Denuded hillsides with gully erosion of soil on Kalymnos.

(continued overleaf)

Different invaders had different impacts on the landscape, and in some areas forests began to regenerate themselves. The forests of the western Mediterranean—southern France and northern Spain—started to recover with the fall of the Roman Empire and the arrival of the Visigoths. Large-scale land clearance here did not begin again until the Middle Ages, especially in upland areas.

Today, forests cover just 10 per cent of the land area of the Basin. A general recovery of Mediterranean forests has been under way since the 1939–1945 war, at least on the northern shores, but natural forest has retreated at an increasing pace in North Africa. There is now little or no natural forest in most of these areas and, regrettably, much of the replanting has been of pine and introduced *Eucalyptus* (Figure 5.27).

Similarly, much of the deciduous forest in the wetter regions has been replaced by sclerophyllous scrub, dominated by the evergreens adapted to the drier conditions. This is because deforestation brings increasing aridity and soil erosion. Removing trees allows rainfall to strike the soil directly and run off without percolating to any depth. Higher surface velocities loosen and wash away soil particles, eroding the fertile upper layers, especially on the steeper slopes. This gulley erosion is a feature of much of the eastern Mediterranean, with its overgrazed upland pasture and abandoned agricultural land. Soil erosion rates are around a hundred times higher in areas of the basin where the trees have been lost.

Animal communities

Some characteristics of these communities clearly result from the interactions between its species. For example, the composition of the chaparral of California is, in part, determined by seedling consumption by small mammals. Generally the seedlings of California lilac (*Ceanothus*) are preferred to those of chamise (*Adenostoma*) so any large-scale herbivory quickly leads to the plant community becoming dominated by chamise. If seedling consumption is reduced, *Ceanothus* dominates. In the Mediterranean Basin, intense grazing activity by voles may allow plants other than grasses to dominate in years of large vole populations.

Obviously, the animal community is closely tied to the collection of plants upon which it relies, directly or indirectly, but its species also have to adapt to the abiotic conditions. In particular, animals resident in the soil have to survive the long summer drought, and much animal activity is timed to match the availability of food and water. This is also one reason why these regions are visited by a large number of migratory species. Much of the insect and bird life of the Mediterranean Basin is shared with the rest of Europe because of these seasonal movements.

Their powers of dispersal and also their ancient lineage probably explain why the insects show the greatest similarities between the regions. Another ancient group, but a far less mobile one—the earthworms—also shows remarkable similarities across the globe. Indeed, the larger community of invertebrate decomposers within the soil is very similar between the five regions (Table 5.3). All are notable for their lack of beetles and their high diversity of woodlice, neither of which, incidentally, is readily explained by the biology of these groups. Similarly, mites are more prevalent than springtails (Collembola). Australia has the most distinct soil fauna, probably as a consequence of its early isolation (Section 8.2).

Nevertheless, there are similarities between the animal communities of the five regions. Not only do the soil invertebrates live to a great depth in each area, they also share the same patterns of seasonal movement up and down the soil profile. This reflects the depths to which some plants send roots in search of water and also the water storage organs that many produce. These, along with their invertebrate community, provide an extensive larder for burrowing mammals. Rodents or their equivalents are important in all regions where soils are easy to excavate.

Rabbits are thought to have originated in the western Mediterranean and were probably instrumental, along with goats and sheep, in preventing trees from dominating its drier areas. A number of equivalent grazers, filling similar functional roles, are found in the chaparral of California (brush rabbit and mule deer) and in the matorral (llama, alpaca).

Similarities

So to what extent are the patterns of community structure repeated in these five locations? Some of the obvious comparisons are summarized in Table 5.3. For the most part, the similarities in appearance are not due to shared plant species but shared adaptive features.

First, we should recognize that some regions have more in common than others. Two groupings stand out as distinct: Chile/California and South Africa/Australia (Figure 5.12). In each pairing, the similarities stem not only from a shared geological history but also from shared latitudinal and altitudinal ranges (Figure 5.1). In the New World, Chile and California both have high mountain ranges close to the coast, spanning several degrees of latitude. South Africa and Australia have a small altitudinal range and also a small latitudinal range. The Mediterranean Basin stands out because of its long history of human disturbance, its large longitudinal extent, and the dominance of its limestone geology.

The main plant species of both Chile and California have evolved equivalent carbon-gaining strategies: growth is confined to the spring, but the plants continue to photosynthesize at a low rate throughout the year. California chaparral has little human disturbance and will restore its dominant species readily (Figure 5.8); today the matorral is heavily grazed and exists in its least disturbed form only on the steeper slopes. Elsewhere, animals both trample the native plants and fertilize the soil, promoting changes in the plant community. Within each region, plant communities change with soil types: in South Africa and Australia the main distinction is between the more fertile soils, now largely used for

---

TABLE 5.3   **Comparing the ecological communities of the mediterranean-type climatic regions of the Earth**

(a) Shared features of *all* mediterranean-type communities

Plant community

Types:	sclerophyllous low srub or open woodland/thorn bush
Major adaptation:	summer drought
Secondary adaptation:	periodic fires
Species uniqueness:	high
Endemism:	high

Animal community

Soil community:	deep, stratified with seasonal migration
Soil invertebrate endemism:	high, rapid speciation since the Pleistocene but less than the plant community
Major adaptation:	summer drought
Herbivores:	insect types associated with sclerophyllous plants
Birds:	distinctive feeding categories
Reptiles:	similar number of species

(continued overleaf)

Table 5.3 *(cont'd)*

Relative features of all mediterranean-type communities

	Med. Basin	California	Chile	S. Africa	S.W. Australia
Plant community					
Species richness:	○	○	○	○	○
Impact of fire:	○	○	○	○	○
Animal community					
Distinct bird niches:	○	○	○	○	○
Soil invertebrate diversity:	○	○	○	○	○

(b) The organization of similar mediterranean communities

	California	Chile
Plant communities		
Response to latitude and altitude:		≡
Seasonal (spring) growth pattern:		≡
Carbon-gaining strategy:		≡
Leaf forms in each habitat:		≡
Easily invaded by foreign weeds:		≡
Productivity:	↑	
Litter production:	↑	
Frequency of fire:	↑	
Dominance by one species:	↑	
Forest rather than shrub as main type:		↑
Developed herbaceous layer:		↑
Soil fertility high:		↑
Animal communities		
Niches occupied by reptiles, birds mammals:		≡
Insectivorous birds feeding in vegetation:		≡
Reptiles associated with herbaceous layer:		↑

	Australia	South Africa
Plant communities		
Tall woodlands with grassland understorey:		≡
Growth confined to summer season:		≡
Heathlands on less fertile soils:		≡
Animal communities		
High termite activity:		≡
Soil invertebrate diversity:	↑	

Arrows indicate which region shows a higher response or higher level, otherwise there are shown as equivalent.

The first part of this table lists those features that are shared by all five regions. Next, the relative significance of a particular feature for each region is shown by the size of the symbol—so plant species richness, for example, is indicated to be very high for S. Africa.

In Table 3b, comparisons are made between the pairings where similar abiotic factors apply to both regions—Chile/California and south-western Australia/South Africa. Where a feature is dominant in one region this is indicated by the arrow. The Mediterranean Basin itself has no direct equivalent.

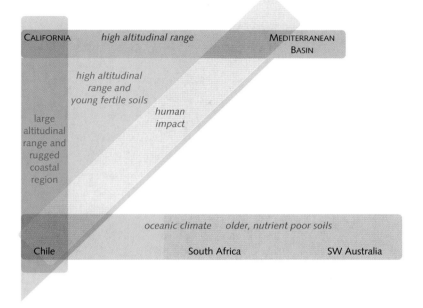

Figure 5.12 A division of the five mediterranean regions according to their shared abiotic factors. South Africa and southwestern Australia represent one distinct group primarily because of their combination of oceanic climate and nutrient-poor soils. Chile and California represent a second group, with the Mediterranean Basin distinct because of it high human impact, its large east–west extent, and the prevalence of limestone rocks within the region.

grazing, and the poor soils that support native fynbos or heathland mallee.

The communities of larger animals, especially birds and lizards, share a number of features between the regions. A key factor for the birds is the vertical development of the shrubs and the density of the plant cover (Figure 5.13). Similar numbers of bird and lizard species are found in each region and comparable population densities occur in equivalent positions in the vegetation. Feeding on seeds and insects also requires similar strategies in each region. Such parallels reflect equivalent species interactions between the dominant plants, insects, and birds. Martin Cody has described the close match between the insectivorous birds of the Californian chaparral and Chilean matorral. Although they contain different species, each niche seems to be occupied by a bird of equivalent size and shape, according to

where they feed in the canopy. The birds are said to comprise a **guild**—a group of species with similar functional roles exploiting the same resource. There is some experimental evidence that the number of species within a guild may be limited (Box 5.3, Section 9.3).

Outside the woodlands, differences start to emerge. The fynbos and its dominant plant group, the proteas, provide opportunities for a guild of nectar-feeding birds, largely absent in the other regions. Likewise, lizard diversity and abundance is higher in Chile because of the well-developed herbaceous layer absent in California. Overall, the complexity of the vegetation in all mediterranean-type regions seems to be a good predictor of bird diversity and also the degree of niche separation of lizards.

Some have argued that this may be a misleading picture for the birds, at least for the Mediterranean

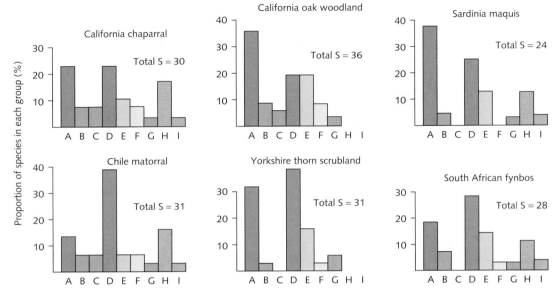

Figure 5.13   A comparison of bird communities in four mediterranean regions and in two very different woodland communities. In each case, the birds are grouped according to their feeding habit (A = foliage insectivores, B = sallying flycatchers, C = nectarivores, D = ground foragers, E = seed/fruit eaters, F = trunk/bark foragers, G = aerial feeders, H = raptors and scavengers, I = crepuscular insectivores). Notice that the proportion of bird species in each category shows the same sort of profile in all communities. However, the mediterranean-type communities do have groups (H and I) which are not found in the other habitats, and there is a greater similarity in the profiles between these regions. This is some indication that the bird community has developed to a similar result in each region, with similar proportions of species in roughly equivalent niches. The most different mediterranean-type community is Sardinia, which has the smallest number of species and an absence of birds in two categories, most probably because it is the one island among this group.

Basin. James Aronson suggests any mature northern temperate woodland in Europe, whether from around the Mediterranean or further north, will tend to have the same number of bird species and more or less the same species list. Moreover, the basin is a crossroads of migrating birds: not only do those from northern Europe stop there to feed on their way to tropical Africa, but so also do a large number of Siberian and Asiatic species that prefer not to cross the Himalaya. Such movements, back and forth, occur through much of the year, so there is a massive turnover of bird species, more than in the other regions.

A rapid speciation has been a characteristic of each region as the modern climate became established. Possibly the higher endemism and higher biodiversity in South Africa and southwestern Australia result from their higher frequency of fires and the higher species turnover this promotes (Section 5.3). Each region started with a different pool of potential colonizing species, but the question is whether these have filled equivalent niches in each location. If they have, we might suspect there were constraints on how these communities were organized, an indication of the rules by which these communities were put together (Box 5.3).

BOX 5.3 **Building a community**

Different species have different roles and make different demands on their environment. We thus expect to find different niches occupied by different species. In fact, being different is a good way of finding space for yourself in a community. As we saw with the scuttle fly (*Megaselia yatesi*—Section 4.3), being adapted to an empty niche may enable a species to establish itself without significant competition. Even so, an invader has to cope with the prevailing abiotic conditions and that may mean reaching adaptive solutions similar to the neighbours'.

These conflicting pressures to conform or to be different are resolved by natural selection, according to the dominant abiotic conditions and the interactions with other members of the community. Species can only coexist over the long term if they can accommodate, and be accommodated by, other members of the community. Some combinations of species may not persist because of competition, parasitism, or predation. It may be, for example, that there are not enough resources to support more than one species of large predator, or that the activity of one successful scavenger prevents others from becoming established. The partitioning of resources will be a key determinant of the structure of all ecological communities.

The relative abundance of species within a community will reflect this partitioning and the interactions between species (Box 9.1). Both interspecific and intraspecific competition are known to determine species combinations in mediterranean-type communities: some plants are excluded from the community of the Chilean matorral if insect herbivory is intense. In the hills above San Diego, James Mills has shown that their preferential consumption of *Ceanothus* seedlings allows chamise (*Adenostoma*) to dominate, regenerating chaparral.

Based on his studies of the bird communities on islands in South East Asia, Jared Diamond suggested there were **rules** of assembly that allowed only particular species to coexist. Only certain combinations of birds were found on these islands because resource utilization and competition (niche overlap) meant that other assemblages did not last. Based on his studies of desert rodents, Barry Fox proposed a 'guild assembly rule' which says that the next species to enter a community is most likely to be from a different functional group. So each category is equally filled—if there are two herbivores and two omnivores present and only one insectivore, the next successful invader is likely to be an insectivore. Within a group (in this case, rodents) such an invader is expected to experience less competitive pressure. This 'rule' has been hotly debated by ecologists, some of whom think it too simplistic and probably only applicable to desert rats.

Another view is that collections of species will self-organize, but according to circumstance, chance, and prevailing conditions. Then there can indeed be different combinations of species as long as key roles are performed. Obviously, species poorly adapted to the abiotic circumstances will not persist, but thereafter the competitive and cooperative associations of the rest, their adaptability, and chance decides community structure. Some outcomes are more likely than others but there is no fixed result—communities can be pushed into various configurations, according to the accidents of history. In this case, convergence towards similar outcomes will be largely a product of the selective pressures exerted by the dominant abiotic factors.

Species will change simply through the processes of migration and local extinctions (Section 9.1). Any ecological community is the current solution to the various interactions between its species and there are often alternative solutions. Because of this, shifting abiotic conditions may allow some communities to cycle between different states.

5.2 Convergence and integration

Our task is to distinguish between a tightly integrated community and a looser collection of species living together because they can survive the prevailing abiotic conditions. For plant communities, especially, Bastow Wilson points to the difficulties in distinguishing common adaptive features from well-defined rules governing coexistence. We see repeating patterns suggesting constraints on community structures, but ecologists have yet to agree on how to distinguish a rule of assembly from a shared adaptation or the 'noise' of chance variation (Box 5.3).

It is certainly not chance that makes California appear similar to Chile and the fynbos comparable to the mallee. Given the climatic regime in each locality, there are only a small number of viable plant growth strategies to cope with the seasonal conditions—especially the summer drought. And animals that exploit such plants will also need to survive the drought and overcome plant defences that protect long-lived leaves.

However, shared characters in plants or animals are some way away from tightly knit communities with matching niches. Differences in detail are likely to create differences in niche space so that the community jigsaw does not map perfectly from one site to another. A fair test would require us to account for all the variables of geology, frequency of fire, history of human disturbance, and so on. Certainly, the closest similarities do occur between regions that share many of the abiotic features in common (Chile and California; South Africa and Australia—Table 5.3), but it seems that even minor variations can become magnified into major structural differences, producing different species assemblages.

Perhaps the simple truth is that no two habitats or regions are sufficiently alike in their abiotic factors. The similarity in mediterranean communities is largely attributable to their plants reaching similar adaptive solutions to long hot summers on poor soils. Thereafter, their different ages or their different degrees of human disturbance mean they are separated today as much by their histories as their ecologies. This natural experiment comes to no conclusion, and, as ever, leaves many questions unanswered. It does, however, provide us with some important insights into community organization. The differences in detail matter, and therefore the resemblance between the mediterranean plant communities is no guarantee of a shared community structure. If there are common rules of assembly, we need more incisive methods to identify them.

The ideal experiment would be to allow communities to assemble themselves, starting with the same abiotic conditions and drawing on the same pool of species. Then, given long enough for species turnover to reach a minimum, perhaps the same collection of species, organized in the same way, would result. Or perhaps, chance would push them in different directions.

What are communities?

We might regard communities as highly integrated if their species assemblages came in bundles—collections which cannot be unpicked. So to have species A, the community must already be occupied by species B and C, or perhaps D must be absent. A tightly knit community, built according to these rules, would be highly predictable for a climate. In fact, such dependences between species are known for most ecological communities, and the presence of some plants, animals, or micro-organisms is critical for the character of the community. We expect to see certain types of plants and animals in temperate woodlands or coral reefs, but in different locations we quickly distinguish them as different species carrying out similar roles.

Such associations can be used to classify community types (Box 5.7). But we also observe that communities grade into one another—grassland gives way to woodland, woodland to high forest—and that they are are continual rather than quantal. Thus, ecologists now ask which configurations of functional roles (niches, rather than particular species) are critical to define the nature of a community (Box 5.4).

Clearly, some species are crucial for the structure of the community and ecosystem to function. Keystone species are those whose presence determines the nature of the community; without them the community would have a very different nature. An example is the African elephant, whose browsing activity on the savanna maintains the open grassland and suppresses encroachment by scrub (Section 3.7). We have also seen how the loss of a keystone species, the ant *Anoplolepis custodiens*, resulted in a dramatic change in the species composition of the South African fynbos (Section 4.2). Yet within any community there will be a number of species that can be lost without an appreciable effect on the community as a whole (Section 9.5). In such cases, the key roles are undertaken by guilds of species in which no single species is important (Box 5.3).

BOX 5.4 | **The right type**

In studying communities we frequently find situations in which no one species is key. Instead, functional groups of several species undertake key roles. Because of this, techniques have been developed to classify species according to functional types, and a community is then described according to the types of organisms that characterize it.

Plant functional types are central to defining communities. These are based on the life-form and physiognomy of plants first described by the Danish botanist Christen Raunkiaer (Figure 5.14). Here, plants are grouped into five categories according to the position of their perennating buds—the buds that give rise to the following season's growth. Phanerophytes (literally 'visible' plants) are trees, shrubs, and other woody plants with buds 25 cm or more above ground. Low-growing woody plants with buds below 25 cm are classified as chamaephytes, whilst those with buds at ground level are hemicryptophytes. Cryptophytes have their buds buried beneath the soil. Finally, there are therophytes—annual species that sit out the unfavourable season as dormant embryos within seeds and do not have persistent buds.

This classification becomes useful when we begin to compare different communities. Often the profile of Raunkiaer's life-forms are characteristic for a community type. Figure 5.15 shows the different life-forms of plants within six mediterranean shrub communities across three geographical regions. Notice how coastal communities have fewer phanerophytes than their inland counterparts, almost certainly a response to the stresses of living by the sea—such as exposure to salt-laden winds.

Jon Keeley and William Bond investigated the degree to which fire-adapted germination occurred in the Californian chaparral and South African fynbos by categorizing the species according to their germination strategies and life-form types (Figure 5.16). Many phanerophytes use fire as a cue to germination. There are also proportionally less cryptophytes and hemicryptophytes but when they do occur, their seeds tend not to be fire adapted.

Equivalent patterns suggest that the plants are responding to similar conditions with similar adaptive strategies. In this way, we can compare communities not on the basis of shared species but on the basis of shared adaptations, indicative of similar responses to the abiotic conditions and perhaps also to the important species interactions within the plant community.

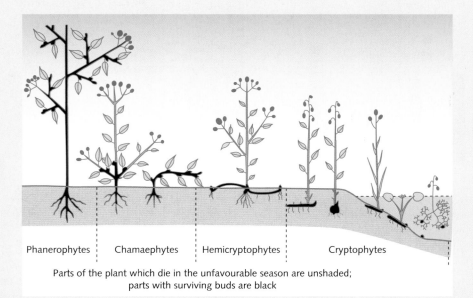

Phanerophytes | Chamaephytes | Hemicryptophytes | Cryptophytes

Parts of the plant which die in the unfavourable season are unshaded; parts with surviving buds are black

Figure 5.14 Raunkiaer's life-form classification in which plants are grouped according to the position of the buds that survive from one year to the next. This classification is a useful way of comparing plant communities not by their species composition but by their shared adaptive forms.

(continued overleaf)

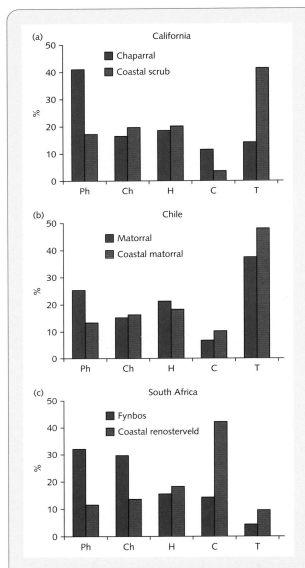

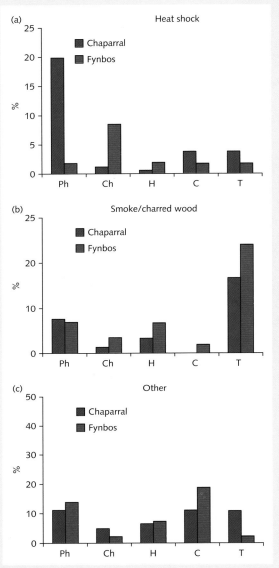

Figure 5.15 Patterns of life-forms in three regions with a mediterranean climate. Raunkiaer's classification of plant forms is shown for coastal and inland zones for each region, with the percentage of each type in the two zones. Drier inland areas tend to be dominated by phanerophytes—tall, perennial trees and shrubs, whereas annuals (therophytes) tend to dominate the moister coastal zones. In South Africa cryptophytes—perennials that protect their buds by keeping them underground—survive the very harsh conditions in the coastal renosterveld. (Ph = phanerophyte, Ch = chamaephyte, H = hemicryptophyte, C = cryptophyte, Th = therophyte.)

Figure 5.16 The pattern of plant life-forms in chaparral and fynbos under different germination regimes. Phanerophytes germinate especially after the heat shock of a fire (a). Therophytes instead require to be primed by the effects of smoke or an abundance of charred wood in the seed bed (b). Most other types require other cues (c). (Ph = phanerophyte, Ch = chamaephyte, H = hemicryptophyte, C = cryptophyte, Th = therophyte.)

5.3 Change in communities

Human disturbance is key to the structure of many mediterranean-type communities and we continue to play an important role in maintaining these assemblages. Yet, even in our absence, species are lost and others invade. The competitive battles and other interactions within a community left alone can quickly lead to a procession of species. Within a region, the stages of development are often highly predictable, hinting that, at least for some habitats, there are only certain ways in which its assemblage can be constructed. This sequence of invasions and replacements is known as a **succession**. A succession ends when colonization and loss are minimal and the community changes very little. Then the plant community assumes a persistent structural form and a stable assemblage of species termed a **climax community**.

Studying succession can tell us which interactions are important to a community and also explain the factors ordering species arrivals and departures (Box 5.5). There are two basic types: a **primary succession** develops on a site where there has been no previous occupation and where colonizing species must initiate ecological processes. These are generally newly exposed sites such as volcanic larva, the stony remains left behind by retreating glaciers, or the shifting sands of developing sand dunes. **Pioneer species** are those able to colonize the available space (Figure 5.17) even though nutrients and organic matter are lacking. Their growth and activity leads to the accumulation of resources that can support later colonists.

Secondary succession is a process of recolonization following disturbance. For example, a site that has been previously occupied may retain some of its original species as a seed bank in the soil. Most importantly, the soil still contains organic matter and nutrients that can be used by invading species. Colonization is thus faster than in primary succession. Which species colonize depends on what remains of the original ecosystem, including the species already present, and also the distance invading species must travel to reach the site (Section 9.1).

Figure 5.17 Plants colonizing lava in Samoa. Pioneer species begin the process of primary succession on the cooled lava flow.

What makes a good colonist? Most obviously, the capacity to disperse and to reach an unoccupied site. Both plant and animal pioneers are typically *r*-selected opportunists, species which can produce large numbers of easily dispersed seeds or offspring. Many annual plants and insects are typical pioneers, able to reproduce prolifically within a short time (Section 4.3). Colonizers also need to survive a wide variety of conditions so that their reproductive success does not depend on a particular resource or on other species. For this reason, early-successional communities typically consist of loose associations of short-lived and highly dispersed species, each

BOX
5.5 **Studying succession**

Ecologists investigating succession are dogged by one problem: time. Gradual, long-term change makes it difficult for one ecologist to follow the successional sequence (known as a **sere**) within a single site. One possible solution is to match successional patterns across several comparable sites but at different stages of development and so obtain a composite picture of the successional sequence. In some cases, an individual site may consist of areas at different stages. An example is the hydrosere—the transition which occurs from open freshwater to dry land, where each stage of community development can be seen over a short distance (Figure 5.18).

Shallow lakes have a relatively short ecological life. The colonizing activity of reeds, rushes, and sedges and the sediment that collects around their roots mean that lakes gradually fill in and shrink. Trees such as willows (*Salix* spp.) grow close to the water's edge, as they tolerate the damp conditions. Their success as facilitators is due to their large roots and fast transpiration rate that pumps water out of the soil. Further back, on drier deposits, willows give way to late-successional species in the form of canopy trees such as oak (*Quercus* spp.) and ground flora that can only survive relatively dry conditions.

Sand dunes also show seral change as a linear sequence. Ecologists can thus substitute change in space for change in time to follow the succession. Going inland from the shore represents a progressively older dune community (Figure 5.19). In the Mediterranean, the stabilizing effect of marram (*Ammophila arenaria*) and the nitrogen-fixing activity of sea medick (*Medicago marina*) mean these are the first to appear on the young dunes. These facilitate the arrival of more demanding species, and a scrubby garrigue will develop on dry ridges. On the more sheltered landward side of the dune ridges, assemblages of xerophytes form, including species of *Cistus* and lavender (*Lavandula stoechas*). On nutrient-poor sites the result is a scrub dominated by juniper (*Juniperus phoenicea*) and mastic tree (*Pistacia lentiscus*). With richer soils and less disturbance, trees

Figure 5.18 The hydrosere: an area of open water gradually being encroached upon by vegetation.

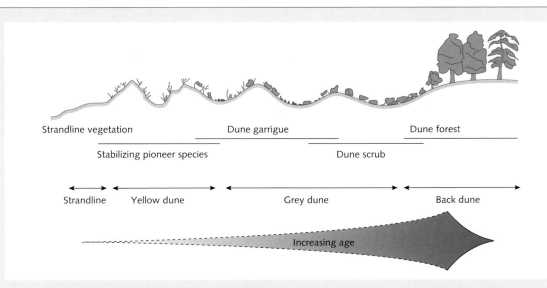

Figure 5.19   Succession on an idealized Mediterranean sand dune. New dunes are colonized at the seaward end, whilst woodland and forest occupy the oldest part of the site, consisting of late successional species that have replaced the earlier colonizers. Note that the changes occur as transitions rather than abrupt zonations, with overlap between the various species assemblages within the succession. The overall trend is one of increasing abundance of woody species over time, and this is reflected in the transition from pioneer communities, through garrigue and scrub into woodland.

colonize the oldest dunes, and a woodland of oak (*Quercus* spp.) and pine (*Pinus* spp.) develops (Figure 5.20). An equivalent pattern of development occurs in other regions with a mediterranean climate. In California, a characteristic sagebrush (*Artemesia californica*) or pine forest may develop, whilst in Australia the soil conditions determine whether a mallee heath or *Eucalyptus* forest forms.

Very few sites have been studied for long enough to give a complete picture of community change. A frequently-used method is to collect fixed-point photographs taken from the same position over many years, to record the advances and retreats of the major species. Ecologists have also used archaeological techniques to uncover the history of sites and to reconstruct the development of their communities over the longer term. Some reveal their past in vertical cores taken from the soil, especially in aquatic or well-layered sediments that can be readily sequenced. The pollen found in layers of a known age can be particularly instructive, indicating which species grew when and their relative abundance (Figure 5.21).

(continued overleaf)

Figure 5.20 Photo-montage of dune succession in the Algarve region of Portugal. Photograph (a) shows sand being trapped by vegetation on the strandline, whilst in (f) pine trees are seen colonizing the oldest dunes. Images (b)–(e) are the intermediate stages of the developing dune system.

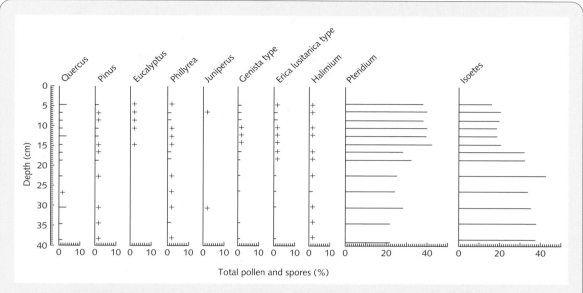

Figure 5.21   Pollen diagrams can assist in the study of succession. Here, pollen and spores from a range of species is plotted according to the depth at which it occurs within a soil core (in this case through a marshy dune—*marismas*—in Cota Donana, Spain). The oldest records are found within the deepest part of the core and come from early successional plants such as quillwort (*Isoetes*) and bracken (*Pteridium aquilinum*). Note that *Eucalyptus* is a relative newcomer as its pollen is only found in the first 15 cm of the soil. (Line length corresponds to percentage of pollen/spores, with '+' indicating presence at less than 1 per cent.)

able to survive a range of habitats and exploit a range of resources.

In most cases, plants have to establish themselves if animal colonists are to survive. Here is one simple rule of assembly for most communities: plants ultimately provide the resources needed by both herbivores and carnivores (Section 6.1). Similarly, the development of the microbial and fungal communities, key components of a living soil, will follow as organic matter is added by the pioneer plant community. This is known as facilitation—one species paving the way for others. Facilitation includes any modification of the environment to create conditions amenable for later arrivals, such as providing shelter or hastening the release of nutrients from rocks. For example, in many coastal dune communities marram grass (*Ammophila arenaria*) is an important pioneer, stabilizing the loose sand blowing landward off the beach. By its growth it adds organic material to the sand, producing a soil that other species can invade. Marram can thrive despite frequent burial, and different species have various adaptations to the shifting sand (Figures 5.22 and 5.23).

Similarly, estuarine saltmarshes are inundated at high tide, when sand and silt are being deposited and moved at rapid rates. Plants growing here have to survive periods of inundation by salt water, followed by exposure to air at low tide. An important pioneer in Mediterranean saltmarshes is perennial glasswort (*Arthrocnemum macrostachyum*) which readily colonizes the open salt pans of the marsh flats. Alfredo Rubio-Casal and his co-workers found that once established *Arthrocnemum* provides the necessary shade and shelter for a series of successors, including annual glasswort (*Salicornia ramosissima*) (Figure 5.24). This pioneer reduces the salinity and increases the nutrient content of the silt, and this allows other plants to establish themselves.

Figure 5.22   Marram (*Ammophila arenaria*) is a colonizer of shifting coastal sands. Here its roots are partially exposed to show the extensive network that helps to bind the sand dune and begin the process of soil formation.

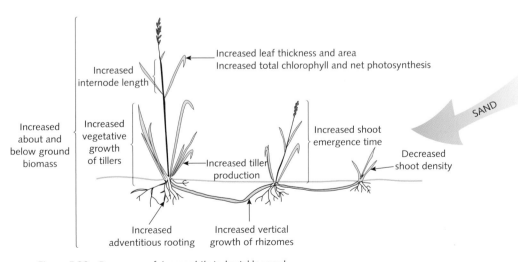

Figure 5.23   Responses of *Ammophila* to burial by sand.

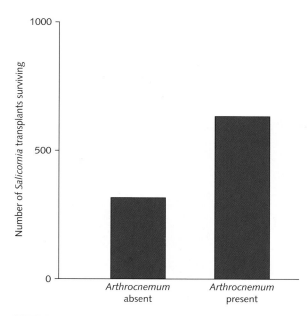

Environmental conditions		
	Salt pan	Canopy
Sunlight (PPDF) ($\mu Em^{-2}s^{-1}$)	1923	250
N (μg/kg soil)	4.68	7.38
P (g/kg soil)	3.4	4.68
Organic matter (g/kg soil)	0.84	2.05

Figure 5.24 The role of the perennial glasswort (*Arthrocnemum macrostachyum*) as a facilitator within Mediterranean salt pans. The graph shows survival of the annual glasswort (*Salicornia ramosissima*) with and without the presence of *A. macrostachyum*. Environmental data for conditions in the open salt pan and under the canopy of *A. macrostachyum* indicate that the shade and shelter provided by perennial glasswort lessens the stresses of life on the open saltmarsh.

Facilitation is most obvious in the early stages of a succession, but not exclusively so. Later in the sequence, well established, possibly keystone species, create conditions or provide resources which much of the rest of the community depends upon. The long-lived oaks of the maquis or chaparral play this role in several mediterranean communities throughout the world, not least in the bird life they support (Figure 5.13).

Early-successional facilitators often sow the seeds of their own destruction and are rarely part of late-successional communities. Species arriving later are typically more competitive and will squeeze out the pioneers. Plants such as *Arthrocnemum* give way to species that become dominant as the community develops. The late arrivals invest for the long term, and have strategies to out-compete their neighbours for space and resources, often inhibiting their growth or establishment. This **inhibition** slows down colonization by other species or excludes them altogether. An example found on Mediterranean coastal dunes is actually a recent invader, *Carpobrotus* (Figure 5.25). Introduced from South Africa, it has within a few centuries become dominant on sheltered early-successional dunes, forming a thick mat that inhibits native colonizers. Although *Carpobrotus* can dominate sites, its hold is eventually loosened by more competitive native species.

Again, we could see this as a general rule of assembly—late-successional communities, without significant disturbance, become dominated by slow-growing competitive species. Succession is thus a process by which the species composition of a community becomes increasingly fixed. Towards the end of a succession, species turnover—arrivals and departures—slows, and the community composition becomes governed less by abiotic factors and more by interactions between its species. Where community development is primarily driven by these internal processes it is termed an **autogenic** succession. Successions principally determined by external, abiotic factors are termed **allogenic**. Ordinarily, the sequence of plants on a sand dune is principally an autogenic succession (Box 5.5), but if there is periodic waterlogging of a dune slack (the hollows formed behind the main ridge), a very different plant community becomes established. The plant and animal species found here are adapted to the wetter conditions, and the community is the result of an allogenic succession.

Changes in the species assemblage are matched by changes in ecological processes. As nutrients and organic matter begin to accumulate, so a reserve of resources and a decomposer community develops (Box 6.2). Again, this facilitates the arrival of other species, and nutrients cycle more tightly within the community. Later on in the succession, nutrients

Figure 5.25 *Carpobrotus* invading a Mediterranean sand dune.

released by the decomposers are quickly utilized by competing plants, so that over time less and less is lost from the system (Section 6.2). We thus expect ecological efficiencies to improve from early- to late-successional communities.

Competition for these resources is often so great in late-successional communities that plants have to be able to survive shortages of nutrients, light, or even water. Plants that survive these conditions are said to show **tolerance** and are typically slow-growing and infrequent reproducers (Section 4.3). Theirs is a waiting game, maintaining themselves in times of shortage, ready to exploit times of plenty. As others fall away, tolerators come to dominate. This again may be a rule of assembly, favouring such strategies in habitats where nutrient supply is limited.

Disturbance and succession

Any disruption that causes the loss of most or all of the resident species will reset the community clock and these processes will start ticking again. The rate of change can be dramatic: volcanic ash and lava can wipe out a community in a matter of hours, whilst the advance of an ice sheet is slower but equally pervasive. In both cases, some species may escape when most others are lost. The likelihood of a particular species remaining or being lost will depend on the availability of refuges.

Colonizing species track the changes in a succession, as pioneers continually advance into new territory, followed by late-successional species that gradually establish a community (Box 5.5). We see this pattern in sand dune formation, where a front line of colonist species advances onto the blown sand, stabilizes it, and makes it part of the dune system (Figures 5.19 and 5.20). A similar sequence follows as a glacier retreats up its valley, uncovering lifeless gravels formerly buried by the ice.

Communities are often adapted to a particular frequency of disturbance and are said to **incorporate** this disturbance (Section 8.1). This is true of mediterranean-type communities that are dominated by plants able to withstand short fires (Section 5.1). Most never reach a stable configuration because fire initiates a cycle of change. This postponement of a climax community is termed an **arrested succession** or a **plagioclimax**. Thus, a

Figure 5.26   The uniform pine forest in the lee of the Pilat dune system near Arcachon, in Les Landes of western France.

garrigue represents a plagioclimax of maquis, with the woodland scrub arrested by a high fire frequency (Figure 5.3b).

The structure and diversity of the whole community may depend on the frequency of such disturbances. In his **intermediate disturbance hypothesis**, Joseph Connell suggests that a community supports its greatest number of species when the disturbance is relatively frequent—not too frequent to cause major extinctions but frequent enough to prevent the competitive dominants from squeezing out other species. Disturbance creates gaps which invasive species may occupy. Some plants, known as **fugitive species**, are able to maintain a population by colonizing one gap after another. The perennial glasswort of the salt-marshes is an example, maintaining a population by colonizing the open pans where new silt is being deposited or existing vegetation has been scoured away.

A community that is largely the result of an autogenic succession is likely to be very elastic, always returning to a similar species configuration after a disturbance. Its species interactions will tend to push it towards a particular endpoint (Figure 5.26). We get some sense of this predictability when we look along the line of a beach and see the same vegetation zones occurring at the same distance from the shore, on the newly forming dunes (Figure 5.20). With less predictability and more frequent disturbance the early stages of a succession have the least stable communities.

The interaction between plants and animals can be equally important. Herbivores that feed on early-successional species can delay the successional process by slowing colonization by other species. If, on the other hand, the plant being grazed inhibits succession, the herbivore may actively promote colonization by other species. Later in the sequence, herbivores have the power either to halt or even reverse a succession. Catherine Bach showed this in her investigation of the role of the flea beetle (*Altica subplicata*) in dune succession on the shores of the American Great Lakes. *Altica* feeds on the dune willow (*Salix cordata*) and Bach compared dune development over a three-year period with and without the activity of *Altica*. Not surprisingly, the density of the willow declined in the presence of the beetle, with a corresponding increase in the density of herbaceous plants. Herbivory by *Altica* seems to facilitate colonization of the dune by plants otherwise inhibited by the willow. This three-way interaction shows how both facilitation and inhibition can result from the activity of a single species.

Few ecologists today would suggest that succession can have a single outcome—a particular climax community for a particular climate. Most now recognize that chance and history, and different starting positions lead to different results. We can, however, see recognizable and repeating patterns that allow us to identify broad community types, invariably shaped, if not defined, by the interactions between their species.

5.4 Communities, change, and conservation

Clearly, some communities have been displaced a long way from their natural state. Today most mediterranean-type communities are plagioclimaxes that have incorporated varying frequencies of disturbance within their limited cycle of change.

The pressure of our numbers and activity has accelerated the pace of disturbance, hastening the loss of plant cover and prompting widespread soil ero-

sion and degradation in these regions (Figure 5.11). Just as Plato described habitat degradation occurring in Ancient Greece, our changes to the Mediterranean Basin today may mean we too will witness paradise lost (Box 5.6). Despite our long history together, the pressures on the landscape, from tourism, agriculture, and industry, may mean that the scale of human disturbance is too great for our partnership with this

---

 BOX 5.6   **The North–South divide**

Jacques Blondel and James Aronson highlight another way in which the Mediterrenean Basin represents a microcosm of the larger world—the clear demarcation of wealth, population growth and economic development between its northern and southern halves . . . and the impact these have on its ecology.

Of the northern shore they say:

> So a gloomy dichotomy emerges: far from the coast there are deserted fields, orchards and pastures, progressively encroached upon by shrublands, and increasingly dense, unproductive and ill managed woodlands. Along the coasts, in the densely urbanized and homogenized industrial zones that continue their inexorable sprawl, all ecological and cultural contact with the Mediterranean past is abandoned and lost.

And of the south:

> A visit to any mountainous areas of North Africa today reveals that demographic pressure, combined

with a highly conservative rural economy still largely disconnected from outside markets, results in very low crop yields and overall productivity. This leads to the all-too-familiar cycle of increasing ploughing and grazing areas followed by soil erosion, and then new clearing elsewhere.

On the northern rim, much of the agricultural land has been abandoned and a scrubby woodland has developed. Without the moderate grazing pressure from large mammals constraining the more competitive species, uplands lose much of their floral and faunal diversity under a scrub of highly competitive sclerophyllous plants. In North Africa, by contrast, the area under cultivation has expanded with the rapid growth of its population. The attempt to feed the people using increasingly infertile and unproductive soils has meant greater encroachment into the upland forest areas, leaving natural woodlands confined to remote, inaccessible pockets.

Today the resident populations of the four major states of the northern shore are not growing, whereas the African countries have annual growth rates around 2 per cent, amongst the highest in the world. Typically these people have one-sixth of the income of those on the European side. In the starkest of contrasts, the Basin is the most popular tourist destination in the world, receiving somewhere between 200 and 250 million visitors each year, most of whom stay on the coast.

The pressure of numbers, either from the resident population on the southern shore or the migratory population that comes to rest on the northern shore in the summer, place major demands on the land. On its northern rim, agriculture, other than viticulture, has been so reduced that most of the economic activity is concentrated in the coastal regions or larger cities—in some places (e.g. Mallorca) half of the coastal fringe has been lost to tourist development.

Meanwhile, the expansion of agriculture on the southern rim of the Basin has caused widespread environmental degradation, especially soil erosion.

The two sides do share one thing in common—an increasing shortage of water—a problem found in all other mediterranean climatic regions. Water supply and waste water treatment are major issues on the border between Southern California and Mexico, where respecting the rights of neighbouring states in both water extraction and pollution is increasingly important. Water rights are also behind much of the political manoeuvrings of the Middle East. Each region has its own peculiar ecological problems and these are likely to get worse if the predicted reductions in rainfall follow with global warming (Section 8.3).

landscape to continue as it is (Boxes 6.6 and 7.1). Along with the threat of increased aridity, some are predicting a major wave of extinctions in the near future, as many of the endemic plants of these regions are lost (Section 9.5).

In recent years, human disturbance has taken a different form. The impact of tourism and its associated development has resulted in the loss or damage of 75 per cent of Mediterranean sand dunes in the last 30 years. The very act of visiting a place involves the use of its resources, adds effluents, and demands facilities, and these combined represent a potent form of disturbance. Visitor pressure, simply walking on the dune, can damage colonization by marram. Research has shown that a mere 500 pedestrian passes on a dune can have a significant, albeit temporary, effect. More intensive visitor pressure can result in permanent change in the vegetation.

Through our travels we have connected the five mediterranean regions to each other, shuffling species between them. Not only do we now cultivate the vine in each region, we have also introduced some less desirable species. *Carpobrotus* from South Africa has become an invasive weed in the Mediterranean Basin and California. Similarly, *Eucalyptus* from Australia is displacing native species (Figure 5.27). Adapted to the nutrient-poor soils of the mallee, its fast growth makes it a formidable competitor as it scavenges

Figure 5.27 *Eucalyptus* displacing native Mediterranean oak/pine forest.

nutrients from richer Mediterranean soils. The result is an impoverished ground flora within *Eucalyptus* plantations, whilst the tree itself is unpalatable to the resident invertebrate community. Its high content of oils both sours soil and increases the occurrence of fires. As a result, communities change with the increased frequency of disturbance and interactions with the newly introduced species.

We are, at least, getting better at following these changes. Using the power of computers, massive databases have been analysed to identify plants likely to occur together. The modern study of phytosociology (Box 5.7) uses these programs to create a classification of plant assemblages. The National Vegetation Classification scheme (NVC) in Britain is used to match a community, electronically, against a database and decide its possible origins, as well as its past, present, and future management. The scheme is also used in the restoration of lost or damaged sites, where it provides information on the characteristic species of an area.

Similar techniques were used by Janet Franklin to characterize plant communities of the Californian chaparral. Aware that conservation managers needed more detailed maps, she developed a system that could predict the occurrence of key species. Using an existing vegetation classification (CALVEG) grouping the region's vegetation into 17 categories, she analysed data from 900 sample plots, incorporating them into the database. A computer model was then used to predict the occurrence of eight key chaparral species—such as *Adenostoma* and *Arctostaphylos*—which are indicators of the principal plant communities.

Understanding communities in terms of their species interactions, and perhaps even generating universal rules of assembly, is invaluable in our efforts to conserve and recreate them. Sometimes, with succession, it may be enough to create the conditions and then stand back to allow the species to assemble themselves; at other times we may need to disturb them on a regular basis to create the desired community. As we seek to protect mediterranean-type communities we will inevitably learn more of how they are put together. Perhaps, for the time being, it is enough to recognize that the concept of the community is a functional definition, rather like the species concept (Section 2.2), more of a device to help ecologists than an ecological reality. Nature, once again, does not come in neat parcels.

BOX 5.7   Classifying communities

Using plant species to classify plant communities has its origins in the eighteenth century. Around this time scientists first documented vegetation patterns and began to recognize them as evidence of interactions between the plants themselves, and with the larger environment. In 1825 Dureau de la Malle carried out a series of felling and regeneration experiments in his woodlands in Normandy and concluded that plant species live socially and that this was '. . . a condition essential to their conservation and to their development'.

In the early years of the twentieth century, the study of plant associations, **phytosociology**, was a peculiarly European pursuit, pioneered by research groups working in Zurich in Switzerland and Montpellier in France and a rival, but nonetheless related, Scandinavian system in Upsalla. Whereas the Zurich–Montpellier approach was descriptive,

its rival was largely numerical, though both classified communities according to predictable species associations. More recent classification schemes include Britain's National Vegetation Classification scheme (NVC). Developed by a team led by John Rodwell and Andrew Malloch, it classifies Britain's vegetation into 300 distinct communities with around 750 sub-communities.

The NVC serves to standardize terminology and survey techniques, allowing comparison between sites at both local and national levels. Each community is given an alphanumeric code, such as W for woodland and H for heathland. Certain categories are identified by two letters such as CG for calcicolous grassland (grasslands of chalk and limestone) and MG for mesotrophic grasslands (grasslands of neutral soils). The communities are numbered and any sub-community is distinguished by a suffix. The codes can

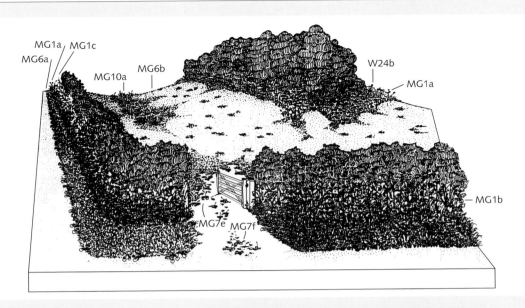

Figure 5.28 A diagrammatic map of a mesotrophic grassland and associated communities. Here, the basic community MG1 (dominated by the grass *Arrhenatherum elatius*) occurs as three sub-communities. One of these (sub-community (a)) is characterized by red fescue (*Festuca rubra*), and the second (MG1b) is a rougher sub-community containing stinging nettle (*Urtica dioica*). A third sub-community (MG1c) is associated with the wetter conditions along the ditch and is typified by meadowsweet (*Filipendula ulmaria*). The diagram shows four other community types, including a scrub community (W24), classified along with the woodlands. A visual representation such as this is useful in showing how communities occur together and grade into one another within the landscape.

be used on maps or within diagrams to illustrate the nature and extent of a community (Figure 5.28).

Sites can also be included in **geographical information systems** (GIS), which use aerial and satellite imaging to map large areas. Community descriptions can also detail threats and community transitions which predict future successional trends, valuable in conservation management. Similarly, it can provide information on a site's past communities, particularly useful in projects that aim to restore lost or damaged habitats.

● SUMMARY

A species assemblage or community represents an association of species adapted to the abiotic conditions of a region, especially climate, but also reflects their interactions with each other. The mediterranean-type community is found in five regions of the world sharing a climate very similar to that of the Mediterranean Basin. Some regions also share similar topographies and geologies, but they differ in the pool of species from which they have been colonized. Disturbance, in particular fire, plays an important role in these communities and, as a result, many sclerophyllous plants, typical of mediterranean habitats, are fire-adapted.

Chile and California have similar plant communities, most probably because of their coastal mountain ranges and high latitudinal extent. A second group, South Africa and southwestern Australia, have flatter landscapes and more ancient, more impoverished soils. The Mediterranean Basin itself has the widest range of habitats, because of its greater area and dominant geology. All show similar plant physiognomies and some equivalence in their animal communities, with high species endemism in all five regions. Even so, the details of their differences confound attempts to find common features by which these communities are organized. Whilst there are indications, especially amongst the bird community, that some patterns are repeated in each habitat, well-defined assembly rules elude ecologists.

Communities change by a succession in which species colonize and are lost with time. In primary succession, a community starts without previous occupiers and organic matter. Secondary succession occurs in disturbed areas recolonizing from the remains of the previous community. Autogenic succession is where the community composition is largely determined by species interactions; allogenic succession is the result of factors from outside the community.

Species interactions through processes such as facilitation, inhibition, and tolerance mean the successional sequence generally moves from rapidly dispersed and quick-reproducing pioneer species to longer-lived competitive species. When species turnover is at a minimum the community is said to be at climax. Disturbance, and its frequency, can also determine the development of the community in an allogenic succession.

Despite having developed alongside humans for much of their history, the mediterranean regions are today threatened by excessive disruption from tourism, increasing industrialization, and changes in land management practices. We have also introduced non-native species that can shift community structure, threatening endemic species. Conservation depends on an ability to recognize and describe communities, along with an understanding of the processes which control them.

● FURTHER READING

Archibold, O. W. 1995. *Ecology of World Vegetation*. Chapman & Hall, London. A comprehensive account of the major biomes.

Blondel, J. and Aronson, J. 1999. *Biology and Wildlife of the Mediterranean Region*. Oxford University Press, Oxford. An accessible review of the ecology and environmental issues of the Mediterranean Basin.

Polunin, O. and Smythies, B. E. 1973. *Flowers of South-West Europe. A Field Guide*. Oxford University Press, Oxford. A flora of the region, with detailed descriptions of its major ecosystems.

● WEB PAGES

A good general-purpose directory which includes links to sources on succession and other ecological principles covered here is:

http://www.biologybrowser.org

The climes, they are a-changing

The Cape Floristic Region of South Africa is home to 8700 plant species, 68 per cent of which are endemic. This flora supports equally unique animal communities, including several endemic vertebrates and the most diverse ant community on Earth. Ants are keystone species in many of the Cape's plant communities, indicated by the 1 300 plant species that depend on them for their survival (Christian 2001; Bond and Slingsby 1984; Keeley and Bond 1997).

As we have seen (Section 4.2), these interactions are now being disrupted by newcomers such as Australian acacia trees (*Acacia saligna*, *A. cyclops*) and the Argentinian ant (*Linepithema humile*). However, global warming may represent a far greater threat to the Cape's ecosystem, with temperature rises perhaps as high as 2 °C predicted in the next 50 years. Unlike ants and acacias, the weather cannot be controlled by chemicals.

The first stage in monitoring environmental change is to measure the ecosystem in its present state. This acts as a baseline against which future change can be compared. When Antoinette Botes and her colleagues did this for the Cape they considered ants to be integral to the functioning of its ecosystems and indicative of the plant communities. Their aim, therefore, was to identify assemblages of ants and plants that, along with other environmental information, would provide a **baseline survey** (Botes *et al.* 2006).

Botes and her team established a transect across a nature reserve—in this case the Greater Cederburg Biodiversity Corridor—collecting ants at various altitudes from sea level to 1926 m (Figure 5.29). They sampled within the six major plant communities of the region—the Strandveld, Succulent Karroo, and four forms of mountain fynbos (Restioid, Proteoid, Ericaceous, and Alpine fynbos: Table 5.4). Botes

Vegetation communities of the Cape Floristic Region

Community type	Vegetation characteristics	Rainfall (mm per yr)	Geomorphology
Strandveld	Low-growing, scattered succulent shrubs (e.g. *Zygophyllum morgsana*, *E. Mauritanica*)	50–300 (coastal influences)	Alkaline, calcium-rich dune sand. Low altitude—sea level—*c.* 300 m
Succulent Karoo	Small, scattered shrubs including Asteraceae (daisies), Crassulaceae (stonecrops), and *Mesembryanthemum* (livingstone daisies)	50–200 (very arid)	Siltstone, shale. Lowland 300–600 m
Restioid Fynbos	Dominated by Cape reed (*Restio* spp.) and sedges (*Carex* spp.), with some *Protea and Leucadendron* spp. (<10%)	200–2 000	Acidic sand with shales. Gentle slopes—generally occurs on lower foothills
Proteoid Fynbos	Dominated by *Protea* spp., (>60%) with some Restionaceae spp.	200–2 000	Acidic sand with shales. Steep slopes—generally mid-montane
Ericaceous Fynbos	Fynbos heathland, dominated by *Erica* spp. (heaths) with some *Protea*	200–2 000	Higher altitudes
Alpine Fynbos	Ericaceous fynbos, with *Erica*. More sparse—occurs at summits	200–2 000	Mountain summits

(continued overleaf)

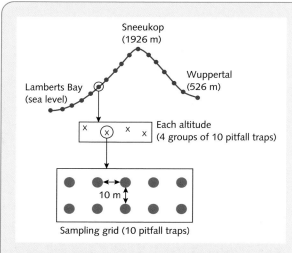

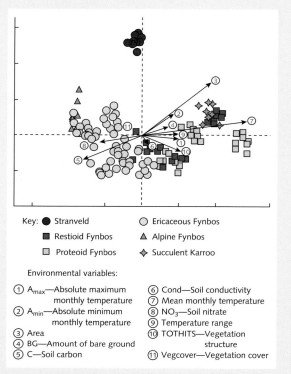

Key: ● Stranveld   ○ Ericaceous Fynbos
     ■ Restioid Fynbos   △ Alpine Fynbos
     ▢ Proteoid Fynbos   ◆ Succulent Karroo

Environmental variables:

① A$_{max}$—Absolute maximum monthly temperature
② A$_{min}$—Absolute minimum monthly temperature
③ Area
④ BG—Amount of bare ground
⑤ C—Soil carbon
⑥ Cond—Soil conductivity
⑦ Mean monthly temperature
⑧ NO$_3$—Soil nitrate
⑨ Temperature range
⑩ TOTHITS—Vegetation structure
⑪ Vegcover—Vegetation cover

Figure 5.29 Diagram of the transect running west to east across the Greater Cederberg Biodiversity Corridor. Botes and her team selected 17 sampling sites along the transect ensuring that each 200 m altitude band was sampled (Botes *et al.* 2006).

had to ensure that the sampling was representative and so set up 17 sampling sites each containing four plots, each with 10 pitfall traps to catch the ants. These were sampled every five days over two separate months to measure seasonal variation. Other data, such as temperature, vegetation cover, and soil nutrients were also collected.

The results were analysed using principal component analysis (PCA), a technique that groups the samples according to their similarity and identifies the principal factors which determine the plant and ant assemblages (Figure 5.30). Significantly, temperature—particularly mean monthly temperature—was the single most important factor in their distributions.

Botes went on to identify species of ant characteristic of their communities, those whose absence might be an early warning of community change. She used a program that calculated indicator values for the ant species based on a combination of the likelihood of them occurring within a particular area—their 'frequency'—and a measure of how characteristic they are of that area—their 'fidelity' (Table 5.5). Only 18 of the 85 ant species found were sufficiently specific to act as **indicator species** for plant communities and altitude (Section 7.2). Using these species

Figure 5.30 A visual representation of ant assemblages in each of the vegetation communities along the transect. The PCA plot is referred to as an ordination as samples are ordered according to the influence of a series of environmental factors. The analysis teases out the 68 samples into groups—some more distinctly than others. For example, the Strandveld assemblages (closed black circles) form a very separate cluster of samples at '12 o'clock' on the plot. This is an indication of their coastal nature—a feature which differentiates them from the other, inland communities. Moving 'clockwise', a cluster of Succulent Karoo assemblages overlap with some of the Restioid Fynbos samples. Further round, the remainder of the Restioid overlaps with the Proteoid Fynbos, which in turn gives way to samples from the Ericaceous and, finally, Alpine assemblages. The environmental variables used to form the groupings are also shown as a series of arrows radiating from the centre of the plot: the longer the arrow the more significant the factor. Here we see mean monthly temperature to be the most important variable.

TABLE 5.5	**Ant indicator species for vegetation types and altitude* in the Cape Floristic Region**

Vegetation type/altitude	Species
Vegetation type	
Strandveld	*Technomyrmex* sp. 1
Succulent Karoo	*Pheidole* sp. #1
	Monomorium sp. #2
	Tetramorium quadrispinosum
	Tetramorium sp. #6
	Messor sp. #1
	Ocymyrmex sp. #1
Restioid and Proteoid	*Camponotus mystaceus*
Fynbos	*Pheidole* sp. #2
Altitude	
0–900 m (W)	*Monomorium fridae*
	Lepisiota sp. 1
500 m (E)	*Pheidole* sp. #1
	Monomorium sp. #2
	Tetramorium quadrispinosum
	Ocymyrmex sp. #1
900–1 700 m (E) & 1 100–1 700 m (W)	*Anoplolepis custodiens*
1 900 m (summit)	*Camponotus* sp. #1
	Meranoplus sp. #1

* Two sets of indicator species were needed as it was difficult to tease out the different ant assemblages—and thus indicator species—on the basis of vegetation type alone (the analysis could only identify characteristic species for the lower mountain fynbos. The altitude analysis provides additional indicator species for summit and near-summit altitudes.

the study has provided a method for monitoring long-term change—the ants are the Cape's equivalent of a miner's canary and changes in their distribution may be a sign of possible trouble ahead.

Identifying temperature as the major factor responsible for the unique communities also allows us to speculate how climate change might alter them. Previous research shows that higher temperatures cause succulent karroo to expand at the expense of other plant communities (Lassau and Hochuli 2004), with the likelihood that mountain fynbos would spread up the mountain, easing out the higher-altitude ericoid and alpine fynbos. The losers would be the ants and many of the plants of the fynbos, especially the proteas. Further, the tight-knit interaction between these two intensely *K*-selected groups could lead to synergistic effects on the larger community, accelerating the impact of climate change in the fynbos. The 'winners' would be the *r*-selected opportunist and generalist ant species that thrive in the hotter drier habitat of the karroo (Lassau and Hochuli 2004).

Investigating climate change with communities that themselves are constantly changing is never going to be easy. It is only through carefully planned baseline surveys and well-chosen indicator species that we can identify and track significant trends and so predict future changes.

References

Bond, W. and Slingsby, P. 1984. Collapse of an ant-plant mutualism: The Argentinian ant (*Iridomyrmex humulis*). *Ecology 65*, 1031–1037.

Botes, A., McGeoch, M. A., Robertson, H. G. van Niekerk, A., Davids, H. P., and Chown, S. L. 2006. Ants, altitude and change in the northern Cape Floristic Region. *Journal of Biogeography 33*, 71–90.

Christian, C. E. 2001. Consequences of a biological invasion reveal the importance of mutualism in plant communities. *Nature 413*, 635–639.

Keeley, J. E. and Bond, W. J. 1997. Convergent seed germination in South African fynbos and Californian chaparral. *Plant Ecology 133*, 153–167.

Lassau, S. A. and Hochuli, D. F. 2004. Effects of habitat complexity on ant assemblages. *Ecography 27*, 157–164.

⬤ EXERCISES

1. Give two reasons why tough leaves may be an adaptation to nutrient-poor soils.

2. Decide which of the following would be most indicative of a community organized principally by its species interactions. Justify your choice.
    (a) The soil invertebrate community undergoes seasonal migration down the soil profile.
    (b) The loss of a key insectivore leads to changes in the plant community.
    (c) The dominant plant species all show similar carbon-gaining strategies.
    (d) The same number of bird species are found in all similar communities.

3. Explain why competition might limit the capacity of a colonizing species to join an existing guild.

4. Group the five mediterranean regions according to their degree of community similarity, and for each grouping, give the main abiotic factors which distinguish it.

5. Describe the advantages of being (a) a cryptophyte, (b) a therophyte in a mediterranean-type environment.

6. Explain why a plant like marram (*Ammophila arenaria*) is such a good pioneer species of sand dune ecosystems.

7. Consider the two descriptions below (a and b) and decide what type of succession they each represent:
    (a) Seeds and propagules in a ploughed field are able to use the existing nutrients and organic matter to establish a community with a wide variety of plant species.
    (b) A few pioneer species colonize wind-blown sand along a seashore. The stability they provide enables organic matter to build up and hold the particles together.

8. Arrange the following successional stages of an idealized Mediterranean sand dune in their correct chronological sequence (youngest to oldest):

Stage	Characteristic vegetation
(a) Dune scrub	Shrubs such as juniper (*Juniperus*) and mastic (*Pistacia*)
(b) Dune woodland	Trees such as oak (*Quercus*) and pine (*Pinus*)
(c) Yellow dune	Herbaceous plants such as marram (*Ammophila*) and sea medick (*Medicago marina*)
(d) Dune garrigue	Sclerophyllous, chamaephytes such as *Lavandula*

Tutorial questions

9. Evaluate some of the experimental techniques you might use to prove whether change in the species composition of a community is due to natural processes or the consequence of human activity.

10. Temperatures are set to rise in the Mediterranean Basin with significant global warming (see Chapter 8). What would be the economic and social implications of the region becoming more arid?

Systems

'All flesh is grass.'
Isaiah 40:6

CHAPTER OUTLINE

- Primary producers and photosynthesis.

- Gross and net primary productivity.

- Metabolic costs—these and other limits to productivity.

- Secondary productivity and its role within the community.

- Food chains and food webs.

- The role of decomposers and the reduction food chain.

- Agricultural systems—their productivity and sustainability.

← Sheep grazing maquis in Portugal.

On this point we have to differ. Although Isaiah is correct in the sentiment, he is wrong in the detail. Certainly, most animals on the planet ultimately depend on the photosynthetic plants for their nourishment, even if these are not actually grasses. Yet there was a time when there were no plants but there were consumers—single-celled organisms that survived by scavenging organic molecules or eating other cells. Before the advent of photosynthesis, life on Earth depended on the energy that could be released from chemical compounds, rather than the radiant energy arriving from the sun.

Such communities still exist today, away from the oxygen-rich atmosphere, at the bottom of deep, dark oceanic trenches where volcanic vents supply both heat and energy-rich compounds. Here, entire ecosystems are driven by the chemical energy captured by bacteria, organisms that would be poisoned by high concentrations of oxygen. Echoing its ancient origins, photosynthesis also needs low oxygen conditions to work. Indeed, the internal life of all cells is protected from this highly reactive element by their rapidly binding its more destructive forms.

Ecological communities are built upon those organisms able to fix free energy in their complex molecules. Whatever its source, the energy stored in their cells and tissues can then be exploited by consumers—the animals, fungi, and bacteria able to break down these molecules, to capture the released energy, and assimilate the raw materials to build their own tissues.

This is the problem facing all organisms—to acquire the resources to maintain themselves, to grow and replicate. Primary producers can fix energy for themselves, but consumers cannot, and so an ecological community assembles around the producers. Today, most of the Earth's production occurs in warm conditions, in moist air or well-lit waters, where the supply of nutrients does not constrain the building of molecules. Because they fix relatively small amounts of energy, the production of bacterial-based communities is just a small fraction of this total, and they do not support large and elaborate communities.

Highly complex and highly productive communities require the arrival of the photosynthesizers and their trick with the light. These producers use sunlight to fix carbon dioxide from the atmosphere and combine it with hydrogen to form energy-rich sugars, releasing oxygen in the process. The abundant radiant energy from the sun, harvested with only minimal efficiency, drives a system that today clothes the Earth and feeds, directly or indirectly, all the consumers that live within its folds.

We begin this chapter by describing energy fixation and the innovation of photosynthesis. We go on to consider how energy moves from one species to another, along food chains and down food webs, and consider what this tells us about community organization. We also describe how this drives nutrient flow through these ecosystems. Finally, we look at the oldest profession, agriculture, as a means of securing energy supplies for ourselves.

6.1 Ecological energetics

Energy is the common currency of the living world. The physical, chemical, and biological worlds are ultimately linked by the laws that govern energy transformations and movement. Energy is stored in chemical bonds and released slowly to drive all biological activity. This maintenance of life is metabolism (Figure 4.1) and the movement of energy through an ecosystem is called ecological energetics.

Energy is formally described as the capacity to do work and is measured as the **joule** (J)—the work done when a force moves 1 kilogram through 1 metre. The more familiar term, perhaps, is the calorie (cal), the amount of heat needed to raise 1 g of water by 1 °C (1 J = 4.2 cal). The laws of thermodynamics describe how energy is conserved in transformations and the properties and exchange of heat. When energy is transformed from one type (say, movement) to another (say, electrical energy) some is inevitably lost. All systems, living or otherwise, operate through a series of transformations with heat dissipating at each step.

Energy doing work is termed **kinetic energy**, whereas stored energy is **potential energy**. Besides synthesizing structural molecules to build new cells, much of the activity of living systems is creating molecules that store energy to be released later to drive metabolism. We detect these transformations when heat is lost through respiration. We can also measure the energy fixed in the tissues, a universal measure of growth to which ecologists refer as production.

6.2 The producers

Energy enters the biosphere principally as sunlight, though it also originates as heat from the Earth itself, from volcanic activity driven by radioactive decay deep within its core. Energy can also be released from the chemical bonds of a range of inorganic minerals. At the surface sunlight is the ultimate source for nearly all biological activity. Each square metre receives an average of 48 million kilojoules (kJ) of radiant energy per year, a small portion of which is captured by photosynthetic organisms.

Because photosynthetic **primary producers** fix their own energy they are known as **autotrophs**, quite literally 'self-nourishers' (or acknowledging their principal energy source, **photoautotrophs**). The most primitive of these, such as the cyanobacteria, have a rudimentary photosynthetic apparatus, some way away from the more elaborate and efficient **chloroplasts** (Figure 6.1) we find in higher plants.

Before the arrival of photoautotrophs, a very different source of energy, the potential energy in

50 μm

Figure 6.1 The leaf, the photosynthetic organ of plants. Electron micrograph of the chloroplast, the cell organelle responsible for photosynthesis. It consists of a series of membranes (thylakoids) on which chlorophyll and its associated proteins are situated.

reduced chemical compounds, drove ecological processes through its release by **chemoautotrophs**. These ancient organisms can still be found in extreme environments where oxygen concentrations are low (Box 6.1), conditions that would have prevailed on the young Earth before the photosynthesizers began to release oxygen, some 3 billion years ago.

BOX 6.1 The extremophiles

There was a time when life on the Earth's surface was so hazardous that ecosystems powered by sunlight were not possible—that is when early life was confined to habitats away from the sun and its hazardous ultraviolet radiation, where chemical energy was abundant and oxygen was largely absent. These simple organisms, principally archaebacteria, are still with us today—extremophiles (literally 'lovers of extreme environments') are found in a range of harsh environments, from ice to boiling water and from salt to battery acid.

Thermophiles and **hyperthermophiles** were among the first extremophiles to be discovered—living in hot springs and thermal vents (Figure 6.2). They obtain their energy by oxidizing hydrogen sulfide dissolved in the volcanic spring waters. Similar archaebacteria are known from 'black-smokers'—underwater volcanoes that release molten magma and sulfurous gases deep within the ocean. Hyperthermophiles live under immense pressure in water close to boiling point, forming the base of a community that includes worms, crustaceans, and fish. One of the strangest animals is *Riftia pachyphila* a 1.5-m-long tubeworm living at the margins of these vents and which concentrates sulfur within its body fluids. Its body cavity is home to symbiotic colonies of sulfur bacteria from which the worm derives its energy.

Methanogens are archaebacteria found in oxygen-poor and waterlogged soils or deep sediments all with a high organic content. Some acquire their energy by combining

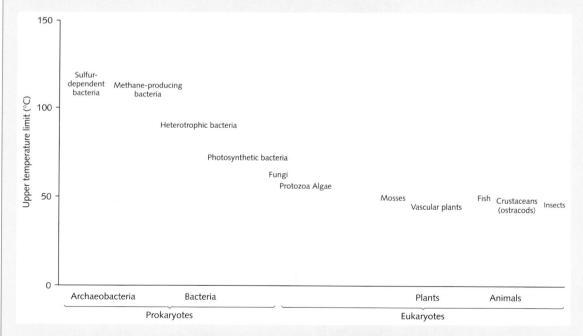

Figure 6.2 Temperature limits for extremophiles occurring within different groups of organisms. Archaebacteria are able to withstand greater temperature than their non-chemoautotrophic counterparts.

hydrogen with carbon dioxide, producing methane and water as wastes, whereas others use acetate as their energy source Research now suggests that methanogens may explain the origins of nitrogen fixation, the process by which some bacteria and cyanobacteria are able to capture nitrogen from the atmosphere (Section 7.1). Mausmi Mehta and John Baross of the University of Washington have described genes coding for nitrogenase (the enzyme responsible for nitrogen fixation) in a methanogen, *Methanocaldococcus jannaschii*, from a thermal vent off the coast near Seattle. These are able to fix nitrogen at temperatures in excess of 90 °C. If nitrogen fixation evolved once, in the archaebacteria, then the gene for nitrogenase must have passed to other micro-organisms by lateral transfer, a process in which genetic code is passed between species of bacteria. Confirmation of this would encourage researchers seeking to transfer this same gene to a wide range of organisms, especially our crop plants.

Salt lakes have colonies of **halophiles**, archaebacteria such as *Halobacter* and *Halococci* that live in salt concentrations three to four times that of seawater. These avoid desiccation by keeping their internal sodium concentration above or equal to that of their surroundings. Eukaryotic organisms can also survive high salt environments, including fish and invertebrates such as the brine shrimp *Ephydra*, whilst the unicellular photosynthetic alga *Dangearidinella altitrix* lives on the surface of the salt crystals.

In the hostile sands of the desert, archaebacteria such as *Metallogenium* and *Pedomicrobium* live off the thin surface layer of iron and manganese known as 'desert varnish'. They obtain their energy by oxidizing these elements, turning the surface of the desert into a rusty mixture of metal oxides. Another group of these 'rock-eating' archaebacteria—**lithoautotrophs**—are found deep below ground in rock fissures. For many years geologists thought the bacteria-like shapes in rock cores were from contaminated drilling equipment. We now know that these archaebacteria play an

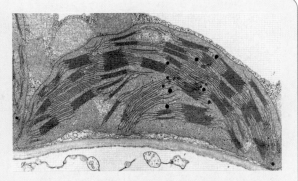

Figure 6.3 Photomicrograph of basalt microbes in volcanic rock from deep below the ocean floor. These were once thought to be artefacts caused by contamination but are now known to be resident in these rocks.

important role in degrading volcanic basalt into a clay-like material (Figure 6.3).

Possibly the most extreme extremophile yet discovered is *Deinococcus radiodurans*. A **radiophile**, it can withstand gamma radiation at levels of 6 Mrad/h, which means it could survive life within a nuclear reactor! This species has attracted attention because its survival is due to an efficient DNA repair mechanism, of great interest to those studying cancer. A recombinant strain of the bacteria is used to the treat organic solvents such as toluene and trichloroethylene from nuclear waste since it can degrade these hydrocarbons whilst unaffected by the radiation.

Enzymes isolated from extremophiles—'extremozymes'—have revolutionized biotechnology. Biological washing powders, 'stonewashed' jeans, contact lens cleaning fluid, and many antibiotics are all products of extremozyme technology. Similarly, DNA polymerase, the enzyme used in genome sequencing and DNA fingerprinting (Box 2.2), was first isolated from the thermophilic sulfur bacteria *Thermus aquaticus*.

Leaves are the energy-collecting surfaces of higher plants, with chloroplasts packed just beneath their surface. A leaf can twist and turn to follow the sun, and may change its shape to control water loss according to the temperature (Figure 6.4). When the leaf is illuminated the chloroplasts use the radiant energy to split water molecules and generate hydrogen, which then combines with carbon dioxide to form glucose (or other simple sugars), releasing oxygen as a by-product:

$$6CO_2 \; + \; 6H_2O \; \longrightarrow \; C_6H_{12}O_6 \; + \; 6O_2$$

Carbon dioxide　　Water　　Radiant energy　　Glucose　　Oxygen

Figure 6.4 A solar array. The solar panels used to capture radiant energy are supported by a structure that supplies the water needed for photosynthesis. Trees grow where water and nutrients allow for large and elaborate arrays, but the shape of any plant and its leaves reflects the conditions in its habitat and its interactions with other species.

Glucose is then used to build more complex molecules either as structural components or as an energy source to fuel metabolic processes.

Chloroplasts contain **chlorophyll**, a group of green pigments arranged within a mass of tightly folded membranes, called **thylakoids**, that pack the pigment within the chloroplast. Chlorophyll appears green because light at the red and blue ends of the spectrum is absorbed and the green wavelengths are reflected (Figure 6.5). There is also a range of **accessory pigments** that absorb energy at shorter wavelengths, to help plants grow in poor light such as the shade of the forest floor or deeper coastal waters. In terrestrial plants, these pigments are usually masked by the dominant chlorophyll but their glorious colours become visible in the autumn when the chlorophyll degrades, just before the leaves fall.

Much of the light striking the leaf passes straight through it and never encounters chlorophyll. Just 44 per cent is at a wavelength that can be absorbed by these pigments. Of the total energy falling on the leaf, typically only 2 per cent will become fixed in a sugar. Despite this low efficiency, the productivity of photoautotrophs is ultimately responsible for nearly all of the living biomass of the biosphere.

The productivity of any species is the amount of living material or **biomass** it produces. This can be expressed either as the amount of its matter within a given area over time ($g/m^2/year$) or its energy content ($kJ/m^2/yr$). The entire photosynthetic output (that is, the energy fixed) of a primary producer is known as its **gross primary production** (GPP). Most of this, often greater than 60 per cent, is used in the plant's metabolism, eventually to be lost as the heat of respiration. It is only the energy fixed in the growth of tissues or the production of gametes that forms the biomass or **net primary productivity** (NPP):

Net primary production (NPP)
= Gross primary production (GPP) − Respiration (R)

The biomass accumulated over a period of time is referred to as the **standing crop**. The wide differences in standing crop between species and between

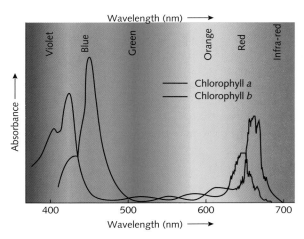

Figure 6.5 Absorption spectra of the two forms of chlorophyll (*a* and *b*). Absorbance peaks at the red and blue ends of the spectrum, wavelengths at which electrons become most excited within the chlorophyll molecule. Green light is not absorbed and is reflected back, giving plants their characteristic green colour.

TABLE 6.1	Global primary production		
Ecosystem type		Mean net primary productivity (g/m²/yr)	Mean biomass (kg/m²)
Continental			
Tropical rainforest		2 200	45.0
Tropical seasonal forest		1 600	35.0
Temperate evergreen forest		1 300	35.0
Temperate deciduous forest		1 200	30.0
Boreal forest		800	20.0
Woodland and shrubland		700	6.0
Savannah		900	4.0
Temperate grassland		600	1.60
Tundra and alpine		140	0.60
Desert and semi-desert scrub		90	0.70
Extreme desert, rock, sand, ice		3	0.02
Cultivated land		650	1.00
Swamp and marsh		2 000	15.00
Lakes and streams		250	0.02
Mean continental		**773**	**12.3**
Marine			
Open ocean		125	0.003
Upwelling zones		500	0.02
Continental shelf		360	0.01
Algal beds and reefs		2 500	2.0
Estuaries		1 500	1.0
Mean marine		**152**	**0.01**
Grand total		**333**	**3.6**

ecosystems (Table 6.1) reflect the factors limiting net primary productivity, principally temperature and water availability. Where it is warm and wet, the standing crop will be high and so the productivity of tropical rainforests stands in stark contrast to that of the deserts and ice sheets of the planet.

Measures of the biomass added each year tell us how rapidly this standing crop is replaced in an ecosystem, that is its turnover (Figure 6.6). Turnover times in the forests are relatively long because their standing crop is much larger than their annual production. In these highly competitive environments, where nutrients are often limiting, there is little scope for rapid growth and annual production is relatively constrained. Heavily grazed grasslands, on the other hand, are subject to continual disturbance (Section 5.3), and a seasonality that creates opportunities for rapid growth. Here, herbivores consume a large

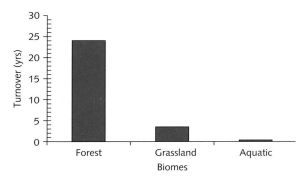

Figure 6.6 Turnover in forest, grassland, and aquatic biomes. This is the time taken to replace the biomass, based on the annual addition as a proportion of the standing crop. Aquatic biomes with abundant nutrients have a high rate of turnover, but a low standing crop. The large standing crop of forests, and to a lesser extent grasslands, is replaced more slowly.

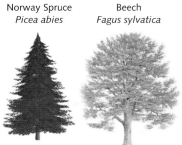

	Norway Spruce *Picea abies*	Beech *Fagus sylvatica*
Life span	100 years	89 years
Leaf shape	Needle	Broad
Annual leaf production	Low	High
Photosynthesis	Low	High
Length of growing season	260 days	176 days
Primary productivity (C fixed)	14.9 tonnes/ha	8.6 tonnes/ha

Figure 6.7 Evergreen species sacrifice short-term productivity for the ability to photosynthesize for longer. The Norway spruce (*Picea abies*) photosynthesizes for almost half as long again as the beech tree (*Fagus sylvatica*) and, despite its slower rate of photosynthesis, the result is higher productivity.

proportion of the annual standing crop, so that turnover is high. Under these conditions nutrients cycle faster and, where water is abundant and temperatures are high, rapid growth is possible.

Most plants have their highest efficiency at relatively low light levels and bright light can inhibit photosynthesis. Plants can only invest in new tissues when the energy fixed in photosynthesis is greater than that used in respiration. When the two are in balance the plant is said to be at its **compensation point**, when all the energy it fixes is used in maintenance. In aquatic systems, the compensation point is reached at a particular depth. Beyond this, photoautotrophs are unable to survive because they use energy faster than they can fix it. In deep waters, light is quickly extinguished both by absorption and reflection by suspended matter. Only certain wavelengths can penetrate any distance, and plants living at depth have to collect energy from whatever light is available. Deep-water kelps use a range of pigments that absorb the wavelengths that penetrate deepest.

Even on land, many plants have to survive extended periods, whole seasons perhaps, below their compensation point, when energy fixation is very slow. These species go into energy-debt for a short time, surviving winters by living off their reserves, but they cannot live beyond their means indefinitely. One strategy is to reduce respiration costs by shedding leaves, tissues

that are costly to maintain and redundant if photosynthesis is only fixing small amounts of energy.

An alternative response is to photosynthesize throughout the winter months but at a lower level. Because water is frozen for much of this season, various adaptations have evolved to limit water loss while photosynthesis continues. Many evergreens retain their leaves, fine needles shaped to conserve water that can continue to fix energy and, of course, avoid the costs of growing new leaves each spring. In this way evergreens can match or even beat the productivity of deciduous trees (Figure 6.7).

Water is the key. Carbon dioxide is absorbed by the leaf through a series of breathing pores (**stomata**), the route by which water is lost through **transpiration**. The transpiration stream delivers water and dissolved minerals, drawn up from the roots, to the leaves, where it can be used to create sugars. Shutting the stomata and slowing transpiration halts photosynthesis by cutting off its supply of carbon dioxide. If water is scarce, photosynthesizing plants, with their stomata open, face the risk of desiccation.

One strategy is to adapt the photosynthetic process itself, so that its water use is more efficient. Some plants have evolved a metabolic pathway (termed the **C4 pathway**) that allows carbon dioxide to be fixed even when the stomata are closed. These have a special enzyme—phosphoenol pyruvate carboxylase—which catalyses the binding of carbon dioxide by phosphoenol pyruvate (PEP), even at the very low concentrations found inside the plant. The result is a series of four-carbon storage compounds such as acetic, aspartic, malic, and oxaloacetic acids, which can later supply carbon dioxide and allow photosynthesis to proceed under hot and dry conditions when the stomata are shut. A number of species associated with such environments use the C4 pathway, including crops such as sorghum (*Sorghum bicolor*), maize (*Zea mays*), and sugar cane (*Saccharum officinale*).

Other plants, including many cacti and succulents, have a further modification, known as the Crassulacean acid metabolism (CAM). These plants keep their stomata shut during the heat of the day, opening them at night to absorb carbon dioxide when less water will be lost. The carbon they fix is stored as malic acid, which supplies their chloroplasts with carbon when they are lit and absorbing energy.

The shape and form of the plant and its leaves can also be adapted to reduce water loss. Besides using CAM, cacti survive in hot deserts using their swollen stems for water storage and photosynthesis. Their leaves are reduced to protective spines. In the cold and dry north, the needle-like leaves of Norway spruce (*Picea abies*) (Table 6.2), have a low surface area and deeply set stomata that together conserve water. Other plants produce leaves covered with downy hairs to trap a layer of still air close to their surface. This not only reduces water loss, it also insulates against excessive cold or heat. Such adaptations are found in both the alpine edelweiss (*Leontopodium alpinum*) and amongst the *Salvia* species of the Mediterranean. In many semi-arid habitats, plants create a damp microclimate next to their leaves by forming a compact shape, such as the low tussocks of chaemaephytes or the upright columns of cypresses (Section 5.1). Oils produced as secondary plant products can also reduce water loss (Section 4.4): as the temperature rises, the oils volatilize to form a layer over the leaf which reduces diffusion from the leaf surface.

Many plants of arid and semi-arid areas shed their leaves during drought (Section 5.1), but again there are a range of strategies. In the Californian chaparral, Black sage (*Salvia mellifera*) loses its lower, older leaves but retains the younger leaves at the shoot tips. The germander (*Teucrium polium*) produces differently sized leaves according to the season: between January and March it has large winter leaves, which are replaced with small leaves for the summer months (Figure 6.8). The plant achieves year-round photosynthesis but can reduce its water loss during the summer by between 50 and 76 per cent.

The schedule of growth and reproduction of a plant also represents part of its adaptive solution to these environmental pressures. All organisms have to balance the allocation of their resources between producing tissues or gametes, according to their life history strategy (Section 3.4). For a plant, this is indicated by its lifespan, generation time, size, and growth form. Some complete their life in a very short time—thale cress (*Arabidopsis thaliana*) lives little more than a month—whilst others do not—the bristlecone pine (*Pinus aristata*) may live for several thousands of years. A plant with a short generation time has to devote much of its resources to reproduction, usually investing in a single reproductive event, with only a small proportion of its biomass devoted to non-reproductive growth. This is a strategy typical of changeable habitats, when reproduction has to be completed before conditions become unfavourable (Box 3.3).

Annual plants complete their life cycle within a year (Section 4.4). Energy is used to produce leaves early in life, and only later are resources directed at reproduction. Eventually as much as 90 per cent of an annual plant's energy is invested in flowers and seeds. Such plants are characteristically *r*-selected species, fast-growing opportunists. Long-lived perennials, which survive over a number of seasons, allocate a greater proportion of their energy to structural tissues or storage compounds, allowing them to grow rapidly in the new season. In the early stages of their life few resources are devoted to reproduction, which is delayed for one or more seasons. Instead they accumulate resources for a large reproductive event, and after short periods of respite, may go on to reproduce repeatedly. This is broadly a *K*-selected strategy, adopted by competitive species that dominate stable and predictable environments (Section 3.4).

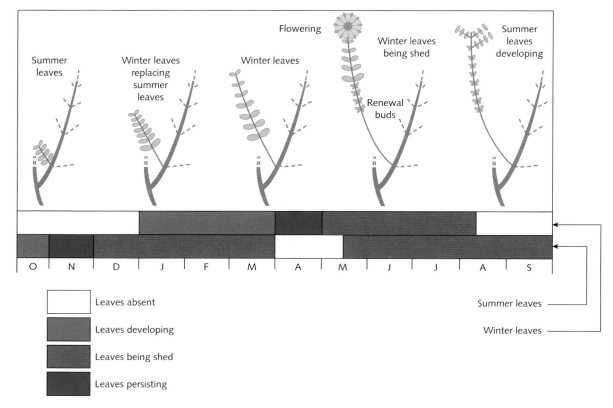

Figure 6.8   Seasonal leaf dimorphism in the Mediterranean germander (*Teucrium polium*), a plant that produces different leaves for different seasons to minimize drought-stress.

Trees and shrubs go a stage further. As they grow, much of their energy goes to produce large amounts of woody tissue, a supportive material that allows for greater height and serves as a plumbing system to deliver water to the leaves. In the early phases of rapid growth, more than half of a sapling's biomass may be in the form of leaves. In contrast, a mature tree may consist of 95 per cent non-living woody tissue with its live biomass confined to the shoots, roots, leaves, and a thin layer of cells beneath the protective bark, wrapped around a superstructure of dead wood. Plants investing in wood are at an advantage where water is abundant and when, because of the severe competition for light, height becomes important. For this reason forests dominate in the wetter parts of the world (Section 8.1).

The productivity of forests is evident from their complex structure. Unlike well-grazed grasslands,

forests are multi-layered, with a series of understorey plants adapted to the conditions and light levels beneath the canopy (Figure 6.9). Their leaves do not need bright sunlight to be effective and, with sufficient moisture, can photosynthesize even on the dimly lit forest floor.

A simple way of comparing the complexity of primary production in different terrestrial communities is the leaf area index (LAI), measuring the number of leaf layers that intercept incoming radiation. This is the leaf area as a proportion of the ground area it covers:

LAI = Total leaf area/Area of ground

Forest ecosystems can have LAIs as high as 9, which means that light passes through nine layers of leaves before reaching the ground. In arid and semi-arid areas the LAI is always well below 1. The

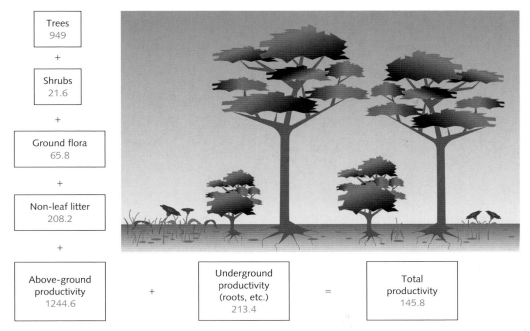

Trees
949

+

Shrubs
21.6

+

Ground flora
65.8

+

Non-leaf litter
208.2

+

Above-ground productivity
1244.6

+

Underground productivity (roots, etc.)
213.4

=

Total productivity
145.8

Figure 6.9 The broken canopy allows sufficient sunlight to support plants growing in the successive layers of a temperate deciduous forest. The plants in the shrub, herb, and ground layers can add substantially to the NPP of the ecosystem (g/m²/yr)—here for a European mixed-oak forest. Notice also that around 15 per cent of the forest's productivity is below the ground in the form of roots, tubers, and other structures.

index is therefore a useful indicator of both the structure and potential primary productivity of an ecosystem.

Grasslands have LAIs somewhere between forests and desert. They have little of the vertical construction of a forest and a much lower standing crop (1.6–4 compared with 20–45 kg/m²). However, grassland plants have very different allocation strategies, adapted to seasonal rainfall, periodic fires, and grazing. Grasses are perennials that invest little in support tissues, but will sprout leaves and continue to grow as long as water is available. When the supply dries up, the leaves simply die. Like other hemicryptophytes (Figure 5.14), a grass protects its growing tissues by keeping them close to the ground, away from grazers and flash fires. This strategy also works well in moister areas, particularly if there is intense grazing pressure.

Algal beds and reefs have the highest rates of primary productivity (Table 6.1). With an NPP of 2 500 g/m²/yr, they can out-perform even tropical rainforests. They too have a multi-layered structure, with algae growing to different heights, extending to different depths, and relying on their accessory pigments (Figure 6.10). Swamps and marshes are also highly productive, because of both their shallow water and abundant nutrients.

Cultivated land appears to have a low net productivity (650 g/m²/yr), but this results from the highly seasonal nature of agricultural activity. Such land supports growing plants for only part of the year, while at other times it is being prepared or awaiting the next growing season. When they are actively growing, crops are among the most efficient primary producers (Table 6.2). In some cases efficiencies close to 10 per cent are achieved, primarily because we remove some of the checks on their photosynthesis by adding water and nutrients. Notice, too, that many crops are C4 plants with photosynthetic processes adapted to drier conditions.

Figure 6.10   Rather like their terrestrial counterparts, the massive vertical development and multi-layering of the kelp forests off the Pacific coast of North America support a large and complex community.

TABLE 6.2	Photosynthetic efficiencies and growth rates of crops			
Crop	Country	Crop growth (g/m²/day)	Total radiation (J/cm²/day)	Light conversion efficiency
Maize *Zea mays**	USA	52	2 090	9.8
Millet *Pennisetum typhoides**	Australia	54	2 134	9.5
Sugar beet *Beta vulgaris*	UK	31	1 230	9.5
Millet *Pennisetum purpureum**	El Salvador	39	1 674	9.3
Sugar cane *Saccharinum* spp.*	Hawai'i	37	1 678	8.4
Tall fescue *Festuca arundinacea*	UK	43	2 201	7.8

* C4 plants

6.3 Links in the chain

A **food chain** describes just one route by which energy passes through a community and the rate of energy transfer between some of its species. When they are consumed, the primary producers pass their energy to the consumers or **heterotrophs** (literally 'nourished by others'). Consumers are collectively called **secondary producers** and are linked together in the food chain in a sequence of **trophic levels** (Figure 6.11).

At the second trophic level are **herbivores**, species which feed directly on plants. Although vegetation is abundant, plant material often represents a poor-quality food requiring considerable investment in time and energy to digest. Consequently, the assimilation efficiencies of herbivores are generally low (Box 6.2). Some concentrate on parts of the plant that are more nutritious or more readily digested than others, such as new shoots or buds. Seed-eaters, for example, consume a food rich in stored carbohydrates and oils. They benefit from an easily digested energy-rich food high in nitrogen which would otherwise have fuelled the germination of the seed. Others have found ways of unlocking the energy contained within the tough cellulose that constitutes the major part of a plant's biomass (Box 6.5).

Living on the energy fixed in the tissues of herbivores or other animals, **carnivores** face other challenges. Flesh is primarily protein and fat—high-energy compounds that are easily degraded and with abundant nutrients. Compared to herbivores, carnivores have a short digestive tract, an indication of the smaller effort required to release the energy and nutrients in their food. However, they incur other costs, most obviously those of catching and killing their prey. These are high when compared to the foraging costs of a non-selective herbivore. Additionally, meals are invariably less frequent and, for some carnivores, whole seasons may pass when their main prey is not available.

Some feeding strategies locate consumers on more than one trophic level. **Omnivores** feed on both plants and animals, and the latter may include both herbivores and carnivores. This means they feed at more than one trophic position, and straddle several levels. Energy flow does not then follow a simple progression along a linear food chain. The pattern becomes more complex if the consumer scavenges dead animals or plants.

We can distinguish two basic routes for energy moving through ecosystems: from herbivores to carnivores (termed a **grazing food chain**), and from decomposers to carnivores (a **decomposer food chain**) (Box 6.3). A scavenger diverts energy heading for the decomposer food chain back into the chain based primarily on herbivores. The two chains are connected every time a blackbird pulls a worm from the soil or a decomposer is eaten by a consumer from the grazing food chain. In most ecosystems, a large proportion of the primary production passes into the decomposer route. On average, only 10 per cent of net primary productivity in terrestrial ecosystems passes to the consumers of a grazing food chain. The rest, 103.5 billion tonnes globally, fuels the decomposer chain. This makes the decomposer chain the most significant route by which energy is transferred through the rest of the ecosystem (Box 6.3).

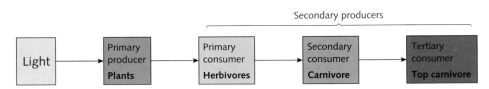

Figure 6.11 A generalized food chain—the transfer of energy down a sequence of trophic levels. Energy fixed by primary producers passes to herbivores, and then to one or two levels of carnivores.

BOX
6.2 **Energy efficiency**

Because it involves a transformation, any movement of energy from one organism to another inevitably leads to some energy loss. Of the energy that is assimilated, a large proportion will be used in metabolic processes, maintaining the organism and supplying energy for its activities. Thus, of the energy assimilated (A), a proportion goes to respiration (R) and the rest can be used in production (P). The latter is divided into the energy fixed in new tissues (growth—Pg) and that fixed in gametes (reproduction—Pr).

Its assimilation efficiency is the proportion of energy in the diet taken up by a consumer:

$$\text{Assimilation efficiency} = \frac{\text{Energy assimilated (A)}}{\text{Energy Consumed}}$$

(An equivalent equation derives photosynthetic efficiency by making the divisor the radiant energy received at the leaf surface.)

By the same principle we work out the efficiency with which energy is fixed in the tissues:

$$\text{Production efficiency} = \frac{\text{Energy fixed in tissues (P)}}{\text{Energy consumed (C)}}$$

The growth efficiency is the proportion of energy assimilated that is used in building new tissues:

$$\text{Growth efficiency} = \frac{\text{Energy fixed in tissues (P)}}{\text{Energy assimilated}}$$

Ecological efficiencies vary amongst animals according to their metabolic costs. Endothermic animals such as birds and mammals have high metabolic costs and may spend over 90 per cent of their energy in maintaining their body temperature. Ectotherms derive most of their heat from external sources and consequently can devote more of their assimilated energy to production (Figure 6.12).

In general, organisms further along a food chain have higher assimilation efficiencies, because they can recover energy more easily from their diet. Herbivores may have a plentiful supply of plant material, but up to a third of this may be cellulose, which they are unable to digest without the help of bacteria in their gut (Box 6.5). Their assimilation efficiencies are low, typically less than 10 per cent, so herbivores need to consume large amounts to meet their energy demands.

Carnivores, on the other hand, receive their energy in a more usable form, as proteins and fats that are both richer in energy and more readily digested. Some carnivores can achieve assimilation efficiencies as high as 90 per cent, but for many it is around 20–30 per cent. This is why carnivores need to eat less, and eat less often, than herbivores.

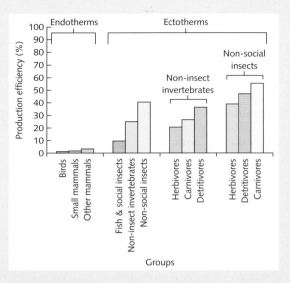

Figure 6.12 Production efficiencies of different animal groups. Endotherms typically have lower efficiencies because they have to meet higher metabolic costs associated with maintaining a constant internal body temperature.

Production efficiency, which takes into account the energy used in respiration, also varies with trophic position (Figure 6.12). Within a group, say the terrestrial invertebrates, herbivore efficiencies are generally lower (21–39 per cent) and carnivore efficiencies are generally higher (27–55 per cent). Across groups, ectotherms have production efficiencies of between 10–55 per cent and endotherms just 1–3 per cent.

These figures would seem to support the argument for always being a carnivore and never being a herbivore. However, most of the biomass of the planet is vegetation and there is consequently more energy available to herbivores than carnivores. The figures also suggest that being an ectotherm is a better strategy when energy is in short supply . . . and this is probably true.

Ecological efficiencies also make it obvious why energy decreases so rapidly along a food chain. A carnivore three or four steps away from the primary producers might have one 10 000th of energy originally fixed at the first trophic level. Figure 6.16 shows how little energy becomes fixed in each successive trophic level of a grassland food chain, so that a weasel contains a mere 0.00026 per cent of net primary productivity. This helps to explain why there are so few weasels and why everybody is not a carnivore.

BOX 6.3 **The decomposers**

Consider a fallen leaf. Having escaped being eaten, it lies, along with other vegetation and dead organic matter, in the layer of litter beneath the trees. What happens to the energy locked within its tissues? How are its nutrients and energy recycled back into the system?

The breakdown of the litter layer to produce **detritus** (the fragmented organic material in the early stages of decomposition) has several stages and involves a series of physical, chemical, and biological processes. The first is a leaching of soluble salts, sugars, and amino acids, reducing its biomass by as much as 30 per cent. When it is consumed by detritivores, the leaf is fragmented into fine particles. On average, earthworms, woodlice, and millipedes assimilate only a small fraction of what they consume and the rest becomes a resource for microbial and fungal decomposers (Figure 6.13). In its passage through their gut the litter is inoculated with bacterial and fungal spores that hastens its decomposition. These micro-organisms release enzymes that break down complex carbohydrates and other macromolecules, releas-

ing minerals and other nutrients that can be assimilated by the roots of primary producers.

The reduction of a leaf to a fine organic debris relies on this combination of invertebrate consumers and microbial and fungal decomposers. We concentrate the activity of such decomposers in our compost heaps. By continually supplying them with plant waste, we can maintain decomposer numbers and (through their waste energy) the high temperatures that speed the process. Decomposition is fastest in warm, wet conditions where oxygen is abundant.

The speed of the process also varies with the quality of the material being decomposed. It is most rapid in a detritus which has a high content of simple sugars, and slower in material dominated by complex macromolecules. For example, the simple compounds of straw are the first to be lost, leaving the more intractable celluloses (Figure 6.14). The complex carbohydrates and those compounds high in tannin and lignin decompose more slowly because these require the enzymes of specialist fungi and bacteria.

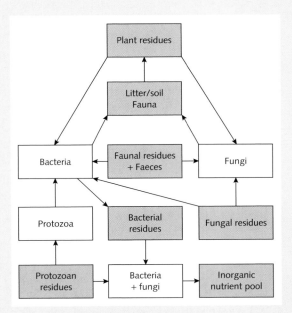

Figure 6.13 A generalized diagram of the flow of material and energy from detritus to a range of detritivores and decomposers.

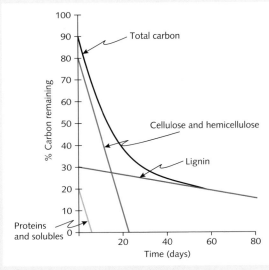

Figure 6.14 The relative rates of decomposition of carbon compounds within straw left on the soil surface. The more easily degraded components—the soluble proteins and carbohydrates—decline most rapidly, whilst the complex macromolecules found in wood (lignin) are more intractable.

(continued overleaf)

Figure 6.15 The familiar sight of the fruiting bodies of fungi. Underground is a large network of fungal hyphae which live by breaking down organic matter.

A community quickly adjusts to the most frequent forms of litter (Section 9.5). Indeed, the balance between the activity of bacteria and fungi is indicative of both the litter and the soil characteristics. Where nutrients are scarce and invertebrate detritivores few, fungi are the dominant decomposers, especially with more intractable wastes (high-cellulose and low-nitrogen, such as wood) and where the soil is relatively acidic. It is only when their caps appear above the litter or sprout from a log that we get some idea of their number and variety (Figure 6.15).

The role decomposers play both in reducing organic waste and making nutrients available for the ecosystem is only now being properly quantified. A single gram of temperate woodland soil may contain as many as 6 million bacteria and 3 000 m of fungal threads (hyphae); a square metre of the woodland floor will be home to a further 200 million decomposers ranging from protozoa through to molluscs. Worms alone process detritus at a rate of between 50 and 170 tonnes per hectare each year. The scale of their activity is indicated by the proportion of the total global biomass they represent, implying that a large proportion of the energy fixed in primary production must pass through the decomposer food chain (Figure 6.16).

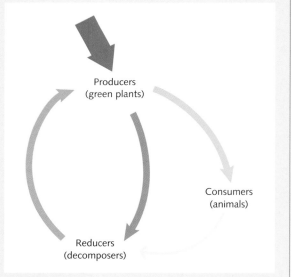

Figure 6.16 Decomposers. In most ecosystems the bulk of the energy fixed by primary producers passes to the decomposer community. This not only fuels the productivity of the decomposers and detritivores, it also drives the recycling of nutrients.

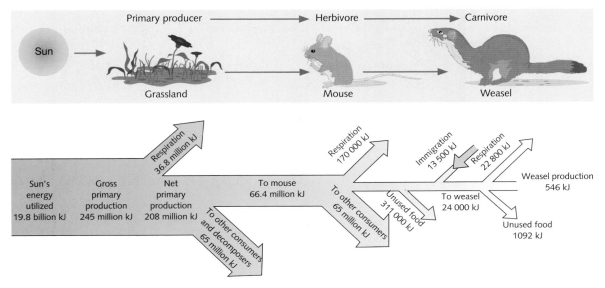

Figure 6.17 The loss of energy (kJ) as it passes along a simple food chain (in this case a North American grassland).

Chain length

Figure 6.17 shows a simple food chain for a grassland ecosystem and the amounts of energy passing from a herbivore (a mouse) to one of its predators (a weasel).

Notice how little energy actually becomes fixed in the tissues of the carnivore. The inefficiency of these transfers and the inevitable losses with each transformation mean that little energy is available to the higher trophic levels. Energy is expended in respiration at each level and only a fraction of that consumed becomes fixed in the tissues. A diagram scaled according to the energy content of each trophic level produces a pyramid (Figures 6.18 and 6.19), demonstrating the small proportion of the fixed energy reaching the higher consumers.

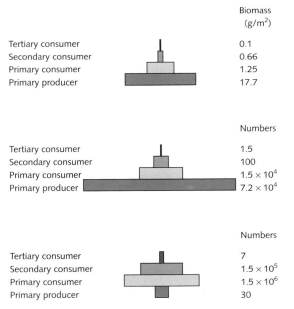

	Biomass (g/m^2)
Tertiary consumer	0.1
Secondary consumer	0.66
Primary consumer	1.25
Primary producer	17.7

	Numbers
Tertiary consumer	1.5
Secondary consumer	100
Primary consumer	1.5×10^4
Primary producer	7.2×10^4

	Numbers
Tertiary consumer	7
Secondary consumer	1.5×10^5
Primary consumer	1.5×10^6
Primary producer	30

Figure 6.19 Some other ecological pyramids. Plots of (a) biomass or (b) numbers of individuals at each trophic level, each showing a decline with successive trophic levels. However, numbers of individuals are a poor indication of the energy at each trophic level: (c) a single tree may support a large number of herbivores, producing an inverted pyramid.

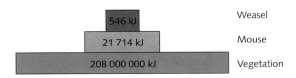

546 kJ	Weasel
21 714 kJ	Mouse
208 000 000 kJ	Vegetation

Figure 6.18 Pyramid of energy (not to scale) based on the productivity of the grassland food chain shown in Figure 6.17.

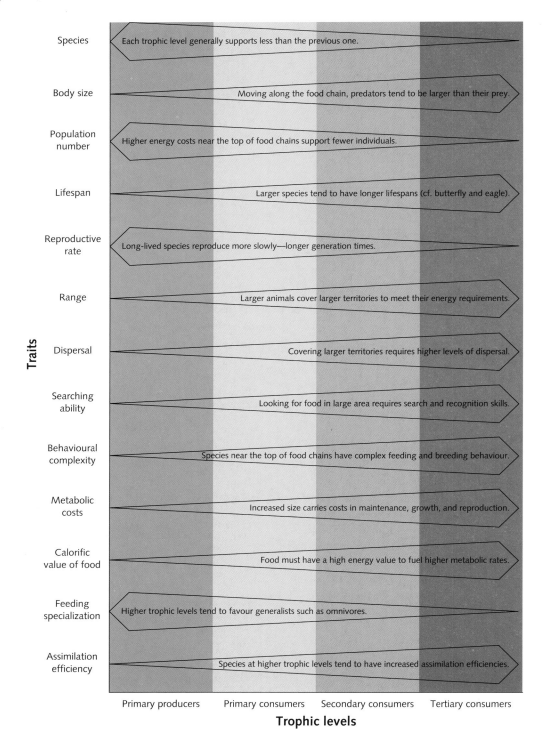

Figure 6.20   General trends along food chains.

This chain has just three links. Long food chains are rare in nature, and rarely extend beyond five links. The rapid decline in energy along the chain (or up the pyramid) was thought to limit the number of links, with too little remaining to support a fourth or fifth level, a consequence of the inefficiency of the energy transfers. If this were the case we should expect longer food chains in those dominated by more efficient ectotherms (such as insects), compared to endotherms (such as mammals). In fact, a number of comparative studies suggest that the assimilation efficiency of its organisms makes little difference to a food chain's length.

Perhaps, then, chain length is controlled by primary productivity, the amount of energy entering the ecosystem. Food chains should then be longer in communities with higher primary productivity. Stuart Pimm and his colleagues compared food chains from highly productive biomes (such as tropical rainforest) with low productivity biomes (such as tundra) and found no difference—each had an average of four levels. Pimm and Kitching also tried boosting the energy input (by adding leaf litter) to some simple communities, but their experimental manipulation failed to extend chain length.

It seems that other features of the food chain must be limiting its length. One is the tendency for consumers to become larger and fewer with each link of the chain (Figure 6.20). Large carnivores require larger territories from which to collect sufficient prey. Perhaps, as Paul Colinvaux has suggested, a predator would have to be so large to feed on the big carnivores at the end of most food chains that it would never secure enough energy to maintain a population. In this case, the hierarchy of the chain becomes self-limiting (Section 8.1).

Predators compete with each other for prey and, together, a collection of predators may check the size of herbivore populations. Despite the abundant vegetation in most ecosystems, herbivore numbers may then be limited by their consumers. In the same way, predators are themselves perhaps limited by the numbers of predators or parasites consuming them (Section 4.4). One possibility, therefore, is that food chains are constrained by interactions between trophic levels, interactions that organize the whole system (Section 9.4).

6.4 **The web**

Perhaps food chains are too much of a simplification. Perhaps the simple linear sequence they portray, grouping species into ill-fitting categories, is part of the problem. Energy does not follow a single route through an ecosystem, but several. A map of all these routes creates a food web—a network of all possible connections, linking several food chains.

Different food chains overlap and connect at different species, and together chains are knitted into the food web. Figure 6.21 shows a web for the community associated with oak trees (*Quercus robur*) in an Oxfordshire woodland. Notice that titmice (birds of the family Paridae) feed at a number of trophic levels, so that both blue and great tits feed on seeds (primary producers), winter moths (herbivores), and beetles and spiders (carnivores). Mice and voles are also omnivorous, eating both insects and plant material. These are all points where food chains connect.

Webs provide a more complete description of the energy pathways in an ecosystem and allow us to locate organisms that do not fit neatly on one trophic level. We can now include detritivores and decomposers, so uniting grazer and decomposer food chains. It is also possible to include parasites and hyperparasites, higher-level consumers that are ignored in many food chains. This is still a simplification, but it is at least a fuller description of the energy pathways within a community.

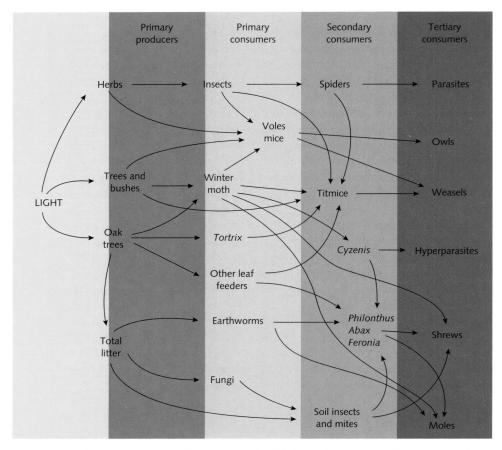

Figure 6.21 A food web as a series of interconnected food chains. In this case, the web shows connections in an oak woodland (Wytham Wood in Oxfordshire).

Describing a web of even simple ecosystems requires a large amount of work and not all the trophic connections may be identified. It also sets somewhat arbitrary boundaries, often by ignoring the energy that leaves the system through migration or transitory predators. Early studies of food webs were limited by the scale and quality of observations, when an occasional or rare feeding event might easily be accorded the same status as a more routine link. Very detailed studies, analysing gut contents and faeces, can provide a more precise profile of the dietary habits of a species and these have now been extended by DNA analyses (Box 2.2). Simon Jarman and his colleagues used this method to identify the number of vertebrate predators feeding on krill in marine ecosystems, including the Adelie penguin (*Pygoscelis adeliae*) and the pygmy blue whale (*Balaenoptera musculus brevicauda*). Approaching the problem from the other direction, Bruce Deagle and Dominic Tollit have used molecular scatology, studying the DNA in faeces. They have been able to show that the proportions of different DNA in the faeces are an accurate reflection of the prey items fed to Steller sea lions (*Eumetopias jubatus*).

Many food webs were first described using **stable isotope analysis**. Isotopes such as ^{13}C and ^{15}N have very long half-lives and their ratios (compared with their common forms—^{12}C and ^{14}N) can be tracked as they pass from one animal to another. Investigators may also introduce a rare isotope and then follow its progress around a food web, or they may distinguish food sources according to natural differences in their ratios. In this way researchers have shown how Pacific salmon (*Onocorhynchus* spp.) transfer key elements (particularly phosphorus) to freshwater rivers when they return to spawn and die.

In establishing feeding connections, these techniques can also cast blame. Cormorants (*Phalacrocorax carbo*) are seabirds that feed on fish in coastal waters. In Britain, some over-winter inland in places where, some anglers believe, they become significant predators of the freshwater fish. Stuart Bearhop and colleagues compared ^{13}C and ^{15}N ratios in feathers from birds found at freshwater sites with those from exclusively coastal colonies. The anglers were right— their differing ratios proved that the inland birds were indeed feeding on the fish targeted by the fishermen.

Any food web takes a snapshot of nature, freezing the dynamic feeding relationships within a community into a static diagram. Species come and go with time, and some even change their trophic position during their life cycle. Most webs ignore this and few give any measure of the strength of the connections, failing to distinguish between rare or routine feeding events. Nevertheless, there are recurring patterns. For example, the ratio of predator to prey species appears to be relatively constant, whereas the proportion of consumers on each trophic level is independent of the total number of species in the web. Additionally, the number of linkages is typically twice the number of species and, as we have seen, food chain length appears to be independent of the web type. Habitat is also an important factor. Food chains are generally shorter in small or frequently disturbed habitats and in ecosystems with little vertical development (a grassland, say, compared to a forest). However, these differences are relatively small and, as Pimm and Kitching observed, the over-all productivity of the ecosystem makes little difference to chain length.

So what causes these patterns? Computer models that incorporate the effects of population sizes and the strength of the connections between species suggest that webs are limited by the number of species they can contain. Established webs are not easily invaded by a new species. For example, an invasive predator attempting to enter a well-developed community must first gain access to a prey species by competing with established predators. If it is successful, interspecific competition must increase for existing species on that trophic level (Section 4.3), and its number of species may therefore stay within a small range. Similarly, interactions between levels will be constraining—in their study of 62 published food webs, Frédéric Briand and Joel Cohen found that the number of predator species was a better predictor of the number of prey species than the other way around.

Whilst aquatic food webs tend to be regulated by top-down forces—constraints exerted by the higher trophic levels—terrestrial food chains appear to be controlled by their lower levels. Terrestrial food webs tend to have less herbivory and a larger detritivore community. They also have a slower turnover time (Figure 6.6) and a tendency to accumulate detritus. Jonathan Shurin points out that the allocation strategies of land plants require them to devote more of their resources to growth, light capture, and defence, to produce a structurally-complex biomass. In contrast, aquatic producers—like phytoplankton—are simpler, frequently unicellular, and short-lived, with less investment in support and protective structures.

As with many ecological concepts, food webs are artificial constructs to aid our understanding and description of important ecological processes, but they may fail to capture the fluidity of the real world. Even so, they suggest ways in which constraint and regulation amongst species are part of the mechanism by which communities are structured. They also help us to manipulate these processes to our benefit and to quantify them. In particular, they can be used as ecological road maps to track the route of pollutants through the ecosystem (Box 6.4).

BOX 6.4   **Bioconcentration, bioaccumulation, and biomagnification**

For many people, it is inevitable that pollutants entering ecosystems will move along food chains, becoming more highly concentrated as one species consumes another. A toxic substance that does not decompose will simply follow the natural trophic pathway, accumulating at each step along the way.

The classic example, and the first pollutant that appeared to show this behaviour, was dichlorodiphenyltrichloroe-thane (DDT). Following its widespread use as an insecticide in the 1950s, DDT caused the population collapse of several predatory birds by reducing the number of eggs they hatched. The realization that spraying insects two or three trophic steps away caused their decline sparked the environmental movement in the west and began the tradition that pollutants were 'biomagnified' along food chains.

Today we realize that the picture is not so simple. **Biomagnification** occurs if the concentration of a pollutant is higher in a consumer than in its diet, that is its **concentration factor** (CF) is greater than 1:

$$CF = \frac{\text{The concentration of the pollutant in the consumer}}{\text{The concentration of the pollutant in the diet}}$$

This equation is equivalent to the assimilation efficiency used to measure energy transfer in Box 6.2, and a trophically mobile pollutant will trace the energy pathways within a food web.

Unless CF > 1, there is no biomagnification and some of the pollutant is lost in the trophic transfer. If both the diet and water are a source for aquatic organisms, the equivalent calculation is called **bioaccumulation**; **bioconcentration** refers specifically to uptake from water alone. In each case, to have a CF greater than 1 the consumer's rate of uptake must exceed its rate of loss.

These are very simple measures . . . and that is part of the problem. First, both terms are time-dependent, yet levels in both the diet and the consumer change continually. A concentration factor only has meaning when the consumer has come to an equilibrium with its diet—only then do we properly measure a consumer's capacity to concentrate the pollutant. This condition is rarely met for specimens collected from the field.

Next, should we measure the whole-body concentration or just concentration in those tissues likely to become part of

the diet of a consumer? Often, much of the pollutant burden remains in the leftovers—in the bones, shell, or hair—and will be lost from the food chain we are mapping. Comparing whole-body concentrations between trophic levels may therefore overstate pollutant availability. Additionally, many predators consume several different prey species, whereas omnivores feed on more than one trophic level. In short, we have to be careful properly to quantify what is assimilated by a particular consumer. Demonstrating biomagnification means placing a species in its appropriate trophic position, and that requires a detailed knowledge of the larger food web.

Pollutants are not necessarily transferred, unaltered, along the food chain. Whilst this may be the case for elemental pollutants, such as metals, many organic compounds are degraded by an organism's metabolism. Even DDT, which is highly insoluble in water, can be degraded at a slow rate by many animals, and its concentration falls within their tissues. Then the body size of the consumer is important: large animals tend to eat more and consume a greater mass of pollutant, but lose it more slowly because they have a slower rate of metabolism per unit of body weight. Thus, the larger animals found at the end of food chains are likely to have a higher concentration simply because they are bigger and excrete the pollutant more slowly—not necessarily because of their trophic position.

Once we begin to address these questions (and consider whether they have been considered in previous studies), we realize that biomagnification does not occur for every pollutant, for every top predator. Take, for example, the polar bear. Susan Polischuk and her colleagues studied the concentrations of various organochlorine (OC) pollutants in bears from Manitoba and showed that DDT (and its derivatives) and PCBs had very different dynamics.

OCs have an affinity for the lipid tissues the bears lay down when they are well-fed. Polischuk and her colleagues measured OC levels in fat, blood, and milk before and after the bears underwent their summer fast. This lasts an average of 56 days, during which time the bears do not defecate, so pollutants could not be lost via this route. As their fat reserves are used up and the bear gets thinner (assuming the pollutant is not broken down by the bear's metabolism) we would expect pollutant concentration to

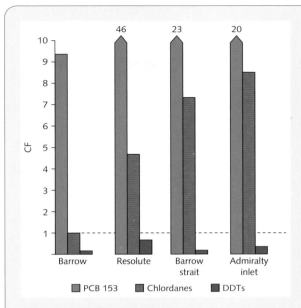

Figure 6.22 The concentration factors (CF) of organochlorine pollutants in the lipid tissues of polar bears from four populations in Alaska, compared to the blubber of its main prey species, the ringed seal. DDT and its derivatives never achieve a concentration factor of 1 whereas PCB 153 is always highly biomagnified. Even so, along with other OCs (such as chlordanes), there is little consistency in CFs between neighbouring populations.

rise. The researchers found that PCB concentrations did rise, but that DDTs declined because they were metabolized. John Kucklick and his colleagues found the same pattern in Alaskan polar bears, where PCBs were also always retained, but at very different levels in different populations (Figure 6.22).

By comparing the concentrations of OCs in the blood and milk with those in the fat tissues, this latter study suggested that these different tissue fractions were close to equilibrium with each other before and after the fast. PCBs are probably lost through subsequent excretion, and in the case of nursing mothers, in the milk. Because these compounds are not metabolized, milk concentrations rise during the fast, just as the cubs are becoming stressed by a lack of food. At this crucial stage, the cubs are being fed increasing concentrations of a pollutant that may affect their development (see First Words).

Interestingly, polar bears have great proficiency in metabolizing DDT and this is one reason why no population of this top predator showed a CF greater than 1 for this pollutant. Biomagnification is far from inevitable, and concentration factors can show considerable variation between pollutants, populations, and species.

6.5 Working the system

Some of the shortest food chains are the ones we use ourselves. As omnivores, agriculture allows us to feed as both herbivores, consuming our crops directly, or as carnivores, feeding on our herbivorous livestock (Figure 6.23).

From the earliest times, human beings have responded to the patterns of productivity in the world around them, but it was the capacity to manipulate primary and secondary productivity that marked the beginning of civilization. For Jacob Bronowski, the change from nomad to **village agriculture** was the biggest single step in the ascent of man. Our ability to work the system has been central to our success, both

in terms of our numbers and the range of nature over which we have influence.

Humankind started out as **hunter-gatherers**—relying on what could be collected or hunted down (Figure 6.24). This way of life is still practised today by many peoples in tropical and sub-tropical areas, among them the San people of the Kalahari and Australian aboriginals.

Although it can never support a large population, at least in semi-arid areas, hunter-gathering is a viable and sustainable feeding strategy. Energetically, at least, it provides a relatively high return on the energy and effort invested—around 10 times as much. By

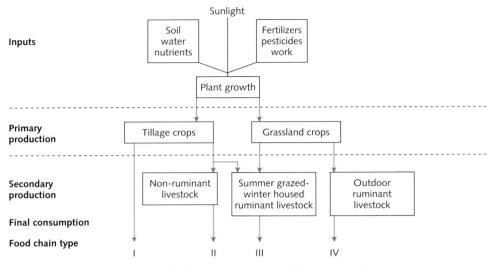

Figure 6.23 A 'web' of agricultural food chains, ranging from (I) humans directly consuming crops and (II) intensive livestock rearing to (III and IV) the grazing of ruminants.

contrast, that epitome of western agriculture, intensive poultry production, returns a small fraction of the energy put in (Box 6.5).

Nomadic-pastoralists are found over a wide range of habitats, following or leading herds of semi-domesticated or domesticated grazers in their search for suitable forage. The Sami of northern Europe track the movements of reindeer to exploit the brief summer of the tundra. The Bakhtiari of Iran and Iraq herd their domesticated sheep and goats over mountain ranges to avoid summer droughts and to find sufficient grazing. Herdsmen of the Alps,

the northern Italian *alpeggios*, Spanish *dehesas*, and *montados* of northern Portugal also move their stock according to season.

The seasonal movement of livestock is termed **transhumance**. Like hunter-gathering, it was practised by some ancient cultures and still has contemporary equivalents. Many ranchers in the Rocky Mountains exploit the high mountain pastures that briefly flourish after the winter snows have melted. Alpine pastures are also used in this way, with herdsfolk moving their stock (and their homes) to the higher slopes during the summer months and returning to the valleys for the winter. This requires a highly coordinated society in which those who herd stock are supported by others who cultivate crops in the valley and make hay for the winter months.

Cultivation probably developed alongside the domestication of the first agricultural animals and almost certainly followed the evolution of a series of heavy-seeded grasses (Box 2.5). Indeed, the sowing, harvesting, threshing, storing, and trading of grain prompted the settled existence that dominates most societies today. As agriculture developed and we learnt to exploit the energetics of ecosystems, so our relations with the living world have become increasingly sophisticated, if somewhat strained.

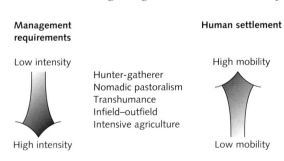

Figure 6.24 A checklist of agricultural systems. The five principal agricultural systems are ranked in terms of their degree of intensity and the amount of mobility in their associated societies. These range from the highly mobile, low-intensity lifestyle of hunter-gatherers to the highly settled, intensive practices of modern agriculture.

Grazing and the cultivation of plants dominate world agriculture. Of the 36 per cent of the Earth's land surface under agricultural production, around one-third is used for cultivation and the remainder for grazing. The oldest established form of animal husbandry is the **infield-outfield** system, characteristic of highly settled village communities. This typically operates on a roughly 10:1 ratio of outfield to infield. The outfield is relatively unproductive land, unfenced and used for grazing by domesticated livestock. The infield surrounds the settlement and is intensively cultivated, growing food for the villagers and for winter fodder, but is primarily used as meadowland for high-quality grazing.

Under this system, farmers obtain a return from the uncultivated outfield, exploiting its productivity and nutrients as feed for their livestock, without much investment in its management. In contrast, considerable effort goes into managing the infield grasslands. While livestock are grazing the outfield, the hay meadows are allowed to grow unhindered for several weeks. After their grasses have set seed, the standing crop is cut for hay and left to dry. Its seeds fall to the ground, providing the next generation for the grassland sward. Once the hay has been collected, livestock may be allowed to graze the infield (aftermath grazing) and are later fed the hay during the winter.

As a result, meadows (and other grazed grasslands) can have very rich floras—a square metre of chalk grassland can support more than 40 species of flowering plants. This diversity follows from the frequency of disturbance and release of nutrients by grazers creating opportunities for a wider range of species. Grazers remove some of the biomass of the faster-growing grasses, recycling their nutrients with their faeces and urine.

New methods of intensifying grass production are bringing about changes in these meadows. Old pastures are oversown with vigorous agricultural grasses, such as perennial rye grass (*Lolium perenne*), supported by the use of inorganic fertilizers. These out-compete other species, especially wild flowers that do not respond as well to the increased nutrients (Section 4.3). Greater changes follow when the meadow is used for silage production. By mowing several times during the year, the farmer collects the production of the meadow and stimulates the further growth of its grasses. The clippings are then fermented to break down part of their cellulose, increasing their nutritional value and reducing their bulk. Unfortunately, frequent cropping means the grassland and its wild flowers get little chance to set seed and, inevitably, floral diversity declines. During the last 60 years in Britain over 95 per cent of flower-rich hay meadows have been lost (Figure 6.25), endangering a large number of wild flower species. Many temperate countries have suffered similar losses and several now have programmes to promote traditional grassland management.

Why grow animals?

The metabolic costs and variable efficiencies of each trophic level mean that energy is lost with each transfer along a food chain (Boxes 4.1 and 6.2). Why then do we not dispense with grazing land and its animals altogether and instead grow crops we could consume ourselves? Surely, we would then recover more of the energy entering the system?

One answer is that, as omnivores, we lack the physiology to assimilate much of the energy available in plant tissues. As carnivores, we can use the capacity of herbivores to release this energy and convert it into a form we can assimilate—their tissues (Figure 6.26). This is particularly true of ruminants which digest cellulose through the activity of symbiotic microorganisms in their gut (Box 6.5). In so doing they unlock otherwise unavailable energy from primary production (cellulose) and make it available in the form of milk and meat.

Grazers also allow us to exploit areas we cannot cultivate. They represent an energy-efficient means of harvesting primary productivity from a wide area. Outfields and rangelands are frequently rough, stony places, with steep hills, deep valleys, and poor soils. This makes them impractical to cultivate, whereas wide-ranging and adaptable grazers, such as sheep and goats, collect some of this habitat's energy in their tissues, which we can then access.

Despite their limited structural complexity, grassland communities can be highly productive. Alpine meadows, for example, can produce as much as 1 200 g biomass/m²/year and this rises to 1 500 g/m²/year for prairies. This is almost double that of crops

Figure 6.25   Well-managed meadows can be both productive and diverse. This traditional meadow in the southeast of England has been managed for centuries by grazing and cutting, and supports a wide range of plants including, in this particular case, green-winged orchid (*Orchis morio*).

grown for our consumption or for fodder (Table 6.1). Because crops are in the soil for only part of the year they cannot match the primary production of permanent pasture.

In contrast to cultivated land, the soil beneath pasture retains its structure and its decomposer community. With minimum disturbance, the grass sward provides year-round protection from severe weather and its roots bind the rich topsoil, preventing erosion. Faeces left by grazers and the remains of the overlying vegetation raise its organic content, enabling the soil to hold water in its upper layers, close to the

---

BOX 6.5   Rumination

The power of bacterial decomposers is exploited by a wide range of animals that maintain bacterial cultures in their gut. Insects, mammals, and even primates have symbiotic relationships that allow them to recover energy from cellulose which would otherwise be excreted. Amongst the mammals, the ruminants—sheep, cattle, deer, antelope, gazelles, and giraffe—have the closest association, with an anatomy and behaviour to suit the bacterial community they host. These are fermentation vessels on legs, operating a highly controlled continuous anaerobic bioreactor as an ecosystem within their digestive tract.

The ruminant stomach is divided into four compartments and contains between 100 and 1 000 million bacteria per millilitre. It also contains roughly half that number of protozoa, predators of the bacteria, which assimilate their protein (making it available to the host) and help to maintain the vigour of this decomposer food web.

The animal maintains a steady environment through its own body heat and, by producing copious amounts of alkaline saliva, maintains the pH of the reaction vessel. The regular supply of vegetation, and the removal of the fermentation products, provides for a rapidly growing bacterial

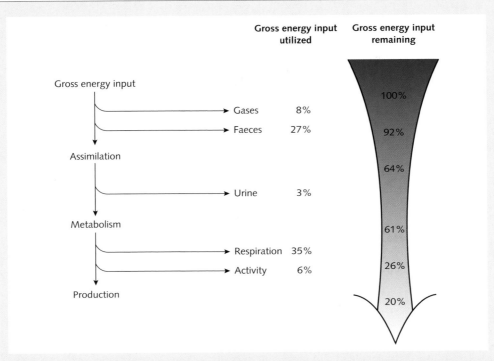

Figure 6.26 Energy budget of a ruminant charting the fate of energy taken in as food (gross energy input). Losses occur at various stages in the process, with the largest being due to metabolic costs. Eventually, 20 per cent becomes fixed in the cow's tissues.

community. By chewing the food, the host creates a well-divided and moist substrate on which the bacteria can grow, promoting fermentation in the rumen. This is later regurgitated to be further reduced in size when 'chewing the cud'. In this way, the ruminant breaks the vegetation into progressively smaller fragments, providing the bacteria with a greater surface area on which to work.

The microbes can be grouped into guilds that specialize in attacking particular components. Cellulolytic bacteria break down cellulose, amylolytic bacteria work on starch. Some are able to deal with both. The main products are volatile fatty acids, such as lactic, butyric, proprionic, and acetic acids, which the host absorbs across the rumen wall into the bloodstream. As with all anaerobic fermentation, the process produces large amounts of carbon dioxide and methane, which the ruminant 'vents'.

Proteins within the food are broken down in the final chamber, the abomasum. This is highly acidic, killing the micro-organisms that pass into it. Eventually, the bacteria

and their predators become part of the diet of the ruminant, an important part of its protein intake.

The benefits of carrying and maintaining a large microbial community in an enlarged stomach are considerable. Pigs and poultry are able to convert between 12 and 14 per cent of their gross energy intake into production, but ruminants achieve around 20 per cent (Figure 6.26). Forage, such as hay, can contain as much as 30 per cent cellulose, and ruminants can tap into an energy source unavailable to most other herbivores. Their partners in the mutualistic association, the micro-organisms, benefit from the benign conditions in most of the animal's stomach.

Domesticated cattle enjoy our protection and husbandry, and this has allowed them to become one of the most populous mammals on the Earth. In turn, we benefit from their capacity to convert low-grade plant material into protein and fat, which we harvest in ways that do not always require the death of the cow.

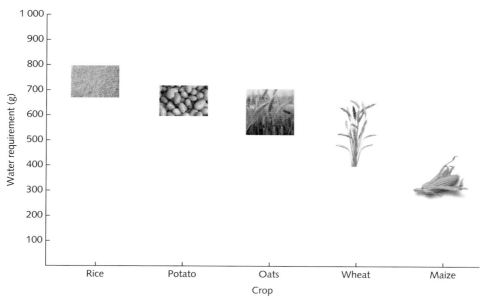

Figure 6.27   The amount of water required (in grams) to produce a gram (dry weight) of foodstuff.

roots of the grasses. Again, compared with other crops, grass has a more efficient water budget and there is a variety of species able to grow under all but the most extreme rainfall regimes. In contrast, the water requirements of crops can be considerable (Figure 6.27) and without irrigation, grasslands may be the only option in some areas.

Getting the grazing right

For many habitats, it is the grazing activity of the livestock that maintains pasture and prevents it from becoming scrub or woodland. Here the grazers are acting as keystone species, but as we saw with the browsers of the African savanna (Section 3.5), at high intensities their feeding can damage the system's capacity to recover.

On the upland pastures of northern Britain, hill-farming has always been a tough existence for farmer and stock alike. To keep sheep-farming viable (and prevent rural depopulation and the loss of open moorland), financial incentives were used to support upland grazing. These took the form of 'headage payments', where farmers were paid a subsidy for each sheep. Not surprisingly, this led to widespread overstocking and considerable environmental degradation.

Sheep prefer bent and fescue grasses (*Agrostis* and *Festuca* species) and these were the first species to decline with overgrazing. Mat grass (*Nardus stricta*), on the other hand, is tough and unpalatable and tends to be avoided by the sheep. Left ungrazed, *Nardus* rapidly spread over large areas. The deterioration in the hill pastures was a concern to both the farming community and conservationists. Hungry sheep headed for moorland and started to overgraze heather. Because heather is crucial to the moorland ecosystem, a decline of this primary producer had severe implications for the other animals it supported further along the food chain. Once the problem was identified, stocking densities were adjusted to the carrying capacity of each area and farmers who stock their land to an optimum density are now rewarded for their good stewardship.

In some cases, undergrazing can be as much of a problem as overgrazing. Moderate grazing controls and rejuvenates the plants within the sward, raising their productivity. It accelerates nutrient cycling and promotes an active decomposer food web. Small patches of bare ground, dung heaps, and other localized disturbance provide conditions for young seedlings to become established. In the absence of grazing, changes start to occur. There is an increase

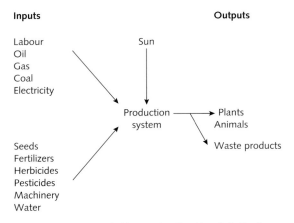

Figure 6.28 Energy subsidies can be direct inputs in the form of fuel or indirect inputs from materials (especially fertilizers and pesticides).

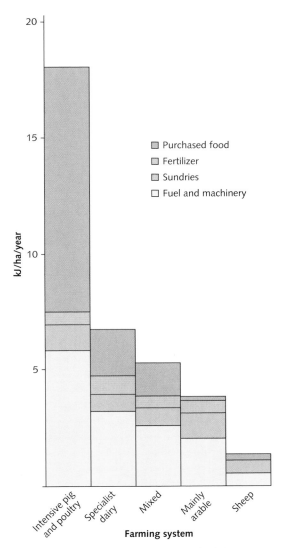

Figure 6.29 An energy subsidy breakdown for five major agricultural systems, from intensive 'factory farming' of livestock through to extensive grazing of sheep.

in the standing crop, a gradual accumulation of biomass, and the community may start to revert to scrub or woodland. Farmers have techniques to compensate for low or no grazing, usually involving some form of biomass removal, such as cutting or burning. Farmers in the American mid-west now acknowledge the key role that wildfire plays in the maintenance of the prairie ecosystem and incorporate it into their grassland management. Mediterranean-type environments also depend on disturbance in the form of grazing and by fire (Section 5.1). Rural depopulation in the north of the Mediterranean Basin and the decline of its farming have resulted in upland areas reverting to scrub (Box 5.6).

Added energy

Agriculture uses energy to manipulate food chains and food webs. We try to prevent some species (weeds and pests) from using the energy entering the system and ensure that this is used by those we are cultivating. Our investment, measured by the energy we apply to the system is its **energy subsidy**. This includes everything from one person turning the soil with a hoe to the waves of combine harvesters that pass over vast wheatfields (Figure 6.28).

We use energy to remove competitive non-crop plants either by tilling the soil or applying herbicides. We can also side-step the nutrient limitations of a soil by using fossil fuels to manufacture and apply

fertilizers. The energy used to manufacture and power agricultural equipment adds to the total subsidy.

Not surprisingly, there are major differences in the size of the subsidy and the returns in different parts of the world (Figure 6.29). An African pastoralist, for example, will expend 21 420 kJ per hectare per year—43 per cent of that of an American maize

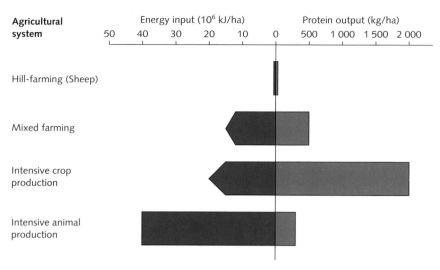

Figure 6.30 Energy inputs and protein yields of four major agricultural systems.

farmer. The returns are greatest in intensive crop production, where the food chain is shortest (chain I in Figure 6.23). Three other systems (chains II–IV) involve a third link—their livestock—which reduces the energy available for our consumption.

The energy subsidy buys us out of some of the constraints, but not all of them. Eventually the costs start to outweigh the benefits and the return on investment or environmental quality declines. For example, phosphate fertilizers allow us to by-pass the limits on productivity imposed by the low availability of this nutrient in most soils, but these are manufactured from phosphate-rich rocks at considerable energy cost. A significant proportion of any fertilizer never reaches the crop to which it is applied and is lost through run-off, enriching other soils and aquatic systems (Box 7.3). This can result in dramatic changes in the species composition of their communities.

Our energy inputs can also determine the protein output of agricultural systems (Figure 6.30), though there is no consistent relationship between energy intake and the amount of protein fixed in the tissues of different consumers. Rather, the energy subsidy is useful as a measure of the cost of production. Intensive animal production is an expensive but reliable means of producing protein. Not only is energy added with the food supply, animals reared indoors frequently require controlled environments so fuel is

burnt to maintain lighting, heating, and ventilation. Western dairy farming also demands a large energy subsidy, both in terms of handling the animals and the milk they produce.

Extensive grazing, in the form of rangeland and pastoralism, remains the most widespread means of producing animal protein. Sheep farming incurs the lowest costs, since the bulk of grazing is confined to the extensive outfield or unmanaged pasture. These animals need only an occasional food supplement, with the main costs being the management of periodic grazing within the infield and its small input of human labour.

Helen Caraveli used changes in the energy subsidy of European agriculture (indicated by the use of tractors and fertilizers) as a measure of intensification in Mediterranean farming (Figure 6.32). Over a 15-year period, Mediterranean countries intensified at a disproportionate rate compared with their northern European counterparts—tractor-use increased by 70 per cent compared to a 7 per cent decline in northern Europe. Given the small size of average Mediterranean farms, and the nature of their soils, the potential for adverse environmental impact is considerable. The intensification of olive oil production has resulted in widespread deterioration in habitats, driven by financial incentives that failed to take into account the environmental implications (Box 6.6).

BOX 6.6

Mounts of olives

Olives are synonymous with the Mediterranean. They invoke its landscape, peoples, and history whilst their cultivation has shaped its traditions, culture, and cuisine.

Today the value we place on olive oil, and the health benefits of a diet in which it is the major fat, has raised its global profile and demand. Olives are grown commercially in other parts of the world including California, Australia, and Argentina, but the Mediterranean Basin remains the main region for production. Helped by subsidies from the European Union (EU), Spain, Italy, Greece, and Portugal dominate the market, producing over 70 per cent of the world's supply from an area of around 5 million hectares.

Farms in southern Europe have traditionally been small, predominantly family affairs (Figure 6.31). In the 1960s and 1970s young people left the farms and the villages to work in the expanding tourist industry on the coast (Box 5.6), leaving an ageing population to contend with the vagaries of an agriculture in which several good harvests might be followed by a disastrous one.

For many years the EU subsidized the farming industry through its Common Agricultural Policy (CAP), providing a guaranteed price for foodstuffs and, through the European Social Fund, welfare assistance to poorer farmers. This helped to guarantee food supplies and provide the financial security the farmers demanded, and to slow the rate of rural depopulation. However, these subsidies also promoted intensification—farmers were now paid for what they could produce, rather than what they could sell. In the case of olive oil production, there were compelling financial incentives to intensify when such cultivation could produce 10 to 20 times more than traditional olive groves (Table 6.3).

Traditionally, olives have been harvested from large-canopied trees, many of which are over 500 years old. This low-input cultivation relies on larger, scattered trees typically planted on ancient terraces banked by stone walls. Management here is minimal, with little or no chemical inputs, and the olives are harvested by hand. Beneath the trees livestock (traditionally sheep or goats) graze, with one or two small plots cultivated for food and fodder.

Needless to say, the productivity of this land can be increased easily with subsidies such as fertilizers, pesticides, and irrigation. These can be financially justified when extra

(a)

(b)

(c)

Figure 6.31 The changing face of olive production. (a) Traditional olive groves with their old trees and rich biodiversity. (b) Intensive olive production—young trees grown in serried ranks in almost clinically tidy conditions. (c) Casualties of change—ancient trees grubbed up to make way for intensification.

(continued overleaf)

TABLE 6.3   **Comparison of three types of olive production**

	Traditional	Semi-intensive	Intensive
Tree characteristics	Big and old	Smaller and younger due to replanting	Dwarf varieties replanted regularly
Tree density	80–150 per ha	150–200 per ha	200–400 per ha
Terraces with supporting walls	Common	Occasional	Rare
Understorey weed control	Harrowed occasionally	Harrowed/cut repeatedly	Controlled with herbicides
Grazing	Rare or common	Rare	No
Chemical inputs	Very low	High	High
Irrigation	Uncommon	Increasing	Common
Harvesting method	By hand	By hand or mechanical	Mechanical
Typical yield	1 200 kg/ha	2 200 kg/ha	5 500 kg/ha
Consistency of annual yield	Very low	Low	High
Soil erosion	Usually low	Often very high	Medium
Biodiversity	High	Low	Very low
Landscape value	High	Low	Low
Other environmental impacts	Fire prevention in marginal areas	Pesticide pollution, irrigation reservoirs	Pesticide pollution, irrigation reservoirs
Subsidy per ha (€)	200	800	1 500

trees are planted to boost olive production. Many farmers have uprooted the older trees and replaced them with closely spaced modern varieties in a monoculture similar to arable farming. Tree densities are increased five-fold with a working life of just 25 years, after which the trees are ripped out and replaced with new stock.

All this has led to a profound change in the landscape and ecology of olive-producing areas. Terraced groves are replaced with plantations filled with little trees (Figure 6.31) and old hill farms are abandoned as their owners move to flat areas in the valleys better suited to intensive production. The environmental costs include soil erosion and, in particular, the loss of biodiversity. The wildflower communities associated with the traditional olive groves are treated as

weeds and are ploughed or sprayed to remove competition. In the absence of the older trees, breeding birds such as the little owl (*Athene noctua*), which nests in tree holes, disappear. Closely spaced trees make life difficult for ground-nesting and ground-feeding birds such as the stone curlew (*Burhinus oedicnemus*), quail (*Coturnix coturnix*), and partridge (*Alectoris rufa*).

The CAP olive oil subsidy regime has now been revised, and the emphasis has shifted away from production towards a more sustainable system. Incentives that reward good environmental practice, such as reduced usage of fertilizers, pesticides, and irrigation, are also planned with incentives for farmers who retain ancient trees, maintain terraced groves, and manage the wildlife habitats on their land.

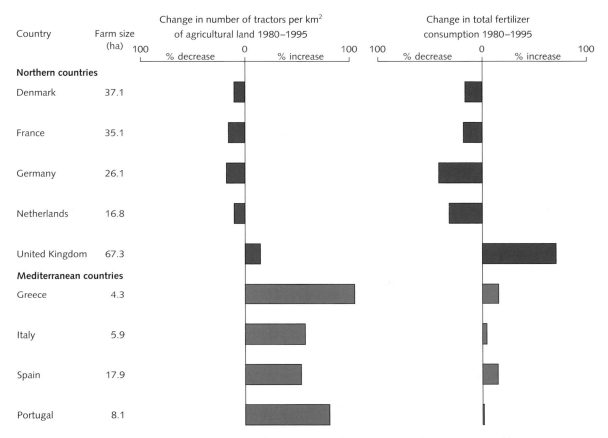

Figure 6.32 Changes in agricultural intensification in the EU between 1980 and 1995 as measured by two components of their energy subsidy: tractors and the application of fertilizers.

Intensive cultivation can damage the soil by altering its structure and disrupting the soil biota (Table 6.4). Frequent ploughing promotes the rapid breakdown (oxidation) of its organic matter. Removing this, the main energy source for the community of decomposers and detritivores, causes their populations to decline, and without their soil-processing activity, the soil texture and its capacity to hold nutrients and water deteriorates.

A frequently turned or disturbed soil has little plant cover. As it dries and its organic content falls, a soil loses its tackiness or crumb structure (Box 7.1) and becomes more mobile. As a consequence, vast amounts of soil are blown or washed away. Estimates suggest that the loss of topsoil amounts to about 1 per cent of the world's cropland each year. In the Andalucian region of Spain, an estimated 80 million tonnes of topsoil are lost annually as a result of intensive olive production (Box 6.6).

Many traditional practices have developed in response to the local conditions of climate, slope, geology, and soils to sustain the productivity of the system and the quality of its soil. For example, techniques rotating crops or shifting cultivation are often followed in places where the soil needs time to recover its nutrient capital. However, around 80 per cent of soil erosion is now attributable to cultivation practices that are unsympathetic to local conditions. Even subsidizing the water supply to agricultural land can cause long-term damage—around one-third of all irrigated land in India has been degraded as a result of salinization (Section 7.1).

TABLE 6.4   Biological functions of soil organisms and the management practices that affect them

Biological function	Functional group	Management practices affecting function
Residue breakdown/decomposition	Residue-borne micro-organisms, soil invertebrates	Burning, soil tillage, pesticide
Carbon binding	Microbial biomass (particularly fungi), macrofauna	Burning, shortening fallow period after slash and burn, soil tillage
Nitrogen fixation	Free and symbiotic nitrogen-fixing micro-organisms	Reduction in crop diversity
Organic matter/redistribution	Roots, fungal hyphae, soil invertebrates	Reduction in crop diversity, soil tillage, fertilization
Nutrient cycling	Soil micro-organisms, soil invertebrates	Soil tillage, irrigation, fertilization, pesticide applications, burning
Soil aggregation	Roots, fungal hyphae, soil invertebrates	Soil tillage, burning, reduction in crop diversity, irrigation
Population control	Predators/grazers, parasites, pathogens	Fertilization, pesticide application, reduction in diversity, soil tillage

Figure 6.33   Monoculture: the familiar face of intensive agriculture.

Even so, improvements in agricultural productivity have followed from the increasing application of technology and innovation (Figure 6.33), especially following the introduction of cheap artificial fertilizers and chemical pesticides in the 1940s and 1950s. In the 1960s and 1970s, improvements in farm management techniques, plant and animal breeding programmes, and pest and disease control all seemed to promise a way of feeding a rapidly growing world population.

The **green revolution**, as it became known, also hinged on the development of new varieties (cultivars) of crop plants. Cultivars were produced that allowed for easier mechanical harvesting and increased productivity. High-yielding varieties were selected for particular characteristics, such as a high leaf:stem ratio and a low carbon:nitrogen ratio: that is, they were bred to produce more leaf and with a higher protein content in their leaves, fruits, or seeds. For example, high-yielding varieties of rice can divert 80 per cent of net production into their seeds, compared to the more usual 20 per cent.

However, the promise of many of these new strains has not always been realized. High-yielding crops in experimental trials often failed to live up to expectations in the field. Sometimes the reason was blindingly obvious: varieties bred by developed countries for use in developing countries were grown without western technologies to support them. Local farmers found the energy subsidy, along with the need for machinery, fertilizers, and pesticides, too costly.

Today's breeding programmes aim to fit crop varieties and cultivation techniques to the environment in which they will be grown. Regarding agricultural land as a habitat like any other allows us to compare it with other, more natural communities. It also helps us understand local traditions and practices—from established patterns of crop rotation to pest control methods. Such techniques and traditions form the basis of **agroforestry**—a land management system used in the tropics for millennia and now being favoured by development agencies. With this approach, farmers working nutrient-poor tropical soils multi-crop and multi-stock their land. Forest trees are managed for food, fuel, and timber and a range of understorey crops like cocoa and coffee are grown in their shelter. This retains part of the original system and its food web, helping to check the growth of pests and diseases. It also protects the soil: with an intact canopy, rainfall has to percolate through several layers of vegetation, entering the soil gradually, with less being lost through evaporation.

Over the last 50 years, around 9 million hectares of productive land have been severely degraded, mainly in over-populated areas (Box 5.6). Perhaps the most significant development at the United Nations Convention on Environment and Development, the Rio Convention of 1992, was Agenda 21, and its 'non-binding blueprint for sustainable development'. This recognizes that future economic development, both agricultural and industrial, needs to be sustainable. For agricultural systems, this means the capacity to maintain productivity without increased species loss or environmental degradation.

● SUMMARY

Primary producers convert sunlight into chemical energy through the process of photosynthesis. This uses radiant energy to produce sugars from carbon dioxide and water, releasing oxygen in the process. Its overall efficiency is around 2 per cent, but the abundance of this source means that photoautotrophs power most of the biological activity of the planet. Some plants have modified the way in which they capture carbon dioxide to sustain their photosynthesis in adverse conditions.

Much of the energy fixed in gross primary production (GPP) is expended in respiration. Net primary production (NPP) is the energy fixed in the tissues and available to higher trophic levels. Herbivores consume part of this biomass and begin a line of consumers that form a food chain. This traces one pathway for energy through a community. Rarely does a food chain have more than four trophic levels, most probably because of species interactions within the larger food web.

Human beings use a range of agricultural strategies to exploit primary production and the energy at different trophic levels. Some of the most ancient, from hunter-gathering to transhumance, are still practised in different parts of the world. Although agriculture seeks to improve the productivity of an ecosystem by applying energy subsidies (to remove competitors, supply nutrients, and so on), these only produce sustainable systems where they are sympathetic to the local ecology.

● FURTHER READING

Tivy, J. 1990. *Agricultural Ecology*. Longman, Harlow. A helpful introduction to agricultural systems and issues such as the energy subsidy.

Salisbury, F. B. and Ross, C. W. 1992. *Plant Physiology*, 4th edn. Wadsworth, Belmont, CA. A standard text on all aspects of plant growth and development—this book is also particularly useful as an introduction to environmental plant physiology.

Vandermeer, H. H. 2003. *Tropical Agroecosystems*. CRC Press, Boca Raton, FL. A useful review of the ecology of tropical agroecosystems. The text provides a classification of farming systems in the tropics and considers these in the context of environmental sustainability.

Whittaker, R. H. 1975. *Communities and Ecosystems*, 2nd edn. Macmillan, New York. A wide-ranging book whose title describes its content exactly. In it, Whittaker reports his original research on the productivity of ecosystems.

● WEB PAGES

The United Nations Environment Program has numerous internal and external links to sustainability issues, including the Earth Summits of 1992 and 2002:

http://www.unep.org

The World Wide Fund for Nature's site is useful for information on agricultural sustainability and conservation. It also provides information on the European Common Agricultural Policy and issues surrounding agricultural intensification in the Mediterranean:

http://www.panda.org/

The real cost of coffee

The coffee bush (*Coffea arabica*) evolved as an understorey shrub, adapted to life in the shade of a tropical forest. Traditionally it was cultivated as an agroforestry crop, grown beneath trees in shaded plantations (Figure 6.34a). This began to change in the 1970s when increasing demand and the rising price of coffee led to widespread replanting with modern high-yielding varieties adapted to life in the sun (Figure 6.35b). In less than 20 years over half of the plantations in Latin America had been converted to 'sun-plantations' (Perfecto *et al.* 1996).

Sun-plantations are expensive to manage, requiring twelve times the energy subsidy, principally because of their intensive management and pest control. With none of the other agroforestry products (citrus fruits, avocados, timber, and fuel-wood) traditionally harvested from shade-plantations to offset these costs, farmers of these plantations were vulnerable to fluctuations in the coffee market, with little to support them when the price of coffee fell.

Not surprisingly, the simplified ecosystem of the sun-plantations also had a major impact on the forest wildlife. Shade-plantations are comparatively species-rich, retaining much of the original native forest and its associated animal species. The 'tidy' sun-plantations have less fallen leaves and dead wood, the litter needed for an active decomposer community. In the 1990s a series of studies demonstrated the impact of reduced cover on the forest bird populations, prompting a campaign to promote shade-grown coffee as both 'environmentally friendly' and a fair-trade option (Komar 2006).

Recent work has sought to quantify the effect removing shade has on the bird community. Cruz-Angon and Greenberg (2005) experimentally manipulated a shade-

(a)

(b)

Figure 6.34 (a) A traditional shade coffee plantation—an agroecosystem rich in species and structural diversity. (b) The more stark, simplified ecosystem of a modern, intensively managed 'sun' coffee plantation.

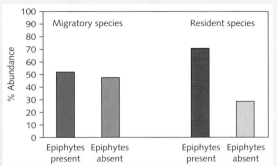

Figure 6.35 The relative abundance of resident and breeding bird species in 'shade' coffee plantations with epiphytes and without epiphytes.

plantation to make it structurally less diverse and potentially less suitable as a habitat for resident and migrant birds. Their change was a relatively small one, removing only one component of cover: the epiphytes (flowering plants, mosses, and lichens) which grow within the trees.

Their experimental area was a large shade-plantation in Veracruz, Mexico, which had already been studied for several years and therefore provided good baseline data. To ensure the experimental plots were comparable, they recorded the tree species present, their height and density, and other vegetation data. They then selected two comparable plots 1 km apart, each of which they divided into

two sub-plots. One of each pair was cleared of epiphytes and the other was left as a control. Between August 2001 and March 2002 bird diversity, abundance, and activity, during and outside the breeding season, were measured in the plots by direct observation. Observers monitored a plot by walking a regular route that wove through it. The walk was timed to last 150 minutes, on mornings of alternate days. All birds seen within 25 m of the observer were recorded and the data was subsequently classified according to the birds' migratory status and feeding guild.

The removal of the epiphytes led to a 41 per cent reduction in the abundance of resident birds (Figure 6.35), probably because of the lack of shelter for roosting and nesting sites. The birds fell into five guilds: fructivores—fruit feeders such as the yellow-throated euphonia (*Euphonia hirundinacea*); granivores—grain feeders such as doves (family Columbidae); insectivores—including warblers (Parulidae); nectarivores—hummingbirds (Trochilidae); and omnivores—such as vireos (Vireonidae). Where the epiphytes had been removed, insectivores, nectarivores, and omnivores all decreased in abundance (Figure 6.36), whereas granivores flourished. This was probably due to the grasses and other seed-bearing plants that now flourished in these opened-up areas.

Whilst Cruz-Angon and Greenberg's study did not involve the total conversion of shade- to sun-plantation, it does show how relatively small changes within an agricultural ecosystem can have a major impact on parts of its animal community. It also gives a measure of the real price we pay

(continued overleaf)

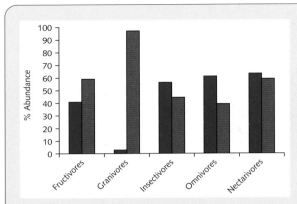

for a popular commodity when intensification is used to reduce the costs of its production.

References

Cruz-Angon, A. and Greenberg, R. 2005. Are epiphytes important for birds in coffee plantations? An experimental assessment. *Journal of Applied Ecology 42*, 150–159.

Komar, O. 2006. Ecology and conservation of birds in coffee plantations: a critical review. *Bird Conservation International 16*, 1–23.

Perfecto, I., Rice, R. A., Greenberg, R., and Van der Voort, M. E. 1996. Shade coffee: a disappearing refuge for biodiversity. *BioScience 46*, 598–609.

Figure 6.36   The relative abundance of major guilds of birds in 'shade' coffee plantations with epiphytes (blue) and without epiphytes (red).

EXERCISES

1. (a) Calculate the efficiency of the energy transfers to herbivores and to carnivores in Figure 6.18.
   (b) Assuming these efficiencies remain fixed, calculate what difference a 10 per cent increase in the energy of the herbivores would make to the energy represented by the weasels.
   (c) If the mice had an efficiency of 0.1 per cent, how much energy would be available to the weasels?
   (d) Decide what kind of efficiencies are represented by your answers to a. (see Box 6.2).
   (e) Why might these efficiencies be underestimated?

2. Examine the life cycle of the malarial parasite given in Figure 4.18 and answer the following questions:
   (a) Identify the consumer and the consumed between the three species involved (definitive host, vector. and parasite).
   (b) What is the source of energy for each consumer at each stage of its lifecycle?

3. Describe how each of the following adaptations allow a plant to reduce its water loss during photosynthesis:
   (a) leaf needle-shaped
   (b) leaf waxy or with volatile oils
   (c) leaf hairy
   (d) special small leaves produced during the summer, winter leaves shed
   (e) leaves reduced to small spines or tiny leaflets and held in a compact plant shape

4. Why should adding artificial fertilizers to temperate grassland reduce the diversity of its plants?

5. The following table gives a balance sheet of the energy in a temperate hardwood forest (Hubbard Brook, New Hampshire USA).
   (a) What proportion of the total energy is held in dead material? (Exclude the soil, where living and dead components are not distinguished.)
   (b) What proportion of the total energy in the system is held in the soil?
   (c) Which part of the food web has the most of the energy in this ecosystem?

Material	Energy values kJ/m² (× 100)	
	Live biomass	Other organic matter
Above-ground		
Trees	2 776	
Dead standing wood		92
Ground level/below ground		
Dead wood		606
Leaf litter		1 003
Live roots	591	
Soil		3 616
Total	**3 367**	**5 317**

6. Construct a simplified food web of a garden lawn ecosystem using the information provided below about the feeding preferences of common garden inhabitants.

Organism	Type of feeder	Diet
Aphid	Herbivore	Live plant material
Blackbird	Omnivore	Worms, slugs, caterpillars, spiders, and centipedes
Caterpillar	Herbivore	Leaves
Centipede	Carnivore	Millipedes, aphids, woodlice, and springtails
Earthworm	Detritivore	Detritus
Ladybird	Carnivore	Aphids
Millipede	Detritivore	Detritus
Slug	Herbivore	Leaves, stems, and roots
Spider	Carnivore	Springtails and caterpillars
Springtail	Herbivore/detritivore	Living and dead plant material
Woodlice	Detritivore	Detritus

7. Grazing animals can be used to produce food intensively (through factory farming) and extensively (through grazing). Draw up a balance sheet of the advantages and disadvantages of both systems.

Tutorial/seminar questions

8. Discuss the ways in which the intensive agricultural techniques practised in temperate areas may not make for a sustainable agriculture in tropical areas.

9. Can we expect to draw general principles about ecological communities when no natural food web has been fully described?

Balances

← The eruption of Mount Ruapehu, New Zealand.

Geologists delight in describing soil as 'rock on its way to the sea'. This is not just evidence of a wry sense of humour but an accurate picture of one long-term process of the planet. Exposed to the atmosphere, rocks crumble and their particles, held briefly in soils, eventually find their way to the bottom of seas. Much later they may be resurrected as new sedimentary rocks, or some of their elements may be released into solution to be diverted through plants and micro-organisms into other parts of the biosphere (Figure 7.1).

In Chapter 6, we described the central role that energy plays in driving living systems and how it fuels metabolism and the building of new tissues. Elements such as carbon, hydrogen, oxygen, nitrogen, and phosphorus are needed to make the macromolecules from which living tissues are built. For plants, these are obtained from the air and soil but for most animals their supply is plants or other animals. Without life, elements move in geochemical cycles of erosion and deposition. With life they move through biogeochemical cycles, diverted through living organisms, on their way to the sea.

The global flux of nutrients is largely determined by the energy available to drive these biotic and abiotic processes. With their greater variety of biochemical pathways, bacteria and fungi in the soil can fix and make nutrients available for the primary producers living above. In turn the photosynthesizers support micro-organisms and the larger decomposer community through the energy-rich material they add to soils or sediments. The energy released during decomposition drives the microbial capture of nitrogen and phosphorus, and the transfer of these, and other nutrients, through the biosphere.

In the last chapter we saw how a major proportion of the energy fixed in terrestrial ecosystems passes through the decomposer food web, and it is this which powers nutrient cycling. Here we describe the role that living organisms play in these cycles. Most communities are adapted to particular nutrient levels in their ecosystem and large changes can follow when these are disrupted. An excess of nutrients may change the species composition of a community and shift it from a diverse and 'balanced' ecosystem. On the other hand, our activities often create ecosystems where nutrients are deficient and which, without help, could not quickly develop a self-sustaining community. We might then attempt to accelerate and manage its development until it becomes self-sufficient, say by recreating a soil and encouraging its decomposer community to develop the capacity to meet plant nutrient demand. Restoring ecosystems in this way becomes a major test of our understanding of ecological processes, challenging us to assemble the parts to make the whole work.

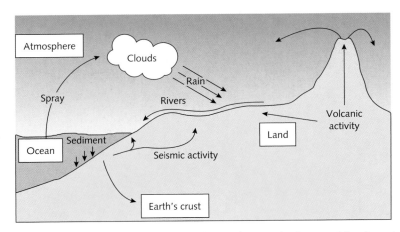

Figure 7.1 The major geochemical processes. Minerals move between land, sea, and the atmosphere. Whilst some elements may be lost to the Earth's crust, others are returned to the ecosystem by volcanic and seismic activity.

7.1 The nutrient cycles of life

We can think of the biosphere as being something like a machine (Figure 7.2), one that builds new tissues and new organisms, driven by the energy captured from sunlight. The life that emerges from these raw materials is not only a product but an integral part of this machine, interacting with the non-living processes to drive part of its mechanism. Through the activity of living organisms, particularly the microbial communities, materials are collected, processed, and made available to different parts of the biosphere. In the process, living tissues are built and ecological communities develop.

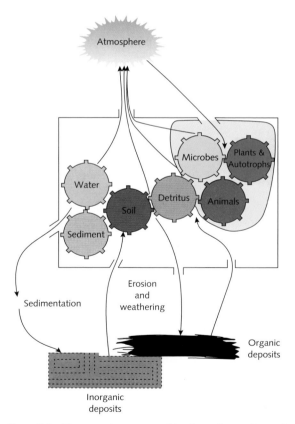

Figure 7.2 The ecosystem as a machine for cycling matter and energy. Minerals move between the atmosphere and the organic and inorganic deposits via the living and non-living processes that drive the system.

When energy is plentiful and conditions are equable, the speed of the machine is governed by the availability of the components needed to build cells. Without these resources, production grinds to a halt and growth ceases. Conditions that are not equable —not warm and wet—slow these processes. The polar regions are bathed in weak sunlight and their low temperatures reduce production and slow decomposition considerably. For example, in the taiga—northern coniferous forests—an atom of calcium passes through the soil and trees in about 43 years, compared with an average of just 10.5 years in tropical rainforests. For this element the taiga is operating at a quarter of the speed of its tropical counterpart (Section 8.2).

The **flux** rate of an element, the speed at which it moves through the biosphere, depends upon its physical and chemical properties and the use to which living organisms put it. Nitrogen passes through the system relatively rapidly, whereas generally phosphorus cycles much more slowly. Phosphorus tends to be relatively immobile, primarily because of the insolubility of its compounds. Nitrogen compounds are highly soluble, highly mobile, and are rapidly taken up by plants or soil bacteria. Most nitrogen compounds are also readily lost in water percolating down through the soil or washing off the land.

These movements also depend on the chemical characteristics of a soil and its biological activity. A high proportion of clay or organic matter will bind charged ions and hold key nutrients in the soil. Equally, soil organisms can fix these nutrients in their cells and tissues. Bacteria and fungi have a variety of biochemical pathways that allow them to capture and fix key elements. Their activity represents the single most important source of both nitrogen and phosphorus in terrestrial ecosystems, without which the rest of the biota would be unable to build new tissues (Box 7.1). When the machine is working effectively the result is a fertile soil, a repository of available nutrients that can support a highly productive community.

BOX 7.1   Profile of a soil

Soils form at the interface between the surface of the planet and its atmosphere, where a fractured and crumbled rock holds enough moisture and heat to support biological activity. When these requirements for life converge a community can form, providing the minerals required by the autotrophs to fix the energy which will power the developing ecosystem.

Much of the biological activity of terrestrial ecosystems is concentrated at this interface: photosynthesizers supply energy below ground as dead and decaying plant material that, with animal wastes and remains, will feed the decomposer community. Water arrives both from above and below, whilst oxygen has to pass into the soil from the air to support aerobic decomposition. This traffic across the air/soil boundary creates a series of gradients, a profile of colour and biological activity that declines with distance away from the surface.

The deep browns of the upper layers reveal where decomposition is concentrated. Without the products of decomposition, soil minerals are often lightly coloured and relatively inert, but the processes and products of decomposition stain these upper layers, and demarcate distinct **soil horizons**. The depth and colour of each layer indicates the nature of its organic material and the microbial community decomposing it, creating a soil profile characteristic for an ecosystem (Figure 7.3, Table 7.2). The roots of trees and other plants cross these horizons in their search for water and nutrients, and create pockets of colour within these layers. Overall, the demarcation of the horizons tell how easily water, minerals and oxygen move down the profile.

Water, oxygen, carbon, and nitrogen arrive at, or are fixed, close to the soil surface. Phosphorus and potassium have to be scavenged from below and are derived from the mineral and organic components of the soil. Water is the main transport medium, moving soluble nutrients through the soil profile, in a process termed **eluviation**. Depending on the climate and soil type, this can cause the leaching of minerals down the profile where they may be deposited, through **illuviation**. Living organisms play an important role in these movements, on a large scale through the activity of burrowing animals, or at the microscopic scale, through the various chemical transformations achieved by bacteria. Roots

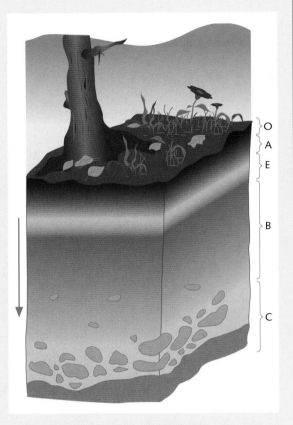

Figure 7.3 A soil profile showing the 'horizons', or layers, that may be found in a temperate forest soil. Here the predominant movement of water is down the profile (eluviation). The surface layer of undecomposed litter is the O horizon. In the layer below it decomposition is in progress and a humus of organic detritus is formed (the A horizon). This is the zone of maximum microbial activity. The leaching of minerals washes out this colour from the next layer, creating a lighter E horizon. These minerals, especially iron, may be precipitated as the soil's chemistry changes further down the profile, producing illuviation in a B horizon. At this depth there is little aerobic microbial or invertebrate activity, and this layer is largely composed of weathered parent material. The parent material, in various stages of breakdown forms the lower C horizon.

and fungal hyphae scavenge scarce or insoluble minerals and also bring water from deep below the surface.

In wet climates, the predominant movement of water is down the profile, though this may be temporarily reversed during drier seasons through evaporation and transpiration (**evapotranspiration**). Where the annual balance of movement is upward, the net movement of minerals is also up the profile and can lead to illuviation close to the surface. Persistent movements of water in this direction lead to soluble minerals being crystallized close to the surface, as various salts. The soils of drier and hotter climates have shallow sub-soils with a poor structure, and are often very saline.

The ease with which a soil drains depends on its parent rock or the deposits from which the soil originated. Coarse particulate soils, dominated by sands and gravels, drain readily but the fine particulates of clays make for soils that are easily waterlogged, that are heavy and cold in the higher latitudes. Very wet soils may never drain properly and then become anaerobic at depth. This promotes a very different microbial community, especially in deeper and oxygen-poor horizons, often dominated by chemoautotrophs (Box 6.1). Closer to the Equator, clays with high concentrations of aluminium or iron can bake to form a hard brick-like material called **laterite**.

The profile of a soil is a close reflection of the vegetation community growing above, itself largely determined by climate (Section 8.1). Soils under temperate woodlands or grasslands have an organic component (or **humus**) rich in calcium and potassium, an alkaline humus or **mull**. Heathlands and coniferous forests have a nutrient-poor acidic or **mor** humus that lacks these elements. Humus and litter are important factors in the chemical nature of a soil because they determine its capacity to hold water and retain its nutrient capital. A soil with a high proportion of either clay or organic matter has a large water-holding capacity. It will also retain more of its mineral nutrients. A soil's **cation exchange capacity** (CEC) is a measure of this binding of charged minerals, especially the major nutrients required by the plant and microbial communities. Negative charges on soil particles will bind positively charged ions (cations), most importantly calcium, magnesium, potassium, and ammonium. In solution, these cations attract the negatively charged anions, most especially nitrates and phosphates, and together these key nutrients are leached from the profile. Thus soils with a low CEC lose their nutrients readily, whereas fertile soils typically have a high CEC (Table 7.1).

TABLE 7.1 **Cation exchange capacity (CEC) of different soil textures**

Soil texture	CEC (meq/100 g)
Sand	1–5
Sandy loam	5–10
Loam	5–15
Clay loam	15–30
Clay	15–100
Organic material	200–400

CEC is measured in milliequivalents per 100 grams and is an indication of a soil's capacity to hold and exchange cations. Soils with fine particles and therefore a large surface area, or with a high organic content, have a higher CEC.

CEC is also a measure of a soil's capacity to bind toxic elements, including aluminium and a range of toxic metals. Many plants cannot establish on acidic soils because unbound ions are available to inhibit root cell division and thus prevent the growth of an extensive root system.

Typically, highly fertile soils have a high organic content and an active microbial community. Organic content is also an important determinant of soil texture. For a good texture, soil particulates need to cling together with a loose and airy structure and form a **crumb**. This requires a good range of particle sizes—a balance of fine and coarser particulates, of clay and sand, giving a **loam**. Much of a soil's stickiness results from the activity of micro-organisms, and the various exudates from plants and fungi. Crumb structure is therefore a good indicator of a soil's biological activity.

Although the parent material of the mineral component may define the chemistry of a soil, its position and local regimes of temperature and moisture are also critical. The major types of soil are consequently classified according to their climate and associated plant community, as well as their rock type (Table 7.2). These same factors determine the nature of the decomposer community, especially the relative role of invertebrates, bacteria, and fungi. For example,

(continued overleaf)

TABLE 7.2   A classification of the major soil types and the conditions under which these occur

Type	Vegetation/climate	Drainage	Texture/structure	Humus/soil pH
Podzol	High latitude or high altitude coniferous forests or heathland	Free, though periodic waterlogging at depth	Sandy	Mor, strongly acidic
Gley	Swamp, wet grassland	Poorly drained	Often heavy clay	
Peat	Bogs	Poorly drained	Very high proportion of poorly-decomposed organic material	Mor
Brown Earth (Cambisols)	Cool temperate deciduous woodland	Free	Loam, good crumb	Mull, going on slightly acidic
Acid Brown Earth	Humid temperate mixed deciduous/coniferous woodland, mountain grasslands	Very free, often highly leached	Sandy loam	Mull, going on acidic
Chernozem (Black soils)	Drier temperate grasslands	Evaporation exceeds annual rainfall	Good crumb	Alkaline upper layers because of calcium deposits
Rendzina (rendolls)	Temperate woodland or grassland on limestone	Highly leached	Thin	Mull
Solonchak (solarthids)	Saline soils of arid areas, dominated by halophytes	Rapid evaporation	Thin	Alkaline
Latosol (Oxisols)	Humid tropics; red latosols dominate under tropical rain forests	Free, with much of the silica leached out, leaving a high concentration of iron and aluminium	Poor; often clays that become brick-like when dry	Acidic, highly weathered

These are very broad categories, and local conditions, such as position, slope, and parent material, will produce variations on each basic theme.

earthworms are largely absent in mor litter beneath coniferous forest where fungi, rather than bacteria, dominate the microbial decomposer community. An open soil, with a good crumb and high nutrient levels can be found under a more balanced mull soil, with high earthworm and bacterial activity. Such soils are associated with grassland and deciduous forest in temperate latitudes, and it is upon these that most western agriculture is now based.

Elsewhere, fierce competition for scarce nutrients can lead to a very different distribution of resources. Most of the nutrient capital of an undisturbed tropical forest is held in the standing crop, its vegetation, and animal life, with only a small fraction in the thin soil. For centuries, the indigenous peoples of these forests have worked with these nutrients cycles, using agricultural practices and traditions that limited their land use according to the nutrient budget of the forest and, crucially, regulating the size of their group through social conventions. As a result, the demands on the forest were relatively light, the cleared areas were soon grown over and the forest recovered quickly when the band move on.

More recent colonizers have ignored these checks on growth at considerable cost. Clear-felling the forest, or burning large areas for intensive agriculture, may produce a bumper harvest for one or two years, but thereafter yields drop. The small nutrient capital of the soil is readily depleted and farmers then have to apply inorganic fertilizers to maintain economic returns. The nutrient capital of the forest was carried away when the above-ground community was cut down.

The nutrient capital stored in the standing crop is a prime factor limiting animal abundance in all ecosystems (Section 8.2). Working on the tropical grasslands of the Serengeti, Sam McNaughton found that the availability of phosphorus, sodium, and magnesium indirectly controlled the density of wildebeest. Areas where the vegetation had higher concentrations of these elements supported greater numbers of large herbivores. Rather like the trees of the topical forests, the wildebeest are part of the nutrient capital of the ecosystem—redistributing some as excretory products, passing some on to their young, and surrendering the balance to the environment on death—either directly through decomposition or indirectly via carnivores and scavengers.

Water

Water plays a key role in liberating and moving many of the important nutrients, but is itself driven through a **hydrological cycle** (Figure 7.4) by solar power. Water enters the atmosphere by evaporation and the transpiration of plants, both of which are driven by solar heat and radiation.

Without life, high temperatures increase evaporation by raising vapour pressures, lifting the moist air to higher altitudes or latitudes. Here it cools, water vapour condenses, and its fine droplets form clouds. The differential heating of land and sea creates regular patterns of air movements (Figure 8.14) and

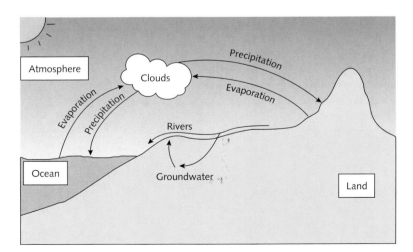

Figure 7.4 The hydrological cycle. Water evaporates from the sea and falls on land, where it enters the ground and eventually returns to the oceans via rivers. The thickness of the lines denote the relative amounts of water moving from one compartment to another.

where moist air cools, its water precipitates as rain or snow. Bands of precipitation form where cold and warm air masses meet. Average precipitation for a region depends on its distance from the sea and the intervening topography (the form of the land). Continental interiors, distant from any large mass of water, tend to have low rainfall. Similarly, land on the leeward side of mountains receives little precipitation because the incoming air sheds its moisture when it is forced to rise and cool over the higher ground. Such areas are said to be in a 'rain shadow'.

Around 97 per cent of the planet's water is held in the oceans with just 0.001 per cent in the atmosphere. Less than 1 per cent occurs on land as ground- or soil-water or as open water in rivers and lakes. The rest (2 per cent) is frozen in polar and glacier ice. Much of the water we use comes from streams and rivers or from groundwater and the aquifers of water-laden rocks.

We may use natural springs or boreholes to draw up water from deep rocks, but when our abstraction exceeds the natural recharge rate the aquifer begins to be depleted. Pumping groundwater, principally for industrial or agricultural use, has reduced once massive reserves in many parts of the world and these would now require considerable time, with no further abstraction, to restore their levels. Our dependence on water that fell as rain many centuries ago means that we can no longer regard aquifers as a renewable resource (Box 7.2).

Even where water is plentiful, its overuse can precipitate ecological disasters. This is especially true in arid and semi-arid areas where evaporation exceeds precipitation. Under these conditions, the supply of excessive water through irrigation systems can draw salts to the surface where they crystallize as a saline crust. A build-up of some elements (especially sodium and magnesium) may reach concentrations toxic to

---

BOX 7.2   The drying of Doñana

Doñana National Park lies on the Atlantic coast of Spain at the mouth of the Guadalquivir river (Figure 7.5). The 50 000-hectare site was designated a national park in 1969, in recognition of its value as one of the most important coastal wetlands in Europe. Home to over 361 different species, Doñana is internationally famous for the 6 million birds which visit on their migration between Africa and Northern Europe. With over 750 species of plants providing food and shelter to a wide range of other animals, it is also an important habitat in its own right.

The original rationale for the setting up of the park was to protect Doñana from the threat of agricultural intensification. However, this has not prevented development taking place along its perimeter and Doñana is now surrounded. To the north lies the intensive farmland of El Rocio—much of it under glass—providing salad crops and strawberries for northern Europe. To the west, the holiday resort of Matalascanas witnesses a seasonal population increase of 200 000. All this has increased the local demand for water, the key resource upon which Doñana depends.

Figure 7.5   Doñana National Park and World Heritage Site is an extensive area of wetlands and coastal dunes that have formed on the delta of the Guadalquivir river on the Atlantic coast of Spain.

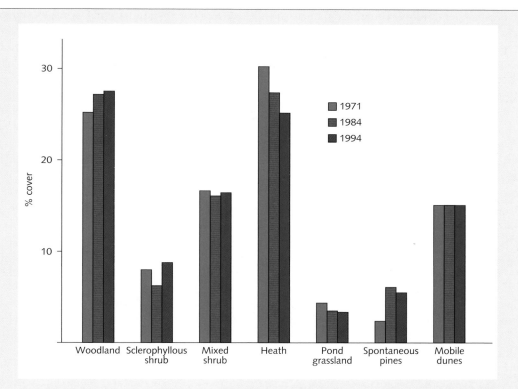

Figure 7.6 Changes in percentage cover of seven vegetation categories in Doñana between 1971 and 1994. Heath and pond grassland have declined as water tables have fallen. Meanwhile, there has been an increase in pines invading areas as they dry, and a smaller increase in juniper woodland and sclerophyllous scrub.

Both agriculture and tourism draw water from a series of boreholes on the perimeter of the park. In excess of 4 million cubic metres are extracted each year, faster than the natural recharge of the groundwater, whilst a sequence of 30 dams upstream on the Guadalquivir river reduces the water entering the site. The water table has been falling for 30 years and, as a result, Doñana has dried—what were wetlands have become coastal dunes. Additionally, commercial forestry has destroyed the native Phoenician juniper (*Juniperus phoenicea*) woodland and replaced it with umbrella pine (*Pinus pinea*) and eucalyptus (*Eucalyptus* species), a further drain on Doñana's water table.

In some places, pine has invaded areas formerly dominated by heather (*Calluna vulgaris*) and green heather (*Erica scoparia*), so the tree's spread closely matches the drying of the soils. Juan Carlos Munoz-Reinso and his colleagues used

specimens of *P. pinea* to pinpoint the time Doñana went into water deficit. They aged specimens growing in dried-up pond beds and concluded that the problem started in 1973–1974. Around this time, conditions became suitable for the pine to grow where previously it had been too wet. Additionally, species that favour a winter water table level of 1–2 m (such as *Calluna* and *Erica*) have progressively given way to new communities dominated by *Halimium halimifolium*, where water tables are now below 3 m in the winter (Figure 7.6).

Doñana's water table continues to fall at rates of 0.1–0.5 m per year, and the nature of the park continues to change—changes which may already be irrevocable. The spread of self-sown trees further dries the soil and so deflects larger areas along a successional trajectory towards woodland (Section 5.3). This is likely to be exacerbated by the impact of climate change.

plants and micro-organisms, rendering the soil useless for agriculture.

The problem of salinization is far from new. Archaeological evidence suggests that a number of ancient civilizations declined when their soils became saline through mismanagement (Section 5.1). Around 6 000 years ago, the people of Mesopotamia (present-day Iraq) built their civilization on the agricultural lands between the rivers Tigris and Euphrates, watering their staple crop, wheat, with intricate irrigation schemes. Evaporation caused salts to accumulate in the upper layers of the soil and within 2 500 years the land had soured. Farmers then began to cultivate barley, a more salt-tolerant crop, but the soil eventually became too saline for this and the fields were abandoned. With no productive lands the people moved north into Babylon where the cooler climate made the soil less susceptible to salinization.

Today over 76 million hectares of salinized land exist across five continents (Figure 7.7). In Western Australia, for example, 11 per cent of the wheat belt is suffering from salinization and this is expected to rise to 30 per cent within the next 50 years. One solution is to farm more sustainably by using species adapted to the semi-arid conditions, avoiding the need for irrigation. In this case, native eucalyptus is being promoted as a suitable crop for drier areas. This makes ecological sense but it would also fill a gap in the economy since Australia currently imports the most Australian of products . . . eucalyptus oil!

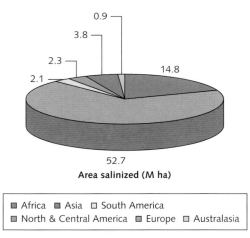

Figure 7.7 The global extent of salinization.

A number of international agencies suggest that water shortages will become severe in several regions over the next few decades. In particular, water shortages and water treatment measures are becoming critical to the economic well-being of regions with a mediterranean climate. Increasing populations and increasing development of these areas are leading to local shortages, and water is now part of the political bartering between neighbouring countries, from southern California and Mexico to the Middle East.

Oxygen

With the advent of photosynthesis about 3 billion years ago, the increasing abundance of oxygen in the atmosphere favoured physiologies that were adapted to its corrosive presence. All higher organisms respire aerobically—that is, they use the capacity of oxygen to gain electrons and bind with hydrogen to pull apart carbon-rich molecules, to release their energy in a step-wise manner. Cells have to protect themselves from the aggressive chemical activity of oxygen and, as we saw previously, descendants of ancient bacteria still flourish where they are sheltered from the oxygen-rich atmosphere. Under such anaerobic conditions, a range of bacteria use sulfate, nitrate, or other elements as the electron acceptor in their respiration. Eventually, their anaerobic respiration leads to oxygen or its compounds being released into the atmosphere.

The oxygen content of the atmosphere is closely linked to the biological activity of the planet, principally through its production by photosynthesis and its depletion by aerobic respiration. As a result, its movement is tightly linked to the carbon cycle and the processes of energy storage and release. However, oxygen forms such a large proportion of the atmosphere (21 per cent) that we do not detect annual shifts in its concentration as we do with carbon (Figure 7.8).

The principal nutrients

Carbon

Carbon makes life on Earth possible. Without its capacity to form molecular chains and rings life would not exist. Dry a human being, and most other

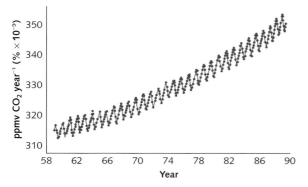

Figure 7.8 Atmospheric carbon dioxide measured at the Mauna Loa Observatory in Hawai'i. The annual oscillations reflect seasonal change in photosynthesis and respiration, and the greater extent of terrestrial ecosystems in the northern hemisphere. Longer term, the trend is for a consistent rise in mean atmospheric concentrations.

living things, and almost half of its weight will be carbon. But the carbon content of the Earth's atmosphere has been far from stable over geological time. In the past, especially the Silurian era, high atmospheric concentrations of carbon dioxide promoted high primary productivity, which eventually became the thick deposits of carbonates and hydrocarbons of the Carboniferous era. As we shall see later (Section 8.3), the rate at which carbon cycles, and its residence time in the atmosphere, also regulates the planet's temperature.

Carbon exists in the atmosphere at very low concentrations—a mere 0.03 per cent for its most abundant gas, carbon dioxide. Most is fixed in sedimentary rocks where it has a slow flux rate, taking perhaps 100 million years to re-enter the biosphere. Over the last 200 years we have been releasing carbon from these sinks at a rapid rate, a rate that is still accelerating. Not only has this changed the carbon balance of the atmosphere, it has also shifted its energy budget (Section 8.3).

Photosynthesis causes a yearly oscillation in atmospheric carbon dioxide of around 5 parts per million. This annual inhalation and exhalation of the planet is due to the disparity in the land area between the northern and southern hemispheres (Figure 7.8). Terrestrial processes in the north dominate, so

carbon dioxide levels are low when its photosynthesis is high, but rise during its winter when respiration here exceeds primary production.

Marine plankton and the forests are the principal routes by which carbon dioxide is removed from the atmosphere. In the absence of large amounts of oxygen, more recently dead plants and animals are held as soil organic matter and peat. Microbial breakdown of soil organic matter is a major source of atmospheric carbon, either through aerobic or anaerobic processes. Agricultural methods using regular soil disturbance and short-lived crops tend to accelerate aerobic decomposition and the release of carbon dioxide. Ploughing destroys the upper layers of a soil profile (Figure 7.3), which leads to changes in the composition of both the invertebrate and microbial communities. Many of the macro-invertebrates in the litter community—woodlice, millipedes, and earthworms—and their predators—beetles, spiders, and centipedes—may then be lost.

As much as 40 000 times the size of the atmospheric pool is fixed in these surface deposits and in the oceans. A large proportion of global carbon is locked away as calcium carbonate in the shells and skeletons of marine organisms, compressed on the sea floor into chalk and limestone. Elsewhere, the remains of other plants and animals exist as hydrocarbons, the gases and oils used to fuel our industrial societies.

Sulfur

The fire and brimstone of the sulphur cycle makes it a truly elemental cycle—linking earth, air, fire, and water (Figure 7.9). Significant amounts are added to the atmosphere through volcanic eruptions, while marine phytoplankton release dimethyl sulfide to the air as a product of their metabolism. This is important in promoting cloud formation, acting as a nucleus for the condensation of water droplets.

The organic matter of sediments and soil are again the main reservoirs. Sulfur forms only a small fraction of living tissues, but plays a crucial role in regulating the structure of many amino acids and proteins. Because its transfers are mediated by a range of specialist bacteria, sulfur has a similar cycle to nitrogen. Both are also added to the atmosphere

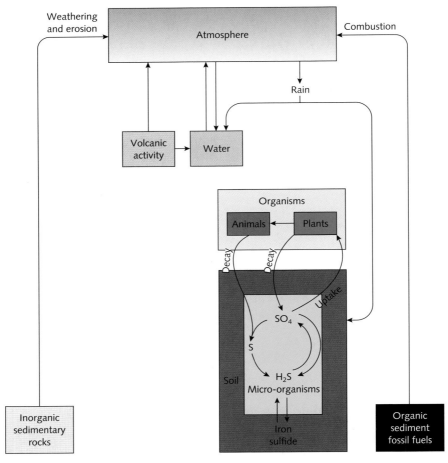

Figure 7.9 The sulfur cycle. Sulphur originates principally from sediments and volcanic eruptions but the burning of fossil fuel has added considerably to localized sulphur pollution and the problem of acid deposition.

through human activity, especially the burning fossil fuels. Sulfur is released from waterlogged soils as hydrogen sulfide through the action of sulfate-reducing bacteria. These anaerobic muds are characteristically stained black by the reduced iron sulfide and give off the characteristic rotten egg smell. Each year, around 100 million tonnes of this gas are produced by species such as *Desulfotomaculo* and *Desulfovibrio* which use sulphate reduction of organic carbon as their energy source. In the upper oxygenated layers other micro-organisms oxidize sulfur compounds back to sulfates (e.g. *Thiobacillus*).

Our production of sulfur compounds far outstrips the contribution from natural sources on land. The result is localized pockets of high sulfur in the air above heavily industrialized and urban areas. As this disperses, sulfur oxides contribute to regional problems of acid rain (Box 7.3). Sulfate aerosols are known to be important in the energy balance of the atmosphere, producing regional cooling by their effects of cloud formation near their source. Making proper allowance for this cooling has been an important factor in improving our modelling of atmospheric energy balances (Section 8.3).

Phosphorus

Phosphorus is an earthbound element that lacks a significant atmospheric component (Figure 7.11). Geological deposits (mainly in the form of a calcium compound called apatite) account for a global resource of 2 000 billion tonnes, but its rate of release by weathering is very slow, just 100 million tonnes per year—less than one 500 000th of the total reserve.

The sea is a rich source of available phosphate and most phosphate-rich rocks are ultimately of marine origin. Intriguingly, the main route back to land for phosphate is through seabirds, feeding on phosphate-rich fish and then defecating at their roosting sites. Their faeces form guano, which, in some locations, has accumulated to such a considerable depth that it can be commercially extracted as a fertilizer.

 BOX 7.3   **Acid rain**

For a long time, the problem of acid rain was thought to be largely a consequence of the sulfur-rich gases from the burning of fossil fuels (especially poor-quality coals) and the smelting of metallic ores. We now know that this picture is far too simple. Acid rain should be more properly described as acid deposition, since it consists of both wet and dry deposition, and its effects are not attributable to sulfur alone.

A significant increase in the acidity of rain was noted in Britain within 100 years of the start of the Industrial Revolution. With the reduction in sulfate emissions from coal-fired power stations over the last three decades, the nitrogen oxides (NO_x) have become increasingly important and now account for much of the acid deposition in Europe. Rainfall is naturally slightly acidic, absorbing small amounts of carbon dioxide from the air to form carbonic acid. Today SO_x and NO_x produced by the internal combustion engine and power stations combine with moisture in the air to form dilute sulfuric and nitric acids. This will produce a rainfall around one pH unit more acidic (that is, a 10-fold increase in its concentration of hydrogen ions) though in some locations it can be as low as pH 3.1.

This is a regional pollution problem and therefore requires international agreements to limit atmospheric concentrations. Beyond the damage it does to buildings and green spaces in urban areas, acid deposition adds large amounts of sulfur and nitrogen to aquatic and forest ecosystems, upsetting their nutrient balance.

Nitrogen and sulfur inputs have a fertilizing effect in some ecosystems and can lead to shifts in the species composition of some plant communities. Many plants growing on nutrient-poor sandy or chalky soils rely on rainfall as their principal source of nutrients (**ombrotrophs**) and are thus adapted to low nutrient levels. Abundant nitrogen allows other species to invade and out-compete the existing flora. Nutrient-poor ecosystems such as grasslands and heathland may change near major sources of nitrogen, such as densely trafficked roads, and become dominated by coarse, rank vegetation.

Acid rain also damages ecosystems from the direct effect of its acidity. Lowering the pH of the soil mobilizes soil nutrients, promoting an initial surge of growth. However, it also displaces key nutrients and cations (such as calcium and magnesium), which limit growth as they are lost. Nitrogen oxides entering through plant stomata can have more direct effects on plants, creating localized acidity damaging to leaves and new shoots (Figure 7.10). The flush of nutrients eventually finds its way into water courses where it causes algal blooms, comparable to eutrophication (Section 7.3).

Much of the pollution originating in northwestern Europe is swept by the prevailing winds into Scandinavia, a region dominated by coniferous forests and thin soils, with little cation-exchange capacity to buffer the incoming acidity or to bind the aluminium toxic to many plants (Box 7.1). Scandinavian lakes have suffered particularly from acidification and again it is the aluminium mobilized from the surrounding soils which causes large-scale fish-kills. As concentrations rise in the water, the mucus on the fish gills coagulates and, because this reduces their oxygen uptake, the fish suffocate. A large number of lakes have been fish-less since the 1960s and Norway, Sweden, and Finland have had to resort to adding considerable amounts of lime to these ecosystems to try to protect or restore them.

(continued overleaf)

Figure 7.10   The damaging effects of acid rain. Acid deposition causes direct damage to vegetation and also makes long-lived species, such as the conifers of this eastern European forest, more susceptible to drought and disease.

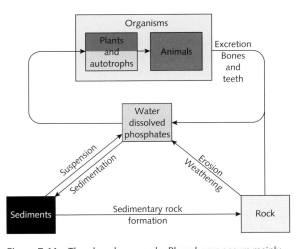

Figure 7.11   The phosphorus cycle. Phosphorus occurs mainly in the form of phosphates, insoluble compounds that are slowly weathered from rocks or resuspended from living organisms. This cycle differs from most others as it has no significant atmospheric phase.

Insolubility is the fate of most free phosphorus and the reason why this element is the key limiting nutrient in many ecosystems. Consequently, plants and animals hold on to their phosphate and concentrate it to many times the level of the surrounding environment. Phosphorus is an essential component of proteins, nucleic acids, cell membranes, teeth, and bones. It plays a central role in a host of cellular processes, not least in the molecule that is the currency of biological energy transfer—adenosine triphosphate (ATP).

Phosphorus enters the food chain primarily through plants and phytoplankton. An atom may remain in an animal for many years and within a tree for centuries. When released, it can be recycled into the living system or become immobilized in the soil or sediments, where it may remain for many thousands of years. Large amounts of oxides of iron and aluminium will immobilize phosphorus in various insoluble compounds. The weathering of apatite from recently exposed rocks is the chief mineral source in a developing soil. Thereafter, the microbial community is the principal source of available phosphates for the rest of the community.

A number of woody plants exploit the capacity of fungi to scavenge phosphorus by forming a symbiotic association with them (Section 4.2), called a **mycorrhiza** (*myco*—fungus, *rhiza*—root). Here, the fungus either grows around (ectomycorrhiza) or around and within (endomycorrhiza) the plant root. In the commonest form of endomycorrhizae, vesicular-arbuscular mycorrhizae, the fungus penetrates the root-forming structures within it (Figure 7.12). The plant supplies the fungus with sugars and other nutrients in exchange for the phosphorus scavenged by the hyphal network. Fungal hyphae secrete organic acids which solubilize a range of phosphates, making these available to themselves and the host plant. The

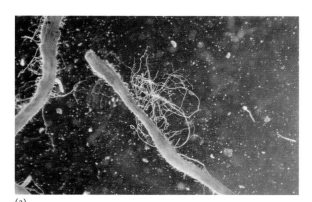

(a)

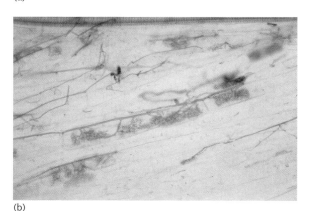

(b)

Figure 7.12 Mycorrhizae: plant and fungus in partnership. Arbuscular mycorrhizal fungi growing (a) around and (b) inside the root.

fungi also supply moisture to the roots, conferring some drought resistance to the host. Our understanding of the importance of these associations is the result of our attempts at establishing plant communities in highly degraded ecosystems (Section 7.3).

Nitrogen

Another important mutualistic association in the soil is based on the microbial capture of the other major plant nutrient: nitrogen. Next to carbon and oxygen, nitrogen is the third most abundant element in biological molecules. Consequently, the flux of nitrogen exerts considerable control on ecosystem productivity and processes (Figure 7.13).

The bulk of the atmosphere is nitrogen (around 78 per cent), the reserve from which living systems ultimately derive their supply. However, it is difficult to coax reactive nitrogen out of the atmosphere. Gaseous nitrogen exists largely in a molecular form (N_2), in which there are three shared electrons, requiring high energies to split the molecule into atomic nitrogen. The massive electrical discharges of thunderstorms do this, though only small amounts of nitrate enter the soil by this route. Most organisms cannot afford the energetic cost of splitting molecular nitrogen and very few micro-organisms have the enzymic equipment.

Some very ancient genera of bacteria and actinomycetes do, and will flourish in habitats devoid of oxygen. A few bacterial genera and the cyanobacteria are responsible for the bulk of biological nitrogen fixation from the atmosphere. They occur in both terrestrial and aquatic environments and can either be free-living or have a symbiotic relationship with certain plants. All possess the enzyme nitrogenase which catalyses the splitting of molecular nitrogen and its combination with hydrogen to form ammonium.

Several groups of higher plants (such as legumes—the pea and bean family) provide low oxygen conditions in their roots where these micro-organisms can survive, and also supply them with an energy-rich carbon source. For their part in this bargain, the fixers provide nitrogen in a form that the plant can readily utilize. Legumes can grow without their symbiont (*Rhizobium*) but they gain an important competitive advantage over competitors when their root nodules are infected with the bacteria (Figure 7.14).

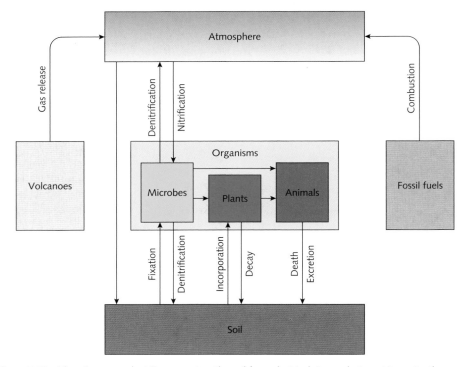

Figure 7.13 The nitrogen cycle. Nitrogen enters the soil from electrical storms, but most importantly through the action of nitrogen-fixing micro-organisms. Additional sources are oxides of nitrogen produced by volcanoes and the burning of fossil fuels.

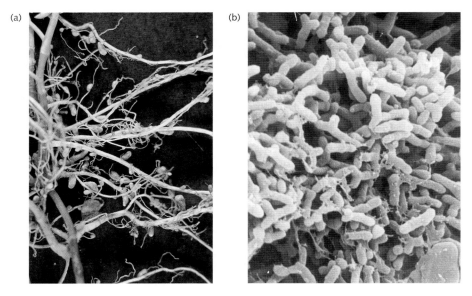

Figure 7.14 Root nodules. (a) Root nodules of clover (*Trifolium* spp.), formed from root epidermal cells as a result of infection by *Rhizobium* bacteria. (b) the y-shaped bacteroids are clearly visible within the nodule.

Rhizobium is the most well studied genus of all the nitrogen-fixing bacteria. These can also survive without their symbiont partner, as free-living saprophytes, using organic material of the soil as their energy source. Most legumes will have a culture in their root nodules but both *Rhizobium* and its host will avoid the high costs of nitrogen fixation whenever possible and will switch to other sources of nitrate or ammonia if these become available in the soil.

Symbiotic nitrogen fixation by is not confined to legumes, and root nodules occur in plants outside the pea and bean family. Alder (*Alnus glutinosa*), for example, hosts a nitrogen-fixing actinomycete bacteria, *Frankia*. There are also free-living bacteria, such as *Thiobacillus*, and cyanobacteria, which fix nitrogen for their own use and are important in maintaining the nitrogen balance of the soil decomposer community.

Needless to say, there have been considerable efforts to introduce such a symbiotic association into major crop plants through genetic engineering, with the prize of reducing our reliance on nitrogenous fertilizers and the massive energy costs of their production. Cereals require particularly large inputs of artificial fertilizers, though when these are applied a considerable proportion will run to waste, creating pollution problems elsewhere, or simply promoting the growth of non-target species.

Nitrate is the main form of nitrogen taken up by plants, but this is highly soluble and readily leached from the soil. In moist climates, soil nitrogen is rapidly depleted and competition for available nitrate governs the productivity of many communities. Some plants have evolved alternative strategies for capturing the nitrogen they need. Carnivorous plants like the Venus fly-trap (*Dionaea muscipula*) and pitcher plants (*Sarracenia* and *Nepenthes* spp.) are found in bogs and swamps where nitrogen concentrations are low (Figure 7.15). The insects caught by these plants are a valuable supplementary source. Directly or indirectly, all animals ultimately depend upon plants for their nitrogen, so this is something like justice, with these plants at least getting their own back!

Plants have also to compete with the soil microbial community for their nitrogen. Different genera of

Figure 7.15 Carnivorous plants like *Sarracenia* have adapted to life in the nutrient-poor conditions of swamps and bogs by obtaining additional nitrogen from insects trapped within the pitcher-like body of the plant.

bacteria use different nitrogen compounds to derive energy from its bonds or produce new cells. Heterotrophic bacteria obtain their energy from mineralizing organic nitrogen compounds, such as amino acids, and produce ammonium in the process (Figure 7.16). The ammonium they do not need for their own growth and reproduction will evaporate unless it is used by other bacteria, especially the nitrifiers. *Nitrosomonas* converts ammonium (NH_4) to nitrite (NO_2), which may then be further oxidized to nitrate (NO_3) by *Nitrobacter*. Together, these two processes are called nitrification. Any nitrate not taken up by plants or micro-organisms may undergo denitrification, in which oxygen is progressively removed from

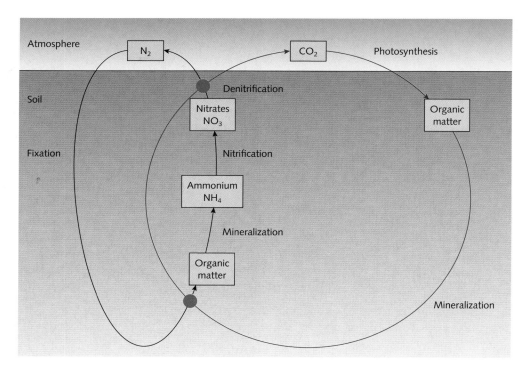

Figure 7.16 Microbial interactions between the nitrogen and carbon cycles. The carbon cycle fuels the nitrogen cycle in both its fixation and denitrification phases.

nitrate, nitrite, and nitrous oxide (N_2O) to yield molecular nitrogen (N_2) that diffuses back into the atmosphere. Denitrification is carried out by a variety of bacteria (including *Bacillus* and *Pseudomonas*) and is an important process in oxygen-poor environments such as waterlogged soils, where it is the major cause of nitrogen deficiency.

The speed of these processes and the access higher plants have to nitrogen depends on the activity of the decomposer community and the balance between carbon and nitrogen in the soil. Micro-organisms need both carbon and nitrogen to build new cells. If carbon is abundant and nitrogen limited, most of the nitrogen becomes locked in microbial cells and there is little available for primary producers. Abundant nitrogen and low carbon levels in the soil cause the decomposer community to run short of the energy needed to drive mineralization, nitrification, and

denitrification. The carbon:nitrogen (C:N) ratio of a soil is then critical, with an optimum between 20:1 and 30:1. Above 30:1, higher plants compete with the microbial decomposers for nitrogen; below 20:1 carbon limits microbial growth and nitrogen leaches from the soil.

Large amounts of nitrogen (around 60 million tonnes) enter the atmosphere from the burning of fossil fuels and also make a significant contribution to acid rain (Box 7.3). Our inorganic fertilizers add a highly concentrated source of nutrients to the soil, in a form that can be readily leached away. These have produced highly localized concentrations of nutrients and, in doing so, upset ecological communities adapted to low-nutrient conditions. A nutrient that becomes abundant in a community of species adapted to low nutrient levels inevitably leads to change in its configuration.

7.2 Eutrophication—too much of a good thing

It is apparent to any farmer or gardener that the fertilizer applied to the soil one year is not there several years hence. For those of us without green fingers it might never have been there in the first place. Runoff and leaching of nutrients mean that the money we spent trying to produce something on our patch of land invariably goes to fertilize something else, somewhere else.

Very often, it is primary production in the waters draining away that takes up these nutrients. Enriched by fertilizers, or other sources of nutrients, rivers, streams, and lakes change their community structure and patterns of cycling through a process termed eutrophication. This is nutrient overload and illustrates how an essential element in the wrong place, at the wrong time can become a serious pollutant.

Eutrophication can be an entirely natural process: many freshwater ecosystems accumulate nutrients as they age, as a succession proceeds (Section 5.3). However, human activity is a major cause of such imbalances in freshwaters and in soils. In both cases, a build-up of fertilizers changes the plant and microbial community, often with a loss of species and the result that one or two species become dominant. The main culprits are phosphorus and nitrogen, with the former being the most serious because ordinarily it is so unavailable in natural systems.

Freshwater eutrophication is most obvious in warm weather, when the water becomes a green soup, thick with phytoplankton and sometimes with floating mats of cyanobacteria. Cyanobacteria will often flourish later in the sequence, when the nitrogen in the water is exhausted. As the nutrient balance changes, they have the advantage of being able to fix nitrogen from the atmosphere. As these mats of cyanobacteria lose their buoyancy they add large amounts of organic matter to the sediments. With few other primary producers, the waters become depleted of oxygen as decomposition begins to dominate. The water rapidly becomes anaerobic and foul-smelling, so that fish-kills are a common sign of serious eutrophication.

The **biochemical oxygen demand** (BOD) is a measure of the amount of oxygen needed for these decomposition processes, and consequently the balance of a freshwater ecosystem. A high BOD implies a high organic content, requiring large amounts of oxygen to decompose this material, depleting the supply for the rest of the biota. If high BOD and nutrient-rich conditions persist for a long time the ecosystem quickly loses it higher plants and animals. This is often seen around a sewage outfall, when untreated human waste is added to a stream or voided to the sea. The abundant nutrients create oxygen-poor conditions close to the discharge and only a small number of species can survive in the turbid waters.

Whilst BOD represents a useful snapshot of current conditions, it says little about past pollution events. The collection of fish and invertebrates living in these waters speaks of the ecosystem's longer history. The presence of these creatures in a habitat is an indication that their abiotic requirements have been met, at least in the recent past (Section 2.5). Plants, animals, and micro-organisms can thus serve as indicator species, sentinels of the long-term conditions in the habitat. For example, stonefly and mayfly larvae are only found in well-oxygenated waters, whereas a freshwater community dominated by oligochaete worms or chironomid fly larvae indicates very low oxygen levels and a persistently high BOD.

A variety of methods of counting and classifying indicator species have been devised, using different species with different weightings to define categories of water quality. Most of these biotic indices have the advantage of being easy and cheap to measure, requiring no elaborate chemical analyses, whilst providing an integrated history of the pollution burden carried by the water. Most importantly, they measure the impact from the community's perspective, critical if we are to restore conditions that allow a sustainable natural community to re-form.

Restoring eutrophic waters

When people think of pollution and effluent they picture industrial plants billowing fumes, and happily forget the number of times they flush more nutrients down the toilet. Likewise, we tend not to connect the food on supermarket shelves with fertilizer runoff or oceans of slurry from livestock housed on the intensive farms used to produce most of the food in industrialized nations. Even free-range livestock can contribute to eutrophication when areas are over-stocked and their wastes are allowed untreated into the drainage system.

Ideally, the best way of dealing with any form of pollution is to control it at its source or to gather waste together for treatment. Domestic and industrial sewage is treated in waste water treatment plants using a series of accelerated and highly controlled decomposer communities. Using a sequence of bio-reactors, or tanks, with rapidly growing bacterial and fungal colonies, a sewage works can reduce BOD and nitrogen levels by over 95 per cent within hours, and produce water that can safely be discharged in aquatic ecosystems.

Sewage that is rich in detergents can account for up to 80 per cent of the phosphate pollution entering such freshwater ecosystems. Unfortunately, this is less easily removed by microbial activity, and can be a major cause of eutrophication. A classic example occurred in Lake Washington near Seattle. As the city expanded in the 1940s and 1950s its sewage discharge led to a deterioration in water quality and a 75 per cent reduction in clarity. During the summer the lake experienced massive growths of the fila-mentous cyanobacteria *Oscillatoria rubescens* which, because of its toxins, represented a major health hazard. Public concern led to a campaign to reduce effluent inputs into 'Lake Stinko', as it was known locally. The phosphorus levels in the water were shown to be the key to the problem and a waste treat-ment programme to reduce these was implemented. It took seven years to reverse these trends. In the last 40 years, a global switch to low-phosphate deter-gents has helped moderate their impact.

Large-scale experiments have examined the causes, effects, and potential cures for eutrophication. In 1968, the Experimental Lakes Area was set up in Ontario where a series of small experimental lakes were deliberately polluted and subsequently moni-tored. In one experiment, a lake (ELA 226) was

divided using a plastic membrane. One-half received carbon and nitrogen and the other carbon, nitrogen, and phosphorus. The latter developed a dramatic algal bloom, confirming the significance of phosphorus. With no check on their growth from a shortage of phosphorus, the cyanobacteria with their capacity to fix atmospheric nitrogen, began to dominate, adding to the nutrient overload in the lake.

Reversing the changes of freshwater eutrophica-tion often requires more than simply halting the nutrient inputs. The Norfolk Broads of East Anglia are low-lying wetlands, the result of peat digging and reed growing in the Middle Ages. Over the centuries these have developed into a community rich in specialist species and the associated fenlands have a number of rare and distinctive plants and animals. However, the area borders some of the richest and most intensively farmed soils in Britain which have received immense inputs of fertilizer. In the 1970s, the Broads began to show signs of eutrophication. Early attempts to address the problem by controlling discharges failed, principally because of the large capital of phosphorus that had accumulated in the sediments. Under anaerobic conditions this was mobilized and released back into the water, again allowing cyanobacteria and other phytoplankton to flourish. This produced cloudy shallow water in which rooted plants (macrophytes) could not grow. Animals adapted to clear waters disappeared, and the fish community became dominated by species able to survive the turbid conditions (especially bottom-feeding bream, *Abramis brama*). One strategy has been to pump out the enriched sediments, using a machine called the 'mudsucker', but even this failed to return clear water to the Broads (Figure 7.17).

Now the Broads Authority uses a management strategy designed to combat eutrophication at several levels. First, by designating Nitrogen Sensitive Areas in which farmers are encouraged to adopt more traditional, less intensive agricultural practices. Second, domestic sewage is being treated to reduce phosphate before discharge. Third, the Broads are being dredged to remove the reserve held in the sedi-ments. Different methods are used on nature reserves according to the severity of the eutrophication. In mild cases, reeds are harvested to 'crop off' excess

Figure 7.17 The mudsucker in action. This machine is used by the Broads Authority to remove nutrient-rich sediments from eutrophic bodies of water.

nutrients. Electric fishing is also used to remove those species that stir up the sediments—and the nutrients represented by their biomass. Clear water species (such as the predatory pike–*Esox lucius*) are re-introduced to encourage the natural fish community to re-establish itself. In the most grossly polluted areas the mudsucker is used in conjunction with phosphate precipitation techniques, adding iron and calcium sulfates to produce insoluble phosphate compounds that sink into the sediments, effectively removing this source of phosphorus.

Other strategies, used with some success on lakes in Holland, involve greater manipulation of the animal community. The idea behind **biomanipulation** is to promote a phytoplankton and rooted plant community typical of clear waters. This typically means introducing certain fish and removing others. Bottom-feeding species are removed and other animals, including filter-feeding molluscs, are introduced to promote clear water conditions, especially the freshwater mussel *Dreissena*, one of the few species able to feed on toxic cyanobacteria. Alongside fish that feed on the zooplankton, and the predatory fish that feed on them, this can promote the growth of

rooted higher plants (macrophytes) characteristic of clear waters. Macrophytes take up nutrients and reduce the supply to the phytoplankton, helping to maintain the high transparency and checking the development of eutrophic conditions.

Seas too can become eutrophic, especially when they are enclosed and slow to recharge with new water from adjoining oceans. The Mediterranean is connected to the Atlantic Ocean by the Straits of Gibraltar and therefore has a long recharge period—around 80 years. Despite its enclosure the Mediterranean is **oligotrophic**; that is, nutrient-poor, indicated by its low productivity (around 150 g carbon/m²/yr.). This is because few rivers discharge their nutrients, sediments, and pollution into it. However, enrichment does occur in a few localized areas, in particular the Adriatic Sea. A third of the Mediterranean's freshwater input arrives in the Adriatic, mainly from the River Po, draining the Alps and northern Italy and the rivers from the Balkans to the east, bringing with it high levels of nutrients and pollutants. The local eutrophication that results can have a very significant economic and ecological impact such as in the Venice lagoon (Box 7.4).

BOX 7.4   **Venice in peril**

Eutrophication in the Mediterranean tends to be localized in coastal waters, close to urban areas with discharges direct to the sea. This is particularly obvious around the Venice lagoon. Extending over 55 000 hectares, the lagoon has an average depth of only 1 m. It is fed on the landward side by a series of rivers and is linked to the Adriatic Sea by three inlets (Figure 7.18). Its waters are warmer, two-thirds less salty, and far more eutrophic than the adjoining sea.

Like any other city, Venice generates liquid waste of various descriptions, most of which ends up in the lagoon. Additional inputs from surrounding agricultural, industrial, and urban sources account for 80–90 per cent of its nutrient load (Figure 7.19). As a result, the Venice lagoon, despite its massive appeal to tourists, often smells during the tourist season. When temperatures are high, the breakdown of its organic matter quickly uses up the available oxygen and the waters of the lagoon become anaerobic.

The excess nutrients have changed the plant community, favouring sea lettuce (*Ulva rigida*), which has gradually displaced the lagoon's natural vegetation. By 1992, dwarf eelgrass (*Zostera noltii*)—a species characteristic of good-quality brackish water—could only be found in small clumps near the edges of inlets. Flourishing with the high nitrate, sea lettuce has become the lagoon's dominant primary producer in the lagoon, adding to its organic loading.

In response, the authorities established the *Consorzio Venezia Nuovo* (CVN) to devise a management programme to reverse the deterioration. Aldo Bettinetti and his col-

Figure 7.18   The Venice lagoon from earth orbit. The city of Venice is at the centre and is linked to the mainland by a causeway. Also clearly visible are the three inlets that connect the lagoon with the Adriatic Sea.

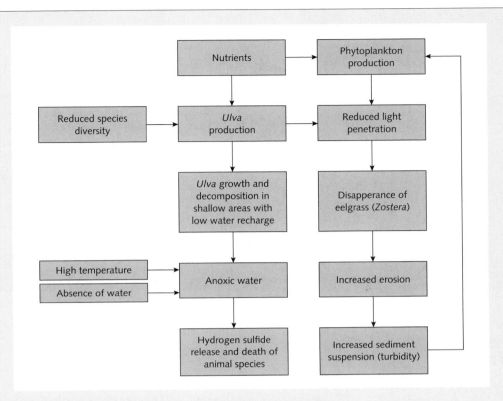

Figure 7.19 The causes and effects of eutrophication in the Venice lagoon.

leagues concluded that the progressive urbanization of the surrounding mainland was a key contributor to the lagoon's problems (Figure 7.19). A natural buffer zone of woodland and marshland had previously helped protect it from sediments and pollution, but this was lost under urban and industrial development.

With the help of a geographical information system (GIS), Bettinetti's team located vast spreads of *Ulva* and the remaining colonies of *Zostera*. Ninety per cent of the lagoon is prime *Zostera* habitat which could easily out-compete *Ulva* in low-nutrient conditions. The plan is physically to remove *Ulva*, to crop off excess nitrate and phosphate, and to provide space for the eelgrass to recolonize. To be effective the *Ulva* must be harvested at the peak of its biomass, but the lagoon waters will also need to be aerated to keep the oxygen concentration above 5 mg/l in order for the eelgrass to establish itself.

Medium-term measures seek to reduce erosion, because shallow waters favour *Zostera*. Sediments are being scraped from the most polluted areas and there are also plans to introduce filter-feeding invertebrates to reduce nutrient loads and the turbidity of the water. Ultimately, better treatment of water before it enters the lagoon will be the most effective way of reducing its nutrient burden.

More enclosed and more eutrophic still is the Black Sea, which is connected to the rest of the world only by the tiny channel of the Bosporus. It was once home to a diverse marine community, but years of pollution and overexploitation have rendered much of its sea floor anoxic (oxygen-depleted). Along with sewage and fertilizer runoff, the Black Sea also receives radioactive waste and suffers frequent oil spills. Its

Figure 7.20 When viewed from space the highly eutrophic Black Sea appears darker than the oligotrophic waters of the eastern Mediterranean.

primary productivity is very high (300 g carbon/m²/year), an indication of its eutrophic state: from space it is visibly darker than the more oligotrophic waters of the eastern Mediterranean (Figure 7.20). For decades it was thought that the damage was irreversible, though there have been international initiatives to measure its nutrient fluxes and various attempts at local mitigation. The result has been a decrease in the volume and incidence of phytoplankton blooms. Because of its geography, the Black Sea is highly changeable and readily affected by activities along its shore. This offers some hope that it can be improved

if concerted action is fully implemented by the nations that surround it.

There can be few pollution incidents more dramatic or distressing than the sight of a broken tanker with oiled seabirds in its wake (Figure 7.21). We transport oil over many thousands of miles and sometimes we spill it—either accidentally or on purpose. Far more frequent are the spillages associated with the loading and unloading, or the surreptitious washing-out, of tanks at sea: an estimated total of 650 000 tonnes of oil is discharged into the Mediterranean annually (Figure 7.22). Around 17 per cent of global marine oil pollution occurs in the Mediterranean, in a sea covering just 0.7 per cent of the surface waters of the planet. Even more deplorable is the deliberate targeting of marine and terrestrial oil installations in times of war (Figure 8.9).

Oil adds to the carbon balance of an ecosystem. Its variety of complex hydrocarbons represents a source of energy to the range of metabolisms present in many bacteria. Indeed, natural oil seepages have their own flourishing decomposer communities based on bacteria and fungi. An oil spill can thus represent a form of eutrophication, causing a community to shift to one dominated by decomposers. Often these decomposition processes slow down as oxygen or other nutrients are depleted. Oxygen is often limiting because the oil itself will limit its passage from the air to the biota (Box 7.5).

With each oil spill we learn more about their causes and consequences. Depressingly, each incident provides us with another opportunity to test our restoration techniques and our methods have evolved considerably over the last 40 years. Today, we seek to contain spills as far as possible and rely on natural degradative processes to deal with them. Human intervention may be best confined to mopping up the oil that comes ashore in the most sensitive areas, or scraping up the most contaminated soils. Otherwise a large range of micro-oganisms already present in the ecosystem can be relied upon to decompose the oil. These are local populations which have been exposed to oil in the past, either from natural seeps or previous spills, and are thus primed to become active and multiply when presented with this carbon source.

Figure 7.21 Seabirds like this male eider (*Somateria mollisima*) are among the first casualties of marine oil spills.

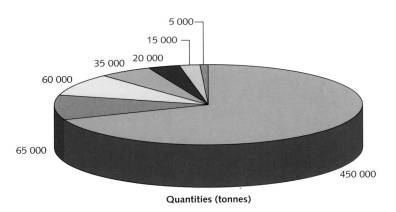

5 000
15 000
20 000
35 000
60 000
65 000
450 000

Quantities (tonnes)

■ Tanker operations (especially deballasting)

■ Navigation accidents

□ Oily bilge waters, sludge, and used lubricating oils from ships

■ Ship repairs (cleaning of tanks and pipes)

■ Pipelines

□ Petroleum handling installations in ports and special terminals

■ Offshore drilling

Figure 7.22 Routine oil pollution within the Mediterranean.

BOX 7.5

Oil pollution—a catalogue of disasters

Faced with a major spill, the obvious response is to contain the oil and to clean it up with whatever means are available. In 1967, when the *Torrey Canyon* broke its back on the Cornish coast, a variety of techniques were used, including bombing the vessel with napalm to set fire to the oil (Table 7.3).

The remainder of the slick was treated with kitchen-sink technology—detergents and dispersants—to break it up. Over 10 000 tonnes of these chemicals were used, but they proved to be more toxic to the wildlife than the oil itself.

(continued overleaf)

TABLE 7.3   Major marine oil spills during the past four decades

Vessel	Date	Location	Extent polluted (km)	Size of spill (tonnes)	Bird deaths
Torrey Canyon	March 1967	Cornish coast, England	225	117 000	30 000
Amoco Cadiz	March 1978	Brittany coast, France	360	233 000	4 600
Exxon Valdez	March 1989	Prince William Sound, Alaska	700	35 000	100 000–300 000
Braer	January 1993	Shetland Isles, Scotland	235	86 000	6 313
Sea Empress	February 1996	Milford Haven, Wales	100	75 000	22 132
Erika	December 1999	Brittany coast, France	400	11 100	150 000–300 000
Prestige	November 2002	Spanish and French coast	>100	77 000	100 000–200 000

As a macropollutant, oil physically obstructs living organisms, smothering and weighing them down. An oil slick on the water surface limits oxygen diffusion into the water, leading to anoxic conditions beneath. Oxygen levels also fall as the microbial community begins to decompose the oil. More direct poisoning follows if an animal ingests it. Oils (and detergents) displace the fat molecules of cell membranes, disrupting their structure and impairing cell function. Because of their high toxicity, detergents and dispersants are today only used as a last resort. Oil is a complex cocktail of hydrocarbons, some of which are volatile and quickly lost. Generally, those components most soluble in water are the least toxic and most readily degraded.

The next challenge was the *Amoco Cadiz* spill of 1978. By far the largest accidental spill, it released a quarter of a million tonnes of oil into the English Channel. Given the reluctance to use detergents, the worst of the pollution was sunk with sawdust and chalk, and the rest was physically removed from the open sea. Detergents were restricted to economically important areas, such as harbours. Elsewhere the strategy relied on the resident micro-organisms. Overall, this was a much more successful approach, since 80 per cent of the main hydrocarbon groups had been degraded within seven months of the spill. A similar operation was used after the *Exxon Valdez* ran aground in Alaska, with the response tailored to the nature of each affected beach. Some only received fertilizers to encourage microbial growth, some were washed down with high-pressure water jets, and others were mopped with an absorbent cloth in an operation called 'rock-polishing'.

Sometimes nature finishes the job without our help. For example, when the *Braer* struck the rocks off the Shetland Isles in winds reaching storm force 11–12, the oil was quickly whipped into an oil–water emulsion (known as a mousse), which assisted in its dispersal. Although the storm conditions meant there was little damage to coastal communities, the wind did carry the oil inland where it damaged sheep pasture.

In 2002, the spill resulting from the sinking of the *Prestige* claimed the record for having contaminated the longest stretch of coastline, from the Galician and Cantabrican coasts of Spain through to the northwestern coast of France (over 1 000 km). Despite the advice of experts to ground the ship and pump out its tanks, it was forced offshore and allowed to sink. Oil issuing from its tanks could not then be contained and spread over a wide area. Later the tanks were partially sealed (using robots) to slow the leakage and eventually they were pumped out.

As with all oil spills, damage to a region's ecology inevitably affects its economy, and the *Prestige* spill was no exception. Fishing was prohibited along 798 km of the Galician coast of which 680 km were badly affected. Within a year, the beaches had recovered with some achieving a higher rating for cleanliness than they had had before the disaster. This was achieved by a combination of collecting the oil as it came ashore and relying on the natural processes of recovery.

Eutrophication in terrestrial ecosystems

Agriculture is the major cause of enrichment in natural and semi-natural terrestrial habitats. This is particularly problematic in areas with nutrient-poor soils, where the community is adapted to intense competition for scarce nutrients. Such assemblages change readily when nutrients become abundant, often with the result that a few species become more abundant and crow most of their neighbours. Typically, nutrient-rich habitats are dominated by fast-growing *r*-selected species. The more competitive *K*-selected species, adapted to nutrient-poor conditions, are those most likely to be lost (Section 4.3).

Changes of this kind were seen in the Park Grass Experiment, a long-standing experiment measuring the impact of nitrogen fertilizers on grassland composition and productivity. Using 150 years of data, Jonathan Silvertown was able to show how a few coarse grass species (such as *Holcus lanatus*) came to dominate the above-ground community. This was associated with a corresponding decline in soil microbial activity. An over-supply of nutrients leads to a build-up of biomass. Similarly, amenity areas such as parks, gardens, and public open spaces can have high levels of nutrients following years of fertilizer applications along with the input from frequent visits by domestic pets. Indeed, research has shown that soils beneath highly manicured golf-course greens are so high in phosphate that they could be bagged up and legally sold as phosphate fertilizer!

Habitats traditionally managed by cutting, grazing, or burning can suffer if this management is suspended and the standing crop is allowed to recycle into the system. Cropping-off excess nutrients by removing this biomass can help, but it may not be enough. Another technique is to bury the rich topsoil or remove it completely. Sometimes, grazers that specifically target nutrient-rich plants can be used to create space for less vigorous plant species. For example, on English lowland heaths highland cattle are being used to browse species such as gorse (*Ulex* spp.), a legume that fixes nitrogen in its roots and thus enriches the soil. The cattle also represent a reserve of major nutrients, but these can easily be removed from the site as either meat or dairy products (Figure 7.23).

Figure 7.23 Biomanipulation in action. In England Highland cattle are used to graze lowland heathland to 'crop-off' the nutrient-rich plant species and open up space for the plants adapted to nutrient-poor soils. Removing the cattle also allows some of the nutrient capital to be taken off the site.

7.3 Restoration—'the acid test of ecology'

With degraded or highly polluted ecosystems we need to use the full range of our ecological knowledge and experience to restore a functional community to a site. Applying these techniques is known as restoration ecology. This is ecology as technology—a science applied to solving real-world problems. It is something of an art too, since restoration often requires creative skills to design entire landscapes and their ecosystems.

Much of modern science is based on a reductionist approach—understanding the system by taking it apart and in some cases, rebuilding it again. Natural ecosystems are not easily deconstructed or reconstructed in this way and are not amenable to such experimentation. However, there are numerous habitats that have suffered degradation, and in attempting to rebuild them ecologists have the chance to test their theories in the real world. For this reason Tony Bradshaw refers to restoration as 'the acid test of ecology': a means of testing theories and hypotheses, as well as our understanding of ecological systems.

The three Rs of restoration

Taken at its most literal, restoration means returning something to the way it was before a change took place. Many projects aim to restore lost habitats and the plant and animal communities they support. In practice, restoration ecology often encompasses several other aims:

- stabilization of land surface;
- pollution control;
- visual enhancement;
- creation of amenity value;
- increase of biological diversity;
- improvement of ecosystem function;
- establishment of productivity.

Much of our expertise has developed with reclamation projects on sites created by extractive or industrial processes. Mining or quarrying often results in large amounts of waste material (spoil) that can be

Figure 7.24 Restoration in progress. Vegetation is used to stabilize chalk spoil produced in the construction of the Channel Tunnel linking Britain and France.

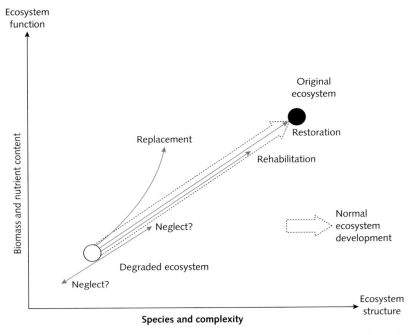

Figure 7.25 The options available in ecological restoration. As degraded sites are reclaimed they show an improvement in both the structure and function of the ecosystems. Complete restoration might not always be the aim of a project. Partial restoration in the form of rehabilitation or alternative replacement habitats may be appropriate endpoints.

both unsightly and a significant source of pollution (Figure 7.24). Here, the aim is use these wastes to produce something more aesthetically pleasing and less of an environmental hazard.

A number of options are available (Figure 7.25). Restoration might involve returning a site to its pre-industrial state, perhaps even to a natural or semi-natural habitat. This can be costly and technically difficult so we may choose partial restoration or rehabilitation, which, with time, might mature to full restoration. Other possibilities include making the land fit for agriculture or housing, the end result more properly described as replacement habitats, a compromise somewhere between the existing degraded state and a restored site. In some situations, it may be best to let nature take its own course and allow the site to recolonize naturally, though this is not an option where a substrate is unstable or a significant source of pollution.

Restoration techniques can change entire landscapes within a relatively short time, though these changes need to be long-lasting. We need the foresight to visualize the consequences of our decisions over ecological time and to consider the history of the site, the existing conditions, and the resources at our disposal. The most complicated option seeks to anticipate future successional changes, by predicting the way an ecosystem might develop, and then to restore the site to a point some way along that natural succession. In this way, restoration becomes a form of accelerated succession.

Restoration methods

Many spoils are devoid of organic matter and nutrients and therefore demand something close to a primary succession (Section 5.3). In other cases, sufficient organic matter and perhaps even some

TABLE 7.4   Problems associated with spoils and their immediate and long-term treatment

Category	Problem	Immediate treatment	Long-term treatment
Physical structure	Too compact	Rip or scarify	Vegetate
	Too open	Compact or cover with fine material	Vegetate
Stability	Unstable	Stabilize/mulch	Regrade or vegetation
Moisture	Too wet	Drain	Drain
	Too dry	Organic mulch	Vegetate
Nutrition			
Macronutrients	Nitrogen	Fertilizer	Legumes
	Others	Fertilizer and lime	Fertilizer and lime
Micronutrients		Fertilizer	
Toxicity			
pH	Too high	Pyritic waste or organic matter	Weathering
	Too low	Lime or leaching	Lime or weathering
Heavy metals		Organic mulch or use metal-tolerant cultivars	Inert covering or metal-tolerant cultivars
Salinity		Weathering or irrigation	Tolerant species
Plants & animals			
Wild plants	Absent or slow	Collect seed and sow, or spread soil containing propagules or plants	Ensure appropriate conditions for colonization
Cultivated plants	Absent	Sow normally or hydro-seeded	Appropriate aftercare
Animals	Slow colonization	Introduce	Ensure appropriate habitat management

topsoil may remain, with a reserve of nutrients to allow a secondary succession to occur. Residual soil, if properly stored, can be used for later restoration. However, its living components are easily lost through poor handling, and the movement and storage of the soil must ensure that its profile is maintained. Eventually this will allow us to return it in the correct order. If there is no risk of the spoil moving or becoming a nuisance, a site near to a source of colonizing species may be left to restore itself naturally.

Spoils vary considerably in their chemical and physical properties (Table 7.4) and therefore vary in their nutrient requirements and site preparation. Often sites need to be landscaped and the spoil broken and mixed to form a substrate with a suitable range of particle sizes. In some cases, ecologists may be working with a relatively inert substrate such as sands, crushed concrete, and brick rubble. Other substrates, such as colliery spoil or smelter wastes, may be excessively acidic or alkaline and rich in a range of toxic metals. Most have few, if any, plant nutrients. Initially, artificial fertilizers can provide these, although without a proper soil structure many spoils are free-draining, and these nutrients are easily washed away. Raising the organic content of the waste helps to create a soil structure and texture, and increase its capacity to bind nutrients. This feeds the microbial community and begins the processes of decomposition and nutrient cycling.

Above ground, ecologists can help the process along by creating a suitable substrate and providing nutrients for the plant community. In its simplest form, this may mean adding soil, fertilizers, and seed. Elsewhere, more drastic measures are used. Limestone quarries in the Peak District of northern England have been restored using 'conservation blasting'. First, quarry faces were blown up to create natural-looking rock falls, after which sewage slurry containing the seeds of lime-loving plants (calcicoles) was sprayed on to the sides of the artificial cliff. The spraying of the slurry dispersed the seed, added nutrients, and provided a substrate in which the seed could germinate.

We can also use plants to improve nutrient capture by exploiting the association of legumes with nitrogen-fixing bacteria and the variety of plants whose mycorrhizal associations make phosphates available. Both associations may require the soil to be inoculated with the microbial or fungal partner, perhaps by introducing topsoil where these are abundant or through the use of commercially available inocula. It is only when the nutrient capital of a soil itself starts to build that species can become established with the result that the ecosystem becomes self-sustaining.

Dealing with the toxic components of a spoil requires a different set of techniques. Many colliery spoils are often rich in iron pyrites (iron sulfide) which readily oxidizes to produce sulfuric acid. This makes the soil acidic and unsuitable for many plants. It is possible to remove large pyritic particles from wastes on some sites, though usually the soil is treated on site by the addition of lime, which 'mops up' the acidity as it develops. Sometimes it is possible to combine acidic and alkaline wastes (such as colliery waste and basic slag), solving two problems at once. Another alternative is to mix the waste with non-toxic material, or to bury it beneath a layer of imported soil. The danger here is that this soil may become soured by acids or metals migrating upwards. The cover needs to be sufficiently deep to stop this

and to prevent deep-rooting plants coming into contact with the waste.

Using resistant species—plants that can tolerate high acidity or high levels of metals—can help offset some of these problems. A range of species are now available commercially and may also be collected from around ancient mineral workings or natural rock outcrops rich in copper, cobalt, lead, and other toxic metals. These ecotypes (Box 2.4) have evolved mechanisms that limit their metal uptake. One example of a metallophyte, which not only tolerates but accumulates a metal, is the 'nickel tree' *Sebertia acuminata* (Figure 7.26). The tree grows on the nickel-laden serpentine soils of New Caledonia and oozes a blue-green nickel-rich latex from its bark. These, and other species, have been used by prospectors as indicators of mineral-rich rocks, an activity known as geobotanical prospecting.

Species selection requires an understanding of a plant's **autecology**, that is how individual species respond to the biotic and abiotic environment. Knowing the characteristics of a plant enables ecologists to anticipate how it might perform in its new habitat with new neighbours. Generally, the ideal plant for a restoration is one that is robust enough to cope with the poor conditions and requires little or no management.

Attempts have been made to assess and quantify species' performance in restoration projects, thereby taking some of the guesswork out of plant selection. Richard Pywell and his colleagues analysed the performance of 58 plants commonly used in British grassland restoration projects in a series of four-year field trials. They measured the performance of each across a series of growth traits (life-form, germination characteristics and so on) and compared these against three well-known classification systems. One was Grime's C–S–R life-strategy classification (Section 4.3). They also used the National Vegetation Classification scheme (NVC) (Section 5.3) to assess a species' role within major plant communities. Finally, species were classified according to the Ellenberg system—a pan-European classification of plants according to their soil and climatic requirements.

The aim was to find traits that were indicators of good performance in restoration situations, such as the ability to colonize gaps, to regenerate vegetatively,

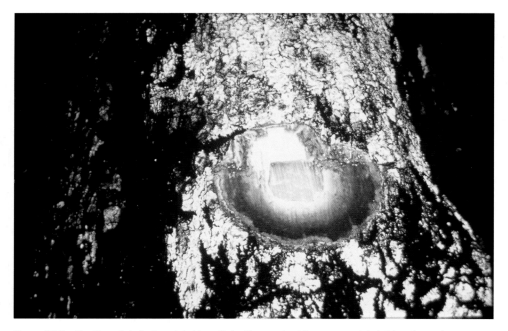

Figure 7.26   The New Caledonian nickel tree (*Sebertia acuminata*) grows on nickel-rich soils overlying serpentine rock and secretes a blue-green nickel-rich latex.

and to compete as well as a generalist rather than a specialist life-history strategy. Not surprisingly, they found that the most successful traits were associated with key stages in the successional process. For example, species that performed well in initial establishments, such as ox-eye daisy (*Leucanthemum vulgare*), had a good colonizing ability which included a high germination rate and a tendency towards a ruderal strategy (*r*-selected 'weediness'). Later in the restoration process, characteristics associated with competitive species, such as vegetative regeneration and the ability to persist within the seed bank, became important.

Extending their findings to the problems of restoring grasslands from former agricultural land, Pywell and his co-workers found that the high residual fertility was a major factor. The nutrient-rich soil conditions tended to favour generalists such as the ox-eye daisy and yarrow (*Achillea millefolium*), rather than more specialist species like yellow rattle (*Rhinanthus minor*). For a species to succeed it had to establish quickly and persist against competitors because the vegetation cover developed very rapidly. Based on these findings, they were able to recom-

mend a 'phased introduction' of species to increase the chances of success.

Perhaps the most extreme form of restoration is **habitat translocation**, in which the principal elements of a habitat—the soil, seed bank, and much of the vegetation cover—are moved from one place to another. Some habitats are easier to translocate than others: grassland, for example, can be lifted as turfs with their topsoil, as can heathland (Figure 7.27). Woodlands too have been moved, although, given the scale of such operations and the species involved, this is more problematic. In some cases partial translocation is possible, with soils, plants, and animals taken from one site to assist the development of another. Habitat translocation is neither easy nor cheap and is only attempted when a valued habitat is under severe threat.

Restoration ecologists have to consider not only the expense of such an operation but also whether the community is likely to survive in its new location given the local geology, topography, drainage, and so on. There is little point in transplanting a nutrient-poor habitat to a place where it is likely to receive nutrient-rich runoff from neighbouring farmland.

Figure 7.27 Habitat translocation in action. Transplanting habitats, such as this patch of the West Pennine Moors in Lancashire, requires speed and accuracy to re-establish the plant material with the minimum of damage.

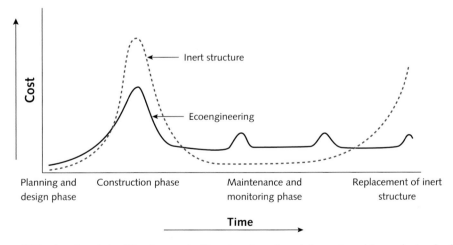

Figure 7.28 A cost analysis of 'hard' versus 'soft' engineering—the relative costs and timescales involved in the planning, construction, and mature operation of ecoengineered and inert structures.

Equally, we need to know what impact the transplanted material will have on the existing communities. Whatever the scale of the restoration undertaken, an established community will need proper aftercare, including site monitoring and a review of its management as the community develops.

Restoration ecology is playing an increasing role in civil engineering, where it offers alternatives to concrete barriers and embankments. Figure 7.28 compares the options for 'hard' and 'soft' engineering, in terms of cost and performance. Inert structures often cost more to construct and have a limited

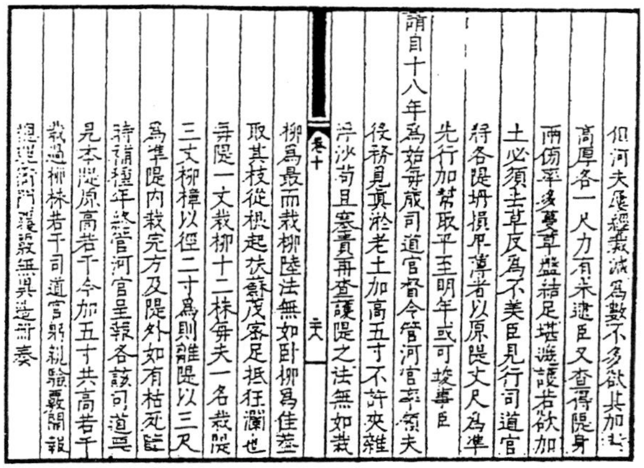

Figure 7.29 Ancient Chinese manuscript, possibly one of the first documented examples of ecoengineering.

It is observed that both sides of the earth dam embankment surfaces are covered with creeping sage grasses, which provide sufficient surface protection. If it is intended to raise and thicken the embankment, the grasses have to be removed and the surface protection condition could become worse . . . It is revealed that the best method for protecting the embankment is planting willows. Among the six planting methods of willow, the lateral planting is best. Since this method permits the willow's branches to grow much closer from the root system, thus it allows the willow to have much blooming branches to resist the impact of impounding water. Every 1 zhang (320 cm) of embankment should be planted with 12 willows . . . The willow should have a minimum girth of 2 cun (3.2 cm) and stick out from the embankment for 3 chi (96 cm). The planting should start from the inner to the outer portion of the embankment. Any dead willows found should be replaced immediately . . .

From a report by Pan, Minister for River Flood Control,
Ming Dynasty (1591)

lifespan. Ecoengineering, on the other hand, makes use of sustainable, self-renewing materials, such as trees, shrubs, and other vegetation. Plant roots are mechanically very strong and provide increased anchorage and stability as they grow through the soil. A mat of vegetation intercepts rain and shelters the ground surface, protecting it from erosion. These habitats are not only cheaper to construct, but can be maintained indefinitely.

People seem to prefer trees to a concrete wall. The acoustic and mechanical properties of vegetation, along with its capacity to trap dust, absorb rainfall, and bind soil, make it preferable to a simple, inert barrier. Despite such innovation ecoengineering is far from new. In the Middle Ages the Chinese used willow in their flood protection schemes (Figure 7.29); then, as ever, using materials that were readily at hand.

Past successes have led to sweeping claims for restoration ecology and there is always the danger that our ability to restore ecosystems could be used to justify their degradation. The truth is that our ability to restore lost and damaged habitats is limited, not least because of our lack of knowledge and the short timescales within which we work. The best we can hope for is an approximation of a natural ecosystem. However, restoration enables us to apply ecological principles in the restoration of damaged and degraded sites, and in the process to learn something about how ecosystems are put together.

● SUMMARY

The physical and chemical processes by which the crustal rocks are degraded and rebuilt are supplemented by the activity of living organisms, creating the major biogeochemical cycles. Organisms use the chemical properties of the different elements in their metabolism and tissues. These properties also determine the range and speed by which these elements cycle through the biosphere.

Beyond carbon fixation by higher plants, bacteria, cyanobacteria, and fungi play a crucial role in scavenging and fixing key nutrients. In the warmer latitudes, shortages of some of these nutrients, especially phosphorus and nitrogen, limit ecological activity. Most ecological communities are adapted to nutrient shortages and a variety of plant species have formed mutualistic associations with soil bacteria or fungi to improve their supply of nitrogen and phosphorus.

Considerable changes can occur in communities when key nutrients become abundant. Nutrient enrichment can lead to eutrophication in both aquatic and terrestrial habitats, with major changes in their species composition. The pollution caused by oil spills is often best treated by containment, leaving nature to restore itself.

At the other extreme, on severely degraded sites recovery happens slowly or not at all, requiring ecological restoration and a range of techniques to accelerate succession. These sites are often short of nutrients and we seek to re-establish the plant and microbial communities that will restore their ecological processes. Other options include partial restoration in the form of rehabilitation or the recreation of an entirely new replacement habitat.

● FURTHER READING

Bradshaw, A. D. and Chadwick, M. J. 1980. *The Restoration of Land*. Blackwell, Oxford. A classic text which considers a wide range of ecological issues concerning restoration.

Jordan, W. R., Gilpin, M. E., and Aber, J. D. (eds). 1987. *Restoration Ecology*. Cambridge University Press, Cambridge. A diverse overview of the subject including some of the socio-economic, aesthetic, and ethical issues concerning ecological restoration.

McCutcheon, S. and Schnoor, J. L. (eds). 2003. *Phytoremediation: Transformation and Control of Contaminants*. John Wiley & Sons, Hoboken, NJ. A useful text on the theory and practice of phytoremediation of organic and inorganic contaminants.

● WEB PAGES

A section of the United Nations Environment Program website provides information about oil pollution and techniques used in cleaning oil spills:
http://www.oils.gpa.unep.org/facts/facts.htm

The Society for Ecological Restoration's website includes a useful online primer on the subject along with links to other resources:
http://www.ser.org

Jonathan Silvertown of the Open University details the origins of the Park Grass Experiment and outlines the results of the first 150 years:
http://www.open.ac.uk/science/biosci/personalpages/j.silvertown/pge.htm

The Woods Hole Oceanographic Institute provides a useful overview of Black Sea nutrient fluxes:
http://www.whoi.edu/science/MCG/cafethorium/website/projects/blacksea.html

CASE STUDY 7

Soiled soil

Diesel oil can be a persistent and toxic pollutant in soils, and a considerable challenge for **phytoremediation**—the use of plants to clean up contamination (Newman and Reynolds 2004). Most of the decomposition of hydrocarbons occurs in the region immediately surrounding the root, the **rhizosphere** (Merkl *et al.* 2005), where microbial activity and numbers are high (Kirk *et al.* 2005). In a normal soil, the roots and their exudates represent a reliable source of carbon upon which a diverse microbial community can develop. Oil contamination restricts the growth and metabolism of roots and may therefore limit the depth of the rhizosphere. However, the associated bacterial community contains species able to use oil as an alternative carbon source, and will switch to these resources when they are available. Because the rhizosphere supports a varied microbial community, establishing a viable above-ground plant community becomes essential to recovering these soils.

Kechavarzi and his colleagues have investigated how plant roots negotiate diesel spillages (Kechavarzi *et al.* 2007), using a fast-growing species, *Lolium perenne* (perennial rye grass), known to remediate oil-contaminated soils. They used 'rhizoboxes'—glass-fronted wooden containers through which they could monitor the growth of the roots. By tilting each box 35° from the vertical, roots growing down intercept the glass, where their development can be observed.

The rhizoboxes were filled with a fertile sandy loam, into which a 15 mm-deep layer of contaminated soil (containing 25 mg diesel/g of soil) was incorporated below the surface. Nine plants were allowed to grow on in each box, so that their roots would eventually encounter the contaminated layer. Five treatments (each replicated three times) were applied, of which two were controls—one with no diesel contamination and one with no plants (Figure 7.30). Treatments 1 and 2 had a contaminated layer 25 mm below the surface; Treatment 1 was watered from above and Treatment 2 was watered from below. Treatment 3 had a breach in its contaminated layer, so that one half of the box had the layer at 25 mm deep and the other at 50 mm. This box was watered from above, as were the two controls. Root growth was measured over 52 days; to follow the root pattern it was traced onto acetate sheets which were then scanned into a computer. Shoot number, shoot height, and root length were recorded at 3-day intervals and at the end

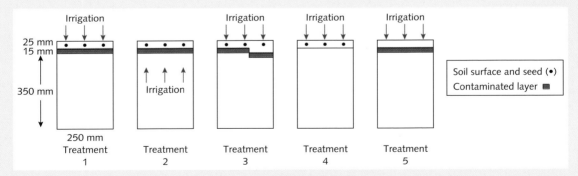

Figure 7.30 The five treatments used to study the establishment of *Lolium perenne* in diesel-contaminated soil by Kechavarzi *et al.* (2007). Each rhizobox was 250 cm wide, 35 cm high, and 2.5 cm deep, creating a narrow section of sandy loam into which the grass could root, and which could be observed through the glass front panel. The pattern of irrigation is shown by the arrows. Treatment 1 used a single 15 mm layer of diesel-contaminated soil 25 mm below the surface, watered from above. Treatment 2 had the same arrangement but was irrigated from below. Treatment 3 had a discontinuous contaminated layer, 25 mm and 50 mm below the surface. Treatment 4 was a control with no contaminated layer, and Treatment 5 was a control with contaminated soil but no plants.

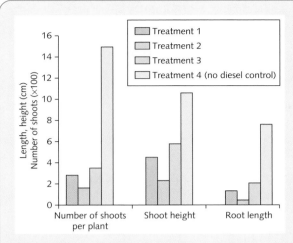

Figure 7.31   The effect of each diesel treatment on the mean values of three parameters measured in *Lolium* at the end of the 52-day experiment.

of the experiments the root and shoot biomass was measured. Levels of hydrocarbons were also measured in the soil above, below, and within the contaminated layer to assess the movement of the pollutant by the end of the experiment.

Not surprisingly, the diesel-contaminated layer had a negative effect on plant growth. All seedlings survived in the diesel-free control, 74 per cent in the staggered layers (Treatment 3), but just 41 per cent in Treatment 2, which had the continuous contaminated layer watered from below. Watering from above (Treatment 1) appears to have ameliorated the effect of a continuous layer, with 67 per cent survival. This pattern, with the greatest impact on plants growing above a continuous layer when irrigated from below, is seen in each measure of performance of the *Lolium* (Figure 7.31).

All dosed soils had poorer root development than their uncontaminated control, but Treatment 2 again had the least developed root system, with only a few roots growing through the contaminated layer. Irrigation from above and a breach in the contaminated layer again allowed the plants in these treatments to develop much fuller root patterns to much greater depths (Figure 7.32).

The upper, uncontaminated layer in which the *Lolium* germinated provided only a shallow rooting zone, and this reduced the survival of plants not able to penetrate the contaminated layer. Those which did were able to establish a much fuller root system beneath the layer, and went on to expand their root network inside the contaminated layer itself as the levels of diesel began to decline. This happened to a much lesser degree with those watered from below. By the end of the experiment, mean levels of diesel in the contaminated layer had fallen to between 16 and 19 per cent of their original concentration, even in the diesel control treatments with no plants. Much of this remediation must therefore have come from the existing microbial community (and from physical and chemical losses), rather than the rhizosphere created by *Lolium*.

Patterns from the staggered contaminated layers provide important clues on how the roots respond to the contamination. First, they develop thicker root patterns in the deeper uncontaminated layer, before they encounter the diesel. Second, the root system appears to grow toward the breach between the two layers, presumably because roots forage away from the contaminant (Figure 7.32). Kechavarzi *et al.* (2007) note that such effects are time-dependent, because eventually the roots penetrate and spread through the contaminated layer.

Irrigating from below is known to speed the degradation of hydrocarbons in contaminated soils by promoting root growth to greater depth, at least when the oils are mixed through the soil profile. As a layer, the diesel-contaminated soil acted as a waterproof barrier, effectively preventing the upward movement of water through the capillary action of the soil particles. In Treatment 2, water that did move up through the layer carried with it small but significant amounts of diesel, and these proved toxic to the young plants.

The experiment enabled Kechavarzi and his colleagues to suggest practical responses to diesel spillage on soils. One strategy might be to provide a deep layer of uncontaminated soil at the surface in which the plants can establish themselves before encountering the contamination. Ploughing helps to reduce the toxic threat by mixing the soil and reducing the concentration of the hydrocarbons, though the roots then have a greater depth of contaminated soil to negotiate. An alternative could be to rip through the contaminated layer with fine tines, to provide channels by which roots might find their way to uncontaminated soil they can colonize.

(continued overleaf)

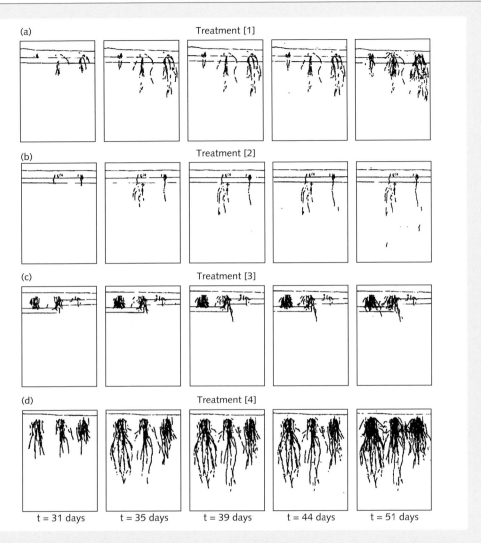

Figure 7.32 The development of the root system in one replicate from treatments 1–4 (a–d) at 31, 35, 39, 44, and 51 days.

References

Kechavarzi, C., Pettersson, K., Leeds-Harrison, P., Ritchie, L., and Ledin, S. 2007. Root establishment of perennial ryegrass (*L. perenne*) in diesel contaminated subsurface soil layers. *Environmental Pollution 145*, 68–74.

Kirk, J. L., Klironomos, J. N., Lee, H., and Trevors, J. T. 2005. The effects of perennial ryegrass and alfalfa on microbial abundance and diversity in petroleum contaminated soil. *Environmental Pollution 133*, 455–465.

Merkl, N., Schultze-Kraft, R., and Infante, C. 2005. Phytoremediation in the tropics—influence of heavy crude oil on root morphological characteristics of graminoids. *Environmental Pollution 138*, 86–91.

Newman, L. A. and Reynolds, C. M. 2004. Phytodegradation of organic compounds. *Current Opinion in Biotechnology 15*, 225–230.

EXERCISES

1. Match the following soil horizons with their descriptions and reorder these in descending order from the top of the soil downwards.

   Horizons: A, B, C, E, O

   Descriptions:
   (i) the bedrock or deposits from which the soil is formed.
   (ii) the horizon from which nutrients leach into the subsoil.
   (iii) a layer containing undecomposed material such as leaf litter and other detritus.
   (iv) the layer where biological activity is at its greatest.
   (v) the subsoil containing minerals, nutrients, and weathered bedrock.

2. Describe three methods by which the nitrogen status of a nitrogen-poor soil can be increased, and set out the merits of each source.

3. Select the most appropriate response to the following statement and comment on your selection.

   Mycorrhizal fungi:
   (a) increase the availability of phosphorus and nitrogen to plants.
   (b) fix nitrogen and increase the availability of water to plants.
   (c) occur as a result of the build-up of salt in the soil.
   (d) increase the availability of phosphorus and water to plants.
   (e) fix nitrogen.

4. The following data were collected from a deciduous forest ecosystem in New England (Hubbard Brook). They give the rate of uptake and loss of nitrogen and calcium by the trees, and the mass of each element in the trees.
   (a) Calculate the turnover time for each element for the trees based on the amount taken up.
   (b) Calculate the turnover time for each element for the trees based on the lost to the soil.
   (c) How do you account for the differences you observe between the two estimates?

	Nitrogen	Calcium
Uptake (kg/hectare/year)	79.6	62.2
Loss (kg/hectare/year)	60.4	43.9
Mass in trees (kg/hectare)	532.0	484.0

5. Examine the map below and consider the following scenario.

    The imaginary fishing village of Safe Haven is situated close to area prized for its wildlife and landscape value. A supertanker has run aground on rocks 6 km offshore and most of its 100 000 tonnes of oil are being driven towards the coast by strong winds. As coordinator of the emergency control centre what are your priorities?

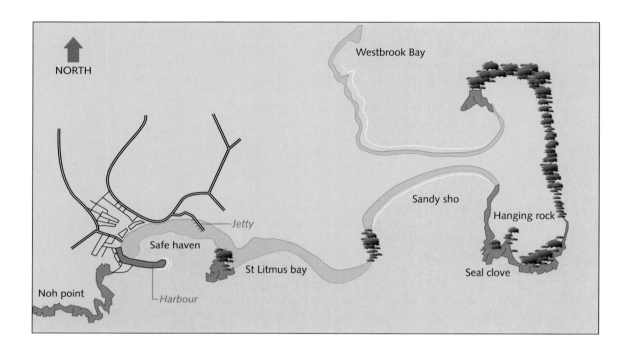

6. Match the following conditions of spoil with an appropriate method of treatment:

    Problem
    (a) High pH
    (b) Salinity
    (c) Low level of macronutrients
    (d) Too dry
    (e) Unstable
    (f) Low pH
    (g) High levels of heavy metals
    (h) Too compact
    (i) Too wet
    (j) Low nitrogen levels

Treatment
(i)   Mulch with organic matter
(ii)   Drain
(iii)   Add pyritic waste and organic matter, and allow to weather
(iv)   Irrigate, allow to weather, use halophytic species to revegetate
(v)   Rip or scarify prior to revegetation
(vi)   Add fertilizer and lime
(vii)   Stabilize, mulch, and possibly regrade prior to revegetation
(viii)   Lime and/or allow to weather
(ix)   Add nitrogen fertilizers/legumes
(x)   Add organic mulch and vegetate with metallophytes

7.   What is the value of removing bottom-feeding fish in our attempts to restore highly eutrophic lakes?

8.   What is meant by success in ecological restoration? Suggest some criteria by which the progress or outcome of a restoration programme may be measured and monitored.

Tutorial/seminar questions

9.   Lakes and ponds tend to become eutrophic as they age. Do rivers and streams? If not, why not?

10.   *Rhizobium* can survive as a saprophyte in the soil. Why should it infect a root nodule and form a symbiotic association with a host plant? (It is worthwhile doing some preliminary research on the nature of this symbiotic association before discussing this in a seminar.)

8

Scales

'Thou art weighed in the balances, and art found wanting.'
Daniel 5:27

CHAPTER OUTLINE

- How ecology encompasses large-scale processes, from the individual to the planet.

- The interaction between small- and large-scale ecological processes.

- Pattern and ecological processes in landscapes.

- How climate governs the distribution of the global ecosystems—tropical, temperate, and boreal biomes, their forests and grasslands.

- The climate and the effects of biological activity on atmospheric conditions.

- The carbon balance of the atmosphere and global climate change.

- Predicting the ecological impact of global climate change.

← Supercell over Nebraska, 28 May 2004.

Travel, it is said, broadens the mind. Choose the right places to visit and the Earth will quickly reveal the immense diversity of its life. Inspired by their visits to the tropics, both Charles Darwin and Alfred Russel Wallace sought to explain this variety. Both were familiar with the ecology of lowland England, but their respective expeditions to the forests of South America and South East Asia spurred them to understand why there were so many species and how new species might arise.

Why should the tropical forests have such a wealth of plants and animals, compared to the forests of the north? Why does the number of species increase from the poles to the tropics in nearly all ecosystems? The damp heat of the tropics certainly promotes a lush and rapid growth, but, as we shall see in the next chapter, a benign climate alone cannot explain their rich diversity of species.

The major terrestrial ecosystems form broad bands that trace the climatic zones of the planet. This is not a perfect match and the pattern is interrupted by discrete communities reflecting local details of landform, geology, or climate. Nor are they in fixed positions—ecosystems advance and retreat with fluctuations in the global climate, across land masses that slide around the planet in geological time. This mobile, three-dimensional puzzle continues to challenge ecologists to explain the current distribution of species and their change through space and time.

In ecology, space and time are linked: ecological processes that operate over large areas operate over long time scales. For example, the ecology of the northern latitudes only begins to make sense when we understand the effects of glaciations over the last two and half million years. Yet, most ecological study has been rooted firmly below the ecosystem level, confined to local communities and relatively short time periods. Until recently, these scales were taken to be the limits of our scientific resolution, and only within this range would we answer questions with any degree of confidence.

Now, ecologists are being asked about long-term change, about the history and the future of life on Earth. Today, ecology uses data from ice cores, the fossil record, and satellites. The spatial and temporal horizons of the science are being stretched. In this chapter, we consider the effect of scale on ecological processes and introduce some of the principles being developed to understand large-scale ecology.

We start with the emerging subject of landscape ecology and the connection between spatial and temporal scales in ecological change. We go on to describe the major terrestrial ecosystems and show how their distribution reflects the prevailing patterns of climate. Finally, we explore the interdependence of the biosphere and the atmosphere. Our understanding of this interaction is crucial if we are to make sensible predictions and adopt policies that anticipate climate change over the next 50 years.

8.1   Landscape ecology

Size and timing are everything in ecology. Ecologists have to be sensitive to the implications of scale because these change from one species to another. An insect that can walk on water exploits a physical property (surface tension) that has no significance for the mobility of much larger animals. Individuals may span short time scales—the adult mayfly has just 12 hours to mate and lay its eggs—and small distances—the damselfly selects particular leaves and lays its eggs millimetres beneath the water surface.

Other animals, including some insects, fly halfway across the world to mate and lay their eggs.

For the fullest picture of the selective pressures operating on a species, we have to see the world from its perspective and the scales to which it responds. For most of its life, the world of a fig-wasp is a fig. Yet the size of an organism is not always a good indicator of the distances over which it ranges. For example, monarch butterflies travel hundreds of kilometres across North America to find the roosting sites from which they emerged. Even the wingless flea can travel the globe attached to the feathers of its host. Ironically, the chances of a bird surviving a migration from one hemisphere to another may depend on the insects that fed on it in the nest or the tiny parasites it now carries south.

The movement of animals, seeds, pollen, and other plant tissues is part of the transfer of energy, nutrients, and DNA occurring within and between ecosystems. Ecologists may like to think they deal with discrete and closed systems, but communities, species, and individuals are leaky and open systems. Genes and resources move across these boundaries, and such inputs and outputs have to be part of our balancing of the accounts for an ecosystem.

The greatest exchanges will be between adjacent ecosystems within a landscape. **Landscape ecology** is a fusion of geography and ecology, a study of the spatial arrangement of ecosystems and the large-scale processes that unite them. A valley will contain a number of different ecosystems, but its overall slope governs rates of drainage, the depth and movement of its soil, the supply of nutrients, and the stability of its surface. Its altitude and forest cover influence how much rainfall it receives, and the chemistry of its waters depends primarily on the soils through which they drain. These soils, in turn, will probably reflect their bedrock and vegetation cover, and those of adjacent communities. The exchange of materials between neighbouring ecosystems means the character of an ecosystem is shaped by the landscape in which it sits (Figure 8.1).

Landscapes range over kilometres and are most obviously defined by natural discontinuities—such as a catchment or some significant change in geology or land use. Ecosystems within a landscape share similar climates, regimes of change, and other large-scale abiotic factors. They become differentiated into different communities because of local conditions—a plant community by the depth of soil or a pond by the water held in a depression, for example. A landscape is thus a mosaic of habitats, each home to a community of species adapted to its particular abiotic conditions. Together the habitat patches within a landscape share a history and are subject to similar geomorphological processes or degrees of human interference.

In some cases the boundaries between landscapes are poorly defined: in the prairies or the Canadian tundra, for example, there may be few obvious discontinuities over large areas, and even the division into catchments becomes almost meaningless. With no sharp boundaries, abiotic change occurs along gradients, and communities grade into each other. In other cases, a community can be isolated by very distinct boundaries—the sharply divided valleys of the Californian or Chilean coastal regions mean that adjacent and very similar ecosystems often have their own endemic plant species, an indication that little genetic exchange occurs between them (Section 5.1).

Within a landscape different ecosystems are most readily defined by their abiotic characteristics and then by their plant communities. Thus, we might distinguish rivers from lakes and ponds, and then lakes by their degree of eutrophication or the plant communities in their shallows. Well-defined ecosystems have flows of energy and materials that typically occur within their boundaries, and exchange much less with other ecosystems. Similarly, we expect more communication between equivalent habitats because particular species favour particular conditions. For example, a species of wildfowl may only roost and migrate between oligotrophic lakes because these have large areas of macrophyte cover close to their shores, in turn avoiding the exposed open-water conditions found on eutrophic lakes (Section 7.2). Natural boundaries represent different sorts of barriers to different species, and ecologists may define 'landscapes' according to the scales relevant to the organism they are studying. The most appropriate spatial scale ranges somewhere between an individual's home range and its species' regional distribution.

Different habitat patches have different degrees of permanence: over hundreds or thousands of years a pond or lake may become dry land through sedimentation and succession (Box 5.5). Such changes mean that a patch's age is often a key determinant of its species composition and diversity. We can distinguish different patch types according to their history (Table 8.1).

The archetypal landscape of lowland England is a patchwork of fields sewn together with hedgerows, roadside verges, and streams, dotted with islands of

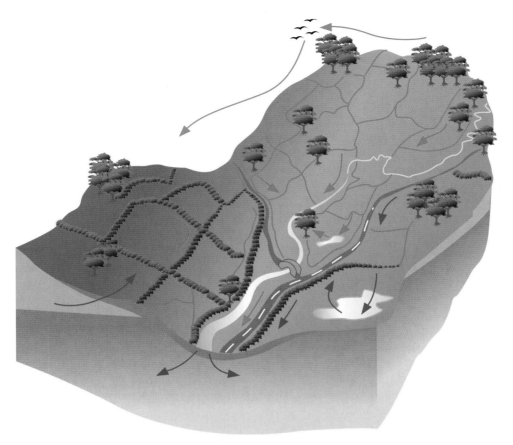

Figure 8.1   Large-scale processes in a landscape, in this case for a temperate agricultural area. Genes, energy, and nutrients move using wind and water, or with plants and animals. Networks of corridors—rivers, hedgerows, or roadside verges—provide routes for some species to move between habitat patches. All biota will be affected by large-scale effects, such as the climate within the valley, or the movement of soils and water down the valley.

water and woodland (Figure 8.2). The vineyards of the Mediterranean are bordered by bramble and scrub, interrupted by rocky outcrops, with woodland confined to steep slopes and upland areas (Figure 8.3). In each case, we can describe a series of landscape elements demarcated by their dominant vegetation.

Landscape ecologists recognize three basic spatial elements—the **patch** (a discrete habitat—say, an area of woodland), the **matrix** (the background ecosystem which dominates the landscape—the fields or vineyards), and the **corridors** that traverse the matrix, connecting the patches (such as hedgerows or streams) (Figure 8.1). Between these elements are more or less

distinct boundaries or **ecotones**. Some organisms exploit the sharp transitions found at ecotones. Many plants favour woodland margins, enjoying the open canopy and the reduced competition for light and nutrients.

Various ways of quantifying these components have been developed to compare landscapes. The length of patch boundaries, the area of the patches, the distance between them, and the extent to which they are connected by corridors can all be measured using detailed maps, aerial photography, or satellite imaging (Figure 8.4). A highly heterogeneous landscape will be fractured into a mosaic of patches,

TABLE 8.1	Forman's patch types
Disturbance patch	Recently created by some form of disturbance, e.g. forest fire
Regenerated patch	Patch recovering from disturbance, e.g. forest clearance in the process of regrowth
Environmental patch	Some sharp discontinuity, e.g. change in soil type that causes a distinct change in the plant community
Remnant patch	Area remaining of the original habitat isolated by an encroaching matrix, e.g. woodland patch amongst agricultural fields
Introduced patch	A habitat created by human activity, e.g. a woodland clearance

Figure 8.2   The patchwork of traditional English agriculture.

Figure 8.3   Typical Mediterranean landscape (Languedoc, France).

(a)

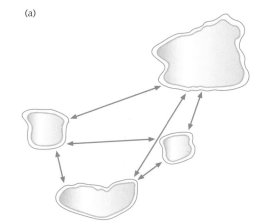

(b)

Figure 8.4   Simple methods of quantifying the ecological properties of a landscape. (a) **Landscape physiognomy** is the physical layout of an ecosystem in the landscape. In this example, the habitat of interest are the ponds, and two measures are indicated here—the length of the boundary (red) of each patch and the distance between pairs (green). (b) **Landscape composition** is a measure of the proportion of the landscape represented by one habitat type. Both of these measures have their counterparts in quantifying the structure of single communities, albeit at much smaller scales.

which will nevertheless be highly interconnected if there is a network of corridors and long boundaries. The large vineyards used in intensive wine production form a very uniform, homogeneous landscape with short (straight-line) boundaries relative to their area (Figure 8.5).

Figure 8.5 The straight lines of intensive viticulture. The monoculture of vines represents the matrix in which habitat patches of woodland or scrub are embedded. The roadsides and streams represent corridors connecting these patches.

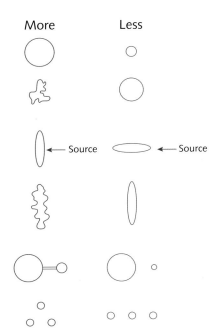

Figure 8.6 The shape and size of habitat patches determines their degree of connectivity. The exchange of species or materials between habitat patches is increased by patch size, by having a more convoluted shape, by providing a larger target, by having a larger patch boundary, by having a connecting corridor, or by having more neighbours.

Connectedness is a feature of the shape and spatial arrangement of patches as well as the distance between them. We would expect large patches to be more readily colonized than small patches, but small patches with a long boundary have a greater exchange with their surroundings. This traffic is larger still if there is a direct connection or corridor that the colonists can use (Figure 8.6).

Species differ in their capacity to cross the matrix, or to make use of any connecting corridors. Often the nature of the matrix serves as a filter, allowing some species or individuals to cross, but not others. Birds fly readily between ponds but fish do not make the same journey quite so easily. Corridors may also be selective (Box 8.1). Isolated patches might be considered habitat 'islands' for some groups, and the collection of species found there will depend on island size and the distance from a colonizing population (Section 9.1).

We can easily measure the degree of connectedness for different agents moving between patches. The simplest method is to count the frequency with which a colonizer reaches a patch, say the number of visits made to a lake by birds from a distant roost. Collecting such data often identifies unforeseen ecological effects of the landscape configuration—an increase in woodland fragmentation, for example, is known to increase nest predation amongst the resident birds. Another effect is that insect parasites found in the nest tend to be higher when the birds are forced to roost at the edge of woodlands or in corridor habitats. Both are a consequence of the increased traffic in predators and parasites moving between patches.

BOX 8.1   Fragmented city

Different species see landscapes in different ways. For the birds and rodents of San Diego, the remnant coastal sage brush and chaparral communities offer different opportunities and challenges. Douglas Bolger and colleagues have studied the responses of these groups to the fragmentation of the natural habitat and the use that different species make of corridors connecting patches of the natural Mediterranean-type scrub vegetation.

Within the urban and suburban areas of San Diego remnant patches of vegetation vary in size from less than 2 hectares to several hundred hectares and have different degrees of connectivity. Some are isolated and some are linked by strips of the original vegetation. Others are connected by roadsides which have been planted with a few of the native shrubs. Bolger and his co-workers recorded rodent and bird species for two years (1992 and 1993) in each of three habitat types—patches, remnant strips, and roadsides. They evaluated how corridors were used by each species, either as a thoroughfare or as a habitat with resources to be exploited. The birds could be divided into two groups—those known to be sensitive to the fragmentation of their native chaparral (and therefore likely to disappear as patch size reduced) and those tolerant of such change.

The vegetation type of the different corridors did not seem to matter for the diversity of either rodents or the birds, either for the remnant strips or in the reduced scrub of the roadsides. However, the extent of shrub cover was important for both groups, and most especially for the

fragmentation-sensitive birds—these avoided corridors with shrub cover of less than 40 per cent and some species were never found on roadsides at all. These same species avoided San Diego and would not over-fly urban or suburban areas. Besides the lack of native shrubs, perhaps the disturbance from the road itself (noise, street lighting, and so on) prevent them using roadside habitats. Only the more tolerant species are found using the corridors—birds that do not appear to demand dense cover comparable to the natural shrub community.

In contrast, all species of rodent recorded in the patches were also found in both corridor types, and most were equally common in the corridors and the patches. The majority were using these as habitats in their own right rather than as passageways, maintaining a reproductive population there. Interestingly, one species of mouse, *Peromyscus eremicus*, was more abundant on the open roadside corridors, perhaps because they benefit from the greater prevalence of herbaceous plants or because here they escape their main predators.

In effect, this same landscape has very different properties for these two animal groups. For many rodent species the corridors are both thoroughfares and habitats to exploit. Several species are able sustain a population in these areas, and they serve as habitat patches for them. For many of the birds, however, these corridors do not increase the connectivity of the San Diego landscape and offer no means of crossing the intervening matrix of urban development.

Beyond their value in comparing landscapes, these measures have considerable ecological significance. The size of any population will be closely tied to the size carrying capacity of its patch. We have already seen how a metapopulation, divided between a number of patches, may be able to sustain itself even if some local populations periodically go extinct (Section 3.7). A highly connected landscape, or a background matrix that is readily crossed, allows most patches to be recolonized quickly. For this reason, the long-term prospects for a species may be determined by the heterogeneity and connectivity of the

landscape it occupies (Box 3.4). Indeed, for an endangered species, it can be crucial, either when trying to find a mate or when disease is spreading through the metapopulation (Section 3.7). Connectivity can cut both ways—isolated patches may provide refuges from an infectious disease or condemn their populations to reduced genetic variation.

The extent to which a landscape is partitioned into distinct patches is measured as its 'graininess'. The 'grain' of a landscape varies with scale and different species resolve different grain sizes according to their specific habitat demands. A small patch of woodland may consist of areas of closed canopy or open glades, each with a particular plant and invertebrate community. Some insects are confined to the glade for the whole of their life cycle. In contrast, a bird over-flying the landscape may distinguish only woodland and open field. Moving through the glade, a butterfly lays its eggs on leaves selected according to the plant species and their nutrient status. At a much larger scale, the bird is seeking out nesting sites according to tree position and the territories of other birds. The two animals are responding to very different grain sizes in the same ecosystem. Favoured microhabitats may also change during a creature's life cycle—eastern massasauga rattlesnakes (*Sistrurus catenatus catenatus*) choose open locations with rock cover when they are carrying eggs but otherwise prefer wetter, open and edge habitats . . . until they are about to hibernate when they seek forest. This range of habitats and, most importantly, their proportions determine the landscape favoured by the rattlesnakes.

At regional scales, enduring changes in abiotic conditions begin to distinguish different landscapes, where processes begin to operate at different rates. Tomas Santos and his co-workers have shown that woodland habitats regenerate from fragments more slowly in the drier areas of Spain. Remnant patches of holm oak (*Quercus ilex*) forest in the centre of the country support far fewer bird species for their size than woodlands amidst the farmland of the temperate north. The central and southern woodlands endure the prolonged summer drought of the Mediterranean region (Section 5.1), and this slows their regeneration. Where rainfall is more consistent and abundant, forest development is faster, so holm oak patches in northern Spain have a greater variety of trees and tree heights. These provide more niches for the birds (Figure 5.13) and, compared to equivalent sized patches in the south, have more bird species per unit area. This demonstrates how features of a landscape's ecology, in this case, bird species richness, are

(a)

Figure 8.7   (a) An ecosystem stranded: abandoned boats at Muynak, formerly a fishing port on the Aral Sea. (b) Wind-blown dust over the Aral Sea seen by the Aqua satellite on 13 June 2006. Maps published as late as 1991 do not show the lake divided into lobes or separate bodies, and the central tongue of land was then just a small island. Today Muynak is 100 km from the shore. The good news is that the damming of the isolated North Aral Sea has allowed this relatively unpolluted portion to grow considerably since 2005, fed by the Syr Darya from the Himalaya.

(b)

Figure 8.7 (*cont'd*)

determined by higher-level and larger-scale factors—here the regional climate and its effect on the plant community structure.

Measuring the connectivity of landscapes in terms of nutrient or material flows also enables us to describe regional ecological processes. Tracing these movements may indicate the routes by which a pollutant moves between patches and between species. Ironically, it has often been the other way around—we learn about ecological processes from following the passage of a pollutant through a landscape. The most spectacular example was the explosion of the Chernobyl nuclear reactor in 1986. This contaminated a swathe of landscapes from Central Europe and northern Italy through Scandinavia into northern Britain. With the major research effort that followed, ecologists were surprised by the ease with which radiocaesium moved though the soils, and quickly learnt that their models of elemental flows through ecosystems were far from complete. One consequence

of Chernobyl has been a better understanding of the ecological mobility of caesium.

Another dramatic case is the pollution and over-exploitation in Kazakhstan and Uzbekistan of the Aral Sea, once the fourth largest lake in the world. For the last 80 years its feeder rivers, the Amu Darya and the Syr Darya, have been diverted to irrigate cotton fields. Since 1960 the Aral has been shrinking and it is now around one quarter of its original size. As its freshwater supply dwindled, the lake became increasingly saline, leading to the loss of its unique fish and invertebrate species. The commercial fishery has collapsed and the large steamers that once used to work the lake are today stranded in desert sands (Figure 8.7).

The impact of the shrinking lake extends many hundreds of kilometres from its original shoreline. Changes in the local climate mean that dry winds blow salt-laden deposits over vast areas, scorching the vegetation and making the land almost useless. Increasing atmospheric pollution and the poor quality of the Aral waters have created major health problems for the local people. Infant mortality rates are among the highest in the world and life expectancy among the shortest. Diverting the Amu Darya and Syr Darya was the start of a slow process of change. It took 20 years for the Aral Sea to show signs of ecological decline. Now the impact covers many thousands of square kilometres. International efforts are being made to restore water levels, but there is no prospect of restoring the ecosystem and its landscape.

One might reasonably argue that no detailed understanding of landscape ecology was required to predict many of the changes that followed the shrinking of the Aral Sea, even at these larger scales. Yet we failed to predict the scope of the Aral Sea tragedy and we are ready to risk making the mistake again. Plans to divert or dam rivers in other parts of the world, from the Euphrates in Turkey to the Mekong in South East Asia, have also raised fears about their climatic, ecological, and economic implications. Nations continue to fund large-scale engineering of landscapes despite the warnings from ecologists, hydrologists, geomorphologists, and history.

The now regular flooding of cities on the Danube has led to calls to restore the forested wetlands alongside the river through the Balkans, to protect lower-lying areas of Rumania and Bulgaria. The increased frequency of floods in central China over recent years has been attributed to the massive engineering of its rivers over a quarter of a century, but such work continues to this day. Rather like the past engineering of the Mississippi Basin, landscape-scale modification to solve local-scale problems may only store up regional-scale problems when the once-in-a-century flood eventually arrives. Across the globe there is a sizeable annual death toll from catastrophic landslips which invariably follow large-scale deforestation of upland areas. Ironically, local peoples, often those most affected, have long understood the role of the trees in holding back mountain soils.

Working down from the level of the landscape we are describing a hierarchy of scales, with the degree of patchiness changing according to the process being studied. Energy, nutrients, and information show greater exchange and connectedness at the smaller scales. To make reasoned decisions and quantified predictions we need to organize the data from these different scales, incorporating the important interactions between individuals, ecosystems, landscapes, and regions. One approach establishes the connections between processes operating at different levels and how these interact to impart stability to a landscape. These ideas have been formulated in more rigorous scientific terms as hierarchy theory.

Hierarchy theory

While a landscape may appear unchanging, its component ecosystems are often much more variable. Even within an ecosystem there will be several habitats—the shoreline, shallows, and deep water of a lake—each with a different inventory of species and each changing in space and time. At smaller scales, measurements taken between habitat patches, or in a single patch over short intervals, are highly variable. Yet when averaged over the whole landscape such numbers change little, even over long periods. The variation we observe depends upon the scale used to measure a particular parameter.

Hierarchy theory suggests this is a general property of many complex systems. It sees a system as being organized into functional levels, differentiated by the rate at which some process operates. The smaller scales and lower levels are 'nested' inside the larger-scale and higher-level processes. Moving from one level to the next in the hierarchy, the rate of a process slows down simply because larger-scale processes run more slowly. Consequently, the variation we observe at the smaller scales has little significance at the higher levels and this 'noise' is gradually lost in successive levels. The whole system is 'buffered'—high-frequency variation is lost with each level and only the larger, more sustained trends transfer to the upper hierarchy. Put at its simplest, hierarchy theory suggests that large-scale systems such as landscapes change slowly because they are big and complex.

Hierarchy theory defines levels not by species but by functional groups, by what they do and the speed at which they do it. Consider the example given in Figure 8.8, where the process is carbon fixation in plant tissues. Within an individual leaf, fixation fluctuates rapidly as carbon dioxide concentrations near the stomata change from one moment to the next. At higher levels in the hierarchy, say the leaves of the whole tree, average rates of fixation, per gram of

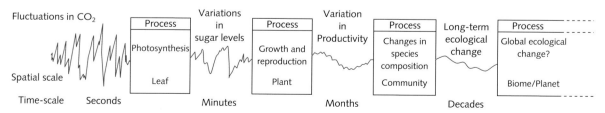

Figure 8.8 A simple example of the effects of a hierarchy of levels governing an ecological process, in this case carbon fixation by photosynthesis.

photosynthetic tissue, vary much less. In the canopy of the forest, over a larger spatial scale, rates are increasingly uniform. As we move up through the hierarchy so the variation decreases further.

At the regional level it may take years, even decades, to detect a significant shift in carbon fixation rates, even following some large-scale change. The general level of fixation for the whole forest ultimately depends upon rates in individual leaves, so that even if the species composition of the forest shifts, change occurs very slowly, if at all. At the level of the forest or the region, the process of carbon fixation is highly stable.

Constraints work in the opposite direction too. Photosynthesis in individual leaves depends upon local humidity which itself depends upon the density of the tree cover. Forests change the climate of their immediate area, inducing rainfall and raising humidity; about half of the rainfall in the Amazon basin originates from the evapotranspiration of the forest itself. So the speed with which water cycles through the system depends, in part, upon rates of photosynthesis and the opening of stomata to fix carbon. In this hierarchy, a high-level attribute, forest cover, governs a process, carbon fixation, operating at the lower levels (the leaves)—an example of a high-level constraint limiting the process at its smallest scales.

Other limits on carbon fixation are imposed by the interactions between species, which have been omitted from our simple hierarchy. For example, very damp conditions on the forest floor favour a particular assemblage of decomposer animals, microorganisms, and fungi. Because the species forming this community determine rates of decomposition and the flow of nutrients between the soil and the canopy, they too will impact primary production (Section 7.1) and rates of carbon fixation.

An ecosystem's processes are thus constrained by the feedback between its functional levels, in the same way that its trophic structure is constrained (Section 7.2). As a result, hierarchy theory posits that an ecosystem organizes itself, forming itself into a stable configuration driven and limited by these interactions. This may be a general property of any complex system with internal checks and balances, one which will inevitably arrive at a state at which it

undergoes little further change. Hierarchy theory may thus go some way to explaining how similarly configured ecosystems appear under equivalent abiotic conditions (Sections 5.1 and 5.3).

Notice here that our hierarchy is only good for carbon fixation and we would need to construct an entirely different one to model some other process. In each case, ecologists have to determine which interactions are important for organizing the system. Take the example of forest regeneration following a fire or a tree fall. At the local scale, regeneration is constrained by the rates of nutrient release as well as the scope for colonizing trees to establish mutualistic associations with soil fungi (mycorrhizae—Section 4.2). At the landscape scale, regeneration depends on the severity and frequency of a disturbance. The latter will determine the distribution and abundance of potential colonizers, the size of regenerating patches, and the ease with which colonizers can find these spaces.

We have learnt much about the role of fire and the effect of scale in natural ecosystems in recent years. Like most of the extensive forested areas in North America, the Yellowstone National Park had been subject to fire-suppression patrols. Prior to 1970, a vigilant forestry service quickly extinguished the smallest fire before it could spread. Much of this forest has extensive tracts of lodgepole pine (*Pinus contorta*), a species that becomes more flammable with age. Without small fires to open gaps and remove some of the older trees, the population in the park became dominated by large combustible trees. At the same time, tinder, in the form of leaf litter, accumulated on the forest floor.

It only needed the extended drought of 1988 and an uncontrolled logging fire to set off the inevitable. Around 70 per cent of the park burnt. The scale of the fires and the dramatic pictures of its destruction led to it being treated as a national disaster. Yet the burn was not uniform and different patches burnt to different degrees. Temperatures high enough to kill seeds were achieved in only relatively small pockets, and trees were killed in only 20 per cent of the park, whilst less than 1 per cent of its elk population was lost. There has since been a rapid regeneration of the trees, the ground flora, and the animals. The removal of old, unproductive wood was followed by a flowering of the forest floor. The nutrients and space released by the conflagration allowed opportunist plants to flourish and resulted in an increase in resources for its herbivorous animals.

The simple lesson is that frequent but small-scale burns are part of the normal economy of coniferous forests, comparable to the same process in mediterranean-type ecosystems. Small burns do not produce dramatic change, but they do create the gaps that prevent some species from becoming dominant (Section 5.1). Change is a part of all ecosystems, and communities are shaped by the most frequent scales of disturbance. If a disturbance exceeds the range of particular species they are lost, and the community shifts to a different configuration. A large-scale disturbance may lead to the loss or replacement of many species, perhaps reducing a community to a plagioclimax (Sections 5.3), even over entire regions. It may be that fire was critical in the human colonization of Australia and that the landscape now reflects its persistent use over an extended period.

Today, the coniferous forests of North America show few signs of the fires in Yellowstone in 1988 and the park is back to a former, younger self. Forest fires are short-term events and hierarchy theory suggests that large-scale ecosystems like Yellowstone are susceptible only to long-term change. If there was persistent change, say, in its patterns of rainfall or temperature, the forest would move its boundaries and occupy a different latitudinal range.

In contrast, the short-term change of a forest fire is absorbed by the system. The fires in the Kuwaiti oil

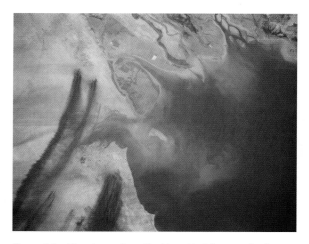

Figure 8.9   The plumes from the Kuwaiti oil fires south of Kuwait City. This photograph was taken by the crew of the space shuttle on mission 37 in April 1991, two months after the fires were ignited.

fields after the Gulf War of 1991 were portrayed as a major ecological disaster at the time (Figure 8.9), yet despite the large amount of carbon dioxide released (240 million tonnes) over the few months they burned, their effect was hardly detectable in the global atmosphere. Any shift in atmospheric composition would need large amounts of carbon to be added over a much longer duration.

The northern coniferous forest may well be moving in response to long-term atmospheric change. As we shall see later in this chapter, climates can change surprisingly quickly. When they do, the ecological regions shift with them.

8.2   The biogeography of Earth

Based on the distribution of large terrestrial vertebrates, the Earth can be divided into six distinct realms (Figure 8.10). First identified by Wallace, the boundaries of these zones pick out the major barriers to animal dispersal that have persisted across the globe. These can be closely matched to the drifting

of the continents: indeed such discontinuities were used by Alfred Wegener to support the theory of continental drift he proposed at the beginning of the twentieth century.

The first mammals appeared around 200 million years ago and one of their fundamental divisions,

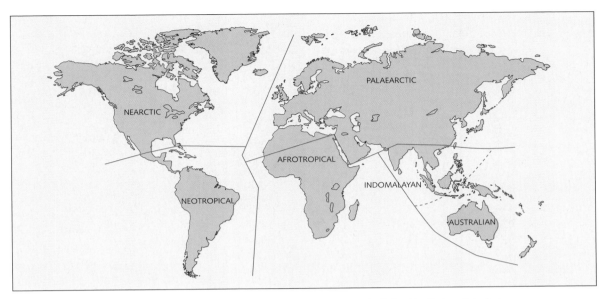

Figure 8.10   The main zoogeographical realms of the planet, primarily defined by their vertebrate fauna. The mammal fauna is particularly characteristic of an area. Before the advent of humans, some orders were entirely excluded from certain realms. The dotted line marks Wallace's line, running through Malaysia and separating the two realms according to their native birds. The flora and fauna of the two realms blend across the whole of this archipelago so the demarcation is not always well defined, even for the marsupials.

between the placentals (Eutheria) and the marsupials (Metatheria), times the break-up of a massive land-mass that once existed in the southern hemisphere. Today the greatest variety of pouched mammals, the marsupials, is found in Australia and New Guinea, land masses which together split away from Antarctica 65 million years ago. Yet the oldest marsupials, the possums, are found on the other side of the globe, in America, the remnants of a diverse metatherian fauna that once occupied South America. The sustained isolation of Australia allowed its marsupials to survive without competition from other mammals until relatively recently—the placentals did not reach Australasia in any numbers before the most recent human colonists. In their isolation, the marsupials had radiated to fill the major niches occupied elsewhere by placentals and have since defended these with varying degrees of success. The northern boundary of the Australasian realm, Wallace's Line, separates these groups, marked by a

deep ocean floor (Figure 8.10). This suggests that there has always been an aquatic boundary dividing the two realms.

Much of the mammalian fauna of the ancient northern continent is common between its realms, simply because there have been no significant barriers between the land masses in the recent geological past. In contrast, mammals that have evolved recently in the tropics show restricted distributions, reflecting the distribution of land and sea during their history. For example, there are no native primates (apes, monkeys, and their relatives) in the Nearctic and *Homo sapiens sapiens* only arrived in the New World in the last few millennia (Box 5.1, Section 1.1).

The distribution of plant communities on the Earth also reflects its more recent past and the prevailing climate. Marking out lines of latitude, major plant assemblages are organized into biomes, collections of species that share adaptations to the climate of each region. Within a biome many plants have a

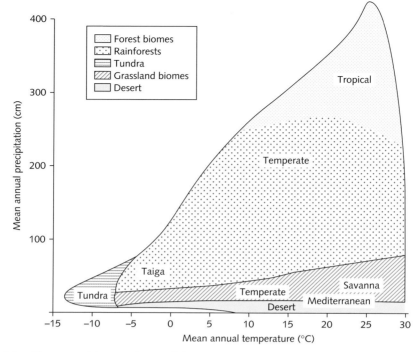

Figure 8.11 The range of the major biomes defined by temperature and moisture. Forests are only found where precipitation is above about 30 cm per year (in the coldest regions) and 120 cm per year (in the warmest regions). Grasslands develop where water is in short supply.

characteristic growth form adapted to the prevailing conditions. The most important factor is the availability of water. Forests dominate where there is sufficient moisture; grassland or low scrub where it is drier.

We can distinguish four major terrestrial biomes—forest, grassland, tundra, and desert—by their regimes of precipitation and temperature (Figure 8.11). For example, deserts are found where evaporation exceeds rainfall, within 30° either side of the Equator. At this latitude, dry air is descending, having shed its rain in the tropics. Deserts also feature on the western edge of several continents, where cold ocean currents induce coastal fogs that deprive inland areas of precipitation.

Local conditions may confound the simple latitudinal pattern of the biomes (Box 8.2). Proximity to an ocean or a mountain range can create local climates that support unique, and isolated, plant communities (Figure 8.12).

Here we limit our task by concentrating on the major terrestrial biomes. This is not to deny that there are characteristic aquatic communities, but most aquatic habitats are defined by their animal community and its adaptations to depth, rates of water flow, salinity, and nutrient source. As we see in Chapter 9, many tropical marine communities show the same latitudinal increase in species diversity as that found on land.

We also concentrate on the most productive biomes, and say little here about the hot and cold deserts. The productive ecosystems fall within three basic climatic zones—tropical, temperate, and boreal, principally defined by their average temperate regimes. For each of these zones, sufficient rainfall will produce a characteristic forest, or where it is drier, a grassland; together this produces a simple classification of two major biomes in each of three climatic regions.

Figure 8.12   The Teleki valley, Kenya. This montane habitat on the equator has a unique plant community reflecting the large temperature range created by the high altitude. Here giant groundsels (*Senecio* spp.) flourish with a range of adaptations to survive very low temperatures. Different species with similar adaptations grow on the equator in the Andes and Hawai'i.

Tropical biomes

Tropical rainforests straddle the Equator, roughly within the tropics of Cancer and Capricorn (Figure 8.16). These are amongst the oldest communities on the planet, with some pockets largely undisturbed, even during the climatic upheavals of the last glaciations. Their diversity is astounding: in Amazonia, there may be as many as 300 tree species in just 2–3 km². Tropical forests have the highest net primary productivity of all terrestrial ecosystems (Section 6.3), yet most grow on some of the most infertile soils.

Tropical forest soils are typically poor because they are old and have been leached of their key minerals. A highly competitive decomposer community quickly scavenges dead and decaying organic matter arriving at the forest floor. This makes for extremely rapid decomposition, promoted also by the high temperatures and abundant moisture. Because its organic matter has such a brief existence, the soil has a poor structure and little capacity for binding nutrients. Consequently, the soil represents a small reservoir of plant nutrients, for which the competition is intense. Some of the forest soils in South East Asian are of volcanic origin and younger, but are also generally lacking in some key nutrients, especially phosphorus. Nevertheless, in upland areas they may build up a significant amount of organic matter to form a thick layer of peat (Box 7.1).

There is also intense competition for light on the forest floor. The plant community is stratified into layers, with groups of species adapted to different degrees of shade (Figure 8.17). The highest layer is the canopy, perhaps 30–40 m above the ground and formed of the upper branches of mature trees jostling for uninterrupted light. Where sufficient light penetrates below this layer a sub-canopy of younger trees develops and further down, an understorey of large-leaved species and palms. At the forest floor, among the buttress roots supporting the taller trees,

BOX 8.2 The global climate

Consider what the atmosphere does. The main source of heat on the planet is the energy arriving as radiation from the sun. Because the Earth is a sphere, spinning on a tilted axis, with an uneven distribution of sea and land, it warms at different rates in different places. Heat disperses to even out these differences. Conduction though the crustal rocks is slow compared with the convection of water, while the atmospheric circulation (also convection) is fastest (Figures 8.13 and 8.14). Weather is simply the movement of air and water vapour caused by these heat imbalances within the atmosphere, and between it and the Earth's surface.

Heat moves in the oceans as currents: warm water rises as it becomes lighter, and colder, denser waters sink to replace it. This creates persistent circulation patterns between the cold polar and warm tropical regions (Figures 8.14 and 8.15). Heat is transferred to the air mainly through the evaporation of water, producing clouds and eventually precipitation. In the atmosphere, winds blow from high-pressure areas with relatively high temperatures to low-pressure areas. The simple atmospheric circulation between the poles and the equatorial regions is complicated by the forces created

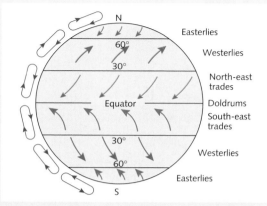

Figure 8.13 A simplified map of atmospheric circulation. Air warmed at the Earth's surface rises and then cools at altitude. Heating is greatest where the sun is directly overhead, somewhere between the tropics throughout the year. Air rises and falls at particular latitudes, generating winds that are then deflected by the spin of the Earth. This creates the characteristic wind patterns of the planet, and combined with seasonal factors and zones of precipitation.

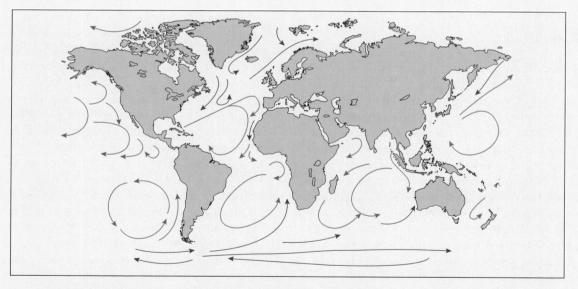

Figure 8.14 Simplified surface currents of the Earth. Warm currents (red arrows) develop either side of the Equator and flow clockwise in the northern hemisphere and counterclockwise in the southern. This movement draws in cold polar currents (blue arrows).

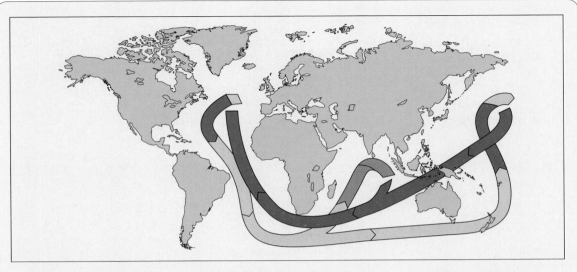

Figure 8.15 The thermohaline is sometimes described as the 'oceanic conveyor belt', a global circulation of deep and surface water currents of different temperatures and salinities. It is a major distributor of heat energy across the planet. Its mixing of water from different levels is important in determining the degree of stratification and the temperature profile of oceanic waters. Cold and saline water moves at depth (blue), replacing the warm and less saline waters toward the ocean surface (red). The heat transfer of the thermohaline is closely associated with the global climate and changes in its circulation mark different phases in an ice age.

by the spin of the Earth, the seasons created by its tilt and the configuration of land and sea. The result is a well-defined, if intricate, pattern of wind and ocean circulation distributing heat around the planet.

These patterns are not fixed over the long term, and oceanic currents in particular will move or cease to flow, changing the heat distribution on the planet. To understand long-term climate change, we need to add in the solar flux (the amount of energy reaching the Earth's surface) and changes in the chemistry of the atmosphere. The solar flux at a particular location varies over thousands of years because of variations in the eccentricity of the Earth's orbit and its relative tilt, as well as variations in the energy output of the sun. This can account for some climatic change in the recent past. However, other factors are important including geological activity and the composition of the atmosphere.

herbaceous plants, particularly ferns, thrive in the semi-dark, dank conditions. Each layer contains seedlings or young trees waiting for sufficient light to join the layer above.

Where the canopy is almost continuous, the interior is relatively open because light levels are too low for much undergrowth. A dense 'jungle' of young trees and climbers can develop only where sunlight penetrates well below the canopy, perhaps where a tree has fallen or the forest forms a boundary with a river. Here, climbers, creepers and 'hangers-on' all use the main trees to gain access to the sunlight. **Epiphytes** attach themselves to the sides of trees and use these for structural support to get higher in the canopy, relying on the nutrients in rainfall and the water draining off their host.

Some tropical forests (especially in South East Asia) are seasonal and have cycles of leaf-fall prompted by a dry season. Most tropical forests are dominated by broad-leaved trees that maintain a canopy throughout

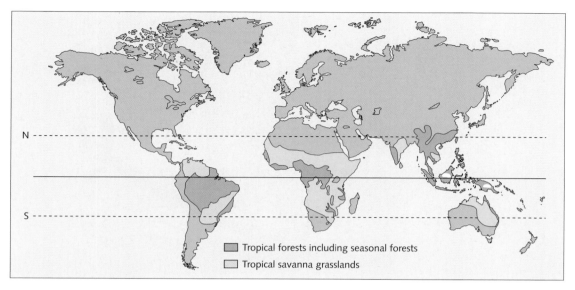

Figure 8.16   The tropical biomes. The dotted lines either side of the Equator (the solid line) represent the tropics of Cancer (northern hemisphere) and Capricorn (southern hemisphere).

Figure 8.17   The stratified structure of tropical rainforests.

the year. The majority of the flowering plants rely on animals for pollination and seed dispersal, using pollen and fruit to attract insects, reptiles, birds, and mammals at various times during the year.

The structural organization of the plant community creates a variety of niches for these animals, which together form distinct assemblages associated with each layer. Many insects and several vertebrates —monkeys, sloths, reptiles, and amphibians—are confined to the canopy or sub-canopy, rarely descending to the ground. A range of carnivores, including birds, feed here as well. Small pools of water held in the leaves of the upper layers or caught in the centre of epiphytic bromeliads support insect and amphibian communities—entire aquatic ecosystems, a few centimetres across, perhaps 30 m above the ground. Equally diverse animal, fungal, and microbial communities wait on the forest floor for material arriving from above.

Rainforest is typical of areas where precipitation exceeds 200 cm per year and where no month has much less than 12 cm of rain. If there is a prolonged dry season, forest gives way to open woodland, thorn scrub, and tropical grassland (**savanna**). The drier conditions favour trees and scrub with small leaves, such as the acacia trees (*Acacia* spp.) of the African plains (Figure 8.18). These protect their leaves with thorns (or aggressive ants—Section 4.2), but even so, browsing is important in limiting the spread of woodland and scrub. The activity of elephants and giraffes, amongst others, are especially important in maintaining an open grassland.

Figure 8.18   The savanna of East Africa.

Grazers are less selective, cropping the above-ground parts of the herbaceous (non-woody) plants. In the absence of grazers, the grasses of the savanna grow high—up to 3 m. Usually tussocked, they die back in the dry season, keeping their living tissues close to the ground, well protected from the grazers who crop closest. Fires are also important here; set by electrical storms, they sweep through the savanna, in quick, short bursts. These, together with the intense grazing, help to maintain an open landscape. Many flowers produce fire-resistant seed that germinate quickly when the rains come, while others sprout from deeply hidden tubers.

Savannas are typical of low, well-eroded flatlands, with depressions that become vital watering holes in the dry season. The soils are again ancient and infertile, with relatively little organic matter so that animal grazing is crucial for nutrient cycling. Their pronounced seasonality means the grass is grazed by successive waves of large mammals, cropping to different levels; in Africa, zebra are followed by wildebeest and thereafter gazelle. In attendance too are the carnivores, the big cats and hyenas and the scavengers who follow them—such as jackals and vultures.

The biomass represented by this hoof and claw testifies to the productivity of this biome. But in terms of biomass consumed, the most important herbivores are insects, primarily grasshoppers, locusts, and ants. Equally important are the inconspicuous termites, obvious only by their sentinel mounds, whose activities are crucial in maintaining soil fertility.

Temperate biomes

The contrast in temperatures between winter and summer has the most profound effect on temperate ecosystems. Cycles of biological activity anticipate and respond to the seasons. This clock times the flowering and fruiting of plants, the feeding, breeding, migration, and hibernation of animals, moving in tempo with each other. This is most obvious in the northern hemisphere which (because of the shape of the continents) has the most extensive temperate forests and grasslands (Figure 8.19).

Again, different biomes reflect the abundance of water during the year. Forests dominate where water is plentiful, grasslands where there is a prolonged dry season. We can distinguish several types of temperate

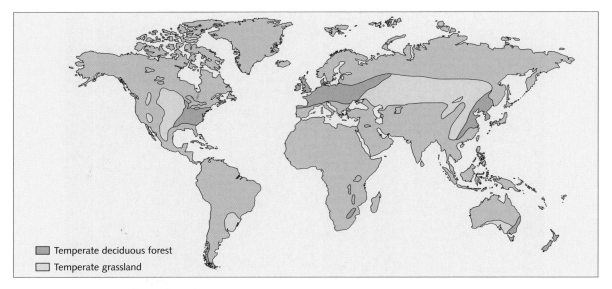

Figure 8.19 The temperate biomes.

forest according to their regimes of temperature and precipitation—from broad-leaved evergreen forest in the warmer latitudes to the conifer-broadleaf mixed forest further from the Equator. Temperate forests have considerably fewer tree species than their tropical counterparts and are typically dominated by just three or four species, principally determined by climate and soil. Often there is a mix of coniferous and broad-leaved trees, with the balance shifting in favour of conifers, especially pine, where soils are nutrient-poor or there is a high risk of forest fires. Many temperate coastal regions, with abundant rainfall or fogs throughout the year, support hemlock (*Tsuga* spp.), fir (*Abies* spp.), and redwood (*Sequoia* spp.), conifers growing in almost continuous stands.

The seasonal contrast in temperatures becomes more marked with distance from the Equator. Where moisture is available all year round broadleaf forest develops (Figure 8.20). Its composition depends upon soil depth and drainage. Beech (*Fagus* spp.) dominates on drier, shallow soils and produces a dense shade with little understorey development. Elsewhere, the forest floor beneath oak (*Quercus* spp.) is comparatively well-lit and undergoes a sequence of herbaceous flowerings each spring before the main canopy develops. There is also a distinct

understorey of saplings and low bushes as well as climbers such as ivy (*Hedera helix*) and honeysuckle (*Lonicera periclymenum*).

At very low temperatures, photosynthesis ceases and respiration is reduced. Average winter temperatures make the balance uneconomic for deciduous trees and they reduce their metabolic costs by shedding leaves. Most conifers can photosynthesize well below 0 °C (Figure 6.7), but have a series of adaptations to reduce water loss from their needles, important when the soil is frozen. Some herbaceous plants over-winter as seeds; others store food in corms or tubers that will fuel growth early in the new season, before the forest canopy reforms.

The annual cycle of leaf-fall makes the soil an important repository of the nutrient wealth of the biome. Much of western agriculture lives off the organic capital remaining in these soils following the clearance of the forest. The soils are deep and rich with a decomposer community stratified down their profile (Box 7.1).

Animals match their life cycles and activity to the times of plenty. Some over-winter as eggs or larvae, some hibernate, reducing their metabolic costs when food is scarce, and others migrate to exploit more abundant resources elsewhere. Once again, insects

Figure 8.20 Temperate broadleaf forest.

are the dominant herbivores and these are, in turn, a resource used by various migratory birds. In the northern hemisphere, mammals include squirrels, wild pigs, and badgers, with deer as the principal browsers. Before humans became the major influence on this biome, wolves, bears, and several species of cat were the larger carnivores.

Where a dry season lasts for most of the summer, the forest gives way to **temperate grassland**. In the centres of large continental masses, away from moist coastal air streams, annual precipitation falls below 60 cm at these latitudes. Australia, South America (pampas), South Africa (veld), Eurasia (steppes), and North America (prairies—Figure 8.21) all have such zones. Once again, the grasslands are typically flat, gently rolling landscapes which, in the northern hemisphere, are associated with the wind-blown sediments from recent glacial activity. In the south, they are old and well-eroded plateaux.

Growth and flowering are distinctly seasonal, with grasses bearing their seed towards the end of the dry summer. Like the savanna, fire and grazing maintain the dominance of the grasses and the legumes, but thorn scrub and trees will appear where grazing pres-

sure is lifted. In the prairies of North America bison once maintained the grassland and the diversity of its flowering plants. In Europe and Asia, various species of horse and antelope were the principal vertebrate grazers, but insects, above and below ground, are major herbivores in all temperate grasslands. Each region also has a range of burrowing animals, both vertebrate and invertebrate. Soils are deep, often with a thatch of undecomposed vegetation overlying a thick humus layer. This is an important protection against wind erosion when the soil is dry. Light grazing helps to speed decomposition processes and adds moisture. Today, most grasslands around the world have few wild grazers because they are intensively farmed for meat and cereals (Section 6.5).

Where temperatures range higher, and rainfall is confined to the winter months, mediterranean-type communities can develop. These have a very restricted distribution (Section 5.1). The largest region, the Mediterranean Basin itself, has lost much of the evergreen woodland that existed before humans arrived and today represents a highly modified landscape. Where rainfall is very low, true desert will form (Figures 8.22 and 8.23).

Figure 8.21   Prairie.

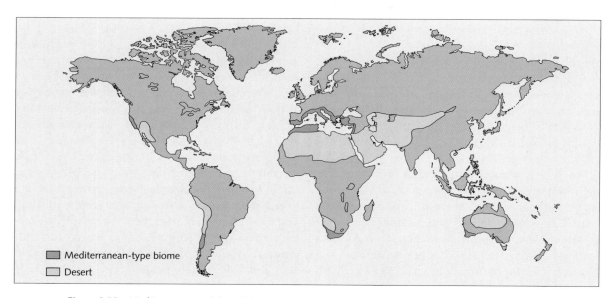

☐ Mediterranean-type biome
☐ Desert

Figure 8.22   Mediterranean and desert biomes.

Boreal biomes

Towards the poles, average temperatures decline further and so does available water. Here, plants are adapted to long periods when little photosynthesis or nutrient uptake is possible.

Again, we can identify a forested region, the northern coniferous forest or taiga and a low-lying plant community where water is frozen for much of the year, the tundra. These communities circle the northern hemisphere where the land mass is almost continuous and together comprise the boreal (northern) biomes (Figure 8.24). Equivalent communities also occur at lower latitudes but at high altitudes, in montane areas, where similarly severe climates exist (Figures 8.12 and 8.27).

Figure 8.23 Desert.

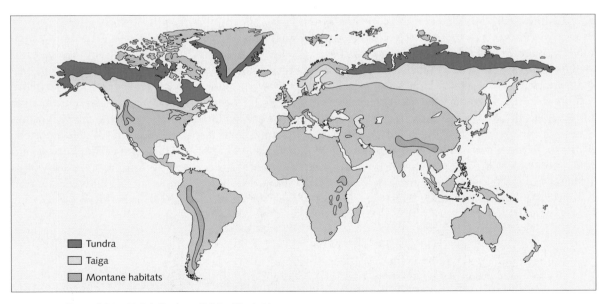

Tundra
Taiga
Montane habitats

Figure 8.24 High latitude and high altitude biomes.

Where the temperate forest gives way to unbroken conifer forests, the taiga starts (Figure 8.25). Vast tracts of this dense forest are dominated by just two or three species—primarily pines, firs, spruces (*Picea* spp.), or hemlock—beneath which is a herbaceous layer of mosses, lichens, and grasses. In some areas deciduous trees are found, mainly birch (*Betula* spp.) and poplars (*Populus* spp.). The fine leaves (needles)

Figure 8.25   Taiga.

of conifers reduce water loss and are efficient for photosynthesis at low temperatures. By retaining their leaves they make full use of the early spring sunlight, but they also have to support or shed heavy loads of winter snow. Accordingly, these trees are shaped with a conical growth habit.

In more northerly areas, the soils are permanently frozen at depth. All are thin, infertile, and covered with a thick layer of partially rotted needles. The nature of this litter and its slow decomposition under cold, semi-waterlogged conditions makes the soil acidic, so favouring fungi as the prime decomposers. A build-up of litter can promote forest fires, import-ant to taiga ecology: fires create gaps which, once again, create opportunities for colonizing species. Some species (such as the jack pine—*Pinus banksiana*) take advantage of the gaps generated by these disturb-ances and shed their seeds only after a fire.

Relying solely on wind pollination, conifers have no direct need for insects and defend themselves against insect attack, using resins and their character-istically pungent terpenes (Section 4.4). Nevertheless, a variety of moths, sawflies, and beetles burrow into the needles, buds, and bark. Seasonal migratory birds feed on these insects, whilst resident birds and squirrels take the conifer seeds. Other herbivores, including

Figure 8.26   Tundra.

elk, caribou, and deer, feed on the low-growing and seasonal vegetation, with dogs (foxes and wolves), mustelids (martens, weasels, wolverines), birds (owls, eagles), bears, and cats as the larger carnivores. Some remain active even when there is deep snow cover but many mammals hibernate for the winter.

In the summer, a number of these herbivores and carnivores move north to feed in the **tundra** (Figure 8.26). This is the zone beyond the taiga, at the edge of the arctic ice sheet. It was once heavily glaciated itself but is now uncovered for two or three months each year. Then the summer temperatures may be high enough to melt the upper layers of the soil, down to perhaps a metre or so. At depth, the soil never thaws and is called **permafrost**. Decomposition under these conditions is slow, and organic matter accumulates as a dark, soggy, and compressed mass known as peat. Similar conditions can exist at lower latitudes but high altitudes (Figure 8.27).

When the thaw comes, water, and the nutrients dissolved in it, becomes available. Plants bloom and grow—grasses, sedges, and low shrubs, such as heathers (*Calluna* spp.), dwarf willows (*Salix* spp.), and birches—and produce seed in quick succession. Many self-pollinate and spread by vegetative growth. Others flower to exploit the short burst of insect activity. Flies (especially mosquitoes) and butterflies that have spent the winter as larvae or pupae emerge to feed and mate, and birds arrive to exploit this pulse of plenty. Elk, reindeer, and other herbivores come to graze exposed mosses and lichens. Some predators, from wolves to hawks, follow and feed especially on rodents, the lemmings and voles.

Across much of North America and Russia, the tundra presents a vast wind-swept and waterlogged

Figure 8.27 Alpine ecosystems at different altitudes in the temperate latitudes.

landscape. Its flat monotony results from its repeated glacial scouring and deposition, as ice sheets have advanced and retreated with the cooling and warming of the climate over the last 2.5 million years.

8.3 Climate change

The position of these biomes is far from fixed. With each glaciation the tundra moved south, in front of advancing ice sheets. During the warmer interglacial periods, tropical conditions extended well into regions that today we regard as temperate. Amongst the sediments deposited over the last 2 million years

we find pollen, mollusc shells, and beetle wing-cases that indicate communities and climates very different from today.

The ice ages are still fresh in the memory of the planet. We still find frozen, well-preserved carcasses of mammoths in Siberia and the bones of rhinos in

Britain. Over the last million years there have been four major glacial advances and retreats, recorded as the remains of cold- and warm-loving species replace each other in the deposits. When the ice retreated last time, the ice age ended abruptly. Temperatures rose extremely rapidly—averages in Europe climbed by about 7 °C in just 50 years and the Earth shifted from ice age to interglacial in the space of a human lifetime (Figure 8.28).

Most likely, this followed an increase in the energy received in the polar regions, perhaps as the position of the Earth relative to the sun changed. Our planet undergoes orbital variations, cycles that vary its distance and angle to the sun, at intervals of tens of thousands of years. These change the amount of radiation hitting different latitudes which, by their effect on circulation patterns, can lead to major change in the global climate.

However, these cycles do not explain the more short-term changes, such as the minor shifts in average air temperatures since the last ice age (including a 'mini-ice age' in the seventeenth century). The last 10 000 years have been relatively mild and today we enjoy a climate substantially warmer than the average for the last 2 million years. Even so, the difference between the ice age and the current interglacial is just 6 °C

Now things seem to be changing again. Since industrialization started 200 years ago, humanity has been slowly altering the chemistry of the atmosphere. Climatologists are trying to model the consequences for the global climate and ecologists are asked to

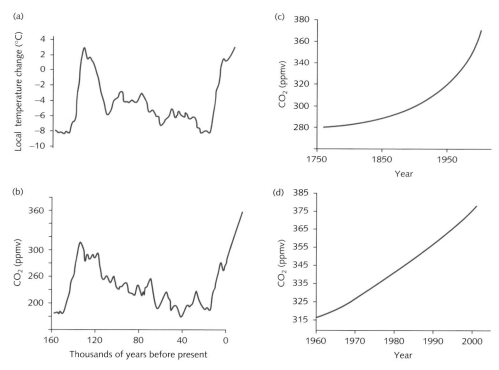

Figure 8.28 Changes in temperature (a) and atmospheric carbon dioxide levels (b) in the last 160 000 years. Atmospheric methane levels show a similar pattern, indicating the rise in carbon-rich gases was due, in part, to increased biological activity as the global climate warmed. A rapid rise in temperatures marks the end of the last glaciation around 10 000 years ago. Carbon dioxide levels have risen consistently since the onset of industrialization in the eighteenth century (c), most rapidly with the increasing fossil fuel consumption and forest clearance in the last 40 years (d). (Note the change of scale in b, c, and d.)

predict the implications for the planet's hierarchy of ecological processes.

Atmospheric composition and mean temperature

Orbital variations may be responsible for the oscillations in the ice advances (though see Chapter 9), but the amplitude of these temperature changes, the scale of their fluctuations, have been magnified by differences in the composition of the atmosphere (Figure 8.28). Samples of ancient atmospheres trapped in the air bubbles of ice cores dating back 160 000 years show that carbon dioxide levels have matched the changes in mean local temperature. Indeed, the same is true of another carbon-rich gas, methane, and since methane is generated by micro-organisms, this

suggests that the level of biological activity has followed these temperature changes.

We know that the composition of the Earth's atmosphere is mediated by biological activity but the problem is to decide the extent to which the current warming is cause or effect. Have carbon dioxide concentrations risen following increased biological activity, causing higher atmospheric temperatures? Or have warmer temperatures prompted more biological activity and greater carbon dioxide release? In fact, it is a combination of both. Along with water vapour, gaseous carbon compounds induce heating of the atmosphere via the greenhouse effect (Box 8.3). The fluctuations in temperature are 'amplified' because of the positive effect temperature has on biological activity. If nothing else changes, a warm climate increases the release of carbon by accelerating

---

BOX 8.3   **The greenhouse effect**

The greenhouse effect is a natural and vital feature of our atmosphere. Without it, the mean atmospheric temperature of the Earth would be −17 °C, rather than its current average of 15 °C. The presence of water vapour, nitrogen, oxygen, and the carbon-rich gases (primarily carbon dioxide) all absorb the heat reflected off the Earth's surface. While the comparison with a greenhouse is inaccurate (there is no physical barrier, like the glass, to heat transfer), the name is at least evocative of the effect of heat capture.

Figure 8.29 shows how this works. Energy arrives from the sun primarily as short-wave radiation warming any surface it strikes. The heat reradiated from these surfaces is in the infrared and these wavelengths are absorbed by different atmospheric constituents (Table 8.2). Besides water vapour, the principal greenhouse gases are CO_2, CH_4, N_2O, and CFCs. The contribution that each makes to global warming depends upon several factors—their concentration, the length of time they remain in the atmosphere, and the wavelength at which they absorb reradiated energy. CFCs are particularly potent because they absorb in part of the spectrum where the atmosphere is ordinarily transparent (Table 8.2).

Any hot body loses its energy through thermal radiation and the hotter the body, the more infrared radiation it emits.

Around 240 W/m² of solar radiation hits the Earth's surface, which must be lost, via the atmosphere, if the atmospheric energy budget is to remain in balance. Air temperatures are high because the reradiated heat is retained by the atmosphere, captured on its return journey by gases not transparent to their infrared wavelengths. Eventually, this energy is lost at high altitude (around 5–10 km), but here the gases are cold and lose their energy slowly. The overall effect of the greenhouse gases is thus to slow the passage of energy from the Earth's surface back into space and in the process raise the mean atmospheric temperature.

A planetary balance between energy received and energy lost can be achieved at any atmospheric temperature, when heat lost matches heat added at the Earth's surface. A higher concentration of greenhouse gases means this balance is reached with more heat held in the atmosphere. Carbon dioxide added to the atmosphere will remain there for about 100 years, so unless we remove carbon at a faster rate than we add it, concentrations in the atmosphere will continue to rise. The excess we add today will continue to have an effect at the beginning of the next century, and stabilization is at least 100 years away.

(continued overleaf)

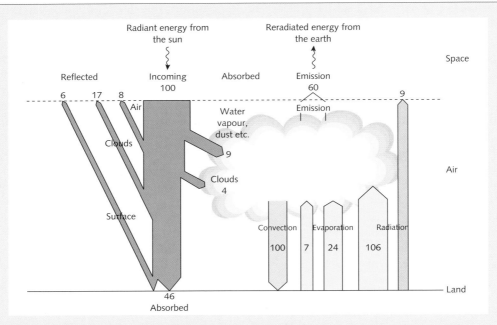

Figure 8.29   An energy budget for the atmosphere and the greenhouse effect using arbitrary units. Solar radiation enters the upper atmosphere, where much of its shorter wavelength ultraviolet frequencies are reflected or absorbed. Longer wavelengths are absorbed as they pass the atmosphere, by water vapour and carbon dioxide. The radiation reaching the ground warms its surfaces and as their temperature rises they emit thermal radiation. The reradiated infrared wavelengths are also absorbed by water vapour and carbon-rich gases in the air. This is the main mechanism by which the temperature of the atmosphere is raised. Increasing the carbon and water content of the atmosphere increases this absorption, and its temperature rises as energy is lost more slowly than it is received.

TABLE 8.2   The principal greenhouse gases

	CO_2	CH_4	N_2O	CFC_{12}
Concentration (ppmv)				
Pre-industrial	280	0.8	0.29	0
Now	378	1.76	0.31	0.00048
Rate of increase (% per year)	0.5	0.9	0.25	4
Lifetime (years)	5–200	12	114	130
Greenhouse effect per molecule over 100 years relative to CO_2	1.0	23	270	7100
Proportional contribution over 100 years*	72	18	4	(Highly variable)
Reduction in emissions needed to stabilize (per cent)	>60	15–20	70–80	75–85

CFCs have a very powerful greenhouse effect but their contribution depends (amongst other things) on their latitude—at higher latitudes they have a reducing effect due to their role in ozone reduction (which also has a greenhouse effect).
* These estimates also allow for the indirect effects of different gases on CO_2 production.

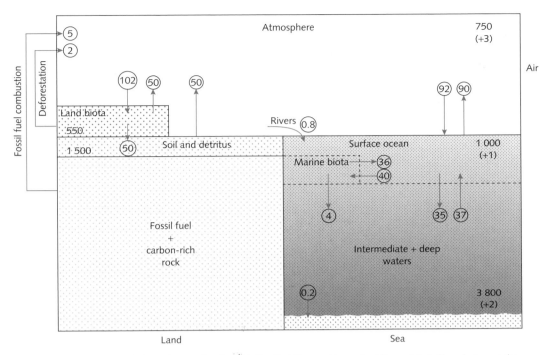

Figure 8.30 The global circulation of carbon. Notice that the most important fluxes (circled) are between the land and air and the sea and air, and each are largely in balance. What has changed in the last two decades is the amount moving through the atmosphere and the reduction of the principal sinks on land, the forests. Fluxes are in 10^{12} kg per year and reservoirs in 10^{12} kg (gigatonnes).

respiration and decomposition and this, in turn, promotes further atmospheric warming.

Carbon-rich gases also originate from non-biological but natural sources, such as volcanic activity. Carbon is removed from the atmosphere through non-biological absorption by the oceans. However, the carbon content of the atmosphere is primarily regulated by the biosphere—fixed by photosynthesis and released by respiration and decomposition (Section 6.1). Carbon fixed in living tissues need not be released again directly, but may instead become incorporated into sediments (Figure 8.30). One important mediator of this process is the marine phytoplankton, whose photosynthesis and productivity rise with temperature. These remove significant amounts of carbon rapidly from the system if they settle out in the sediments. The same happens with the death of marine invertebrates that combine carbon dioxide with calcium to build their shells and exoskeletons.

The scale of this sedimentation can be seen in the vast formations of limestone rocks over the planet.

More significant perhaps, at least as far as our economic activity is concerned, are the coal and oil deposits accumulated from millions of years of photosynthesis, primarily in terrestrial biomes and especially from the compressed forests of the Carboniferous period.

Anthropogenic sources of atmospheric carbon

In the last 200 years, we have been releasing this fossilized carbon at an accelerating rate. On an annual basis, our inputs are relatively small (Box 8.4), but we have been burning coal and oil for a long time now. Overall, anthropogenic (human-derived) sources have led to a 30 per cent increase in carbon dioxide over background levels since industrialization began.

Burning forests (Figure 8.31) not only releases carbon oxides directly to the air, it also prompts an increase in decomposition in the exposed soil. In the

Figure 8.31 Tropical forest fires in the Yucatan peninsula in Mexico, photographed on 20 April 2003 by the Aqua satellite. The red dots mark individual fires. In the past, fires from this region have caused poor air quality in areas of Texas and Oklahoma. The extent of these fires is though to have been due to the dry conditions associated with the intense El Niño event the previous winter.

1980s, about one-fifth to one-fourth of the annual rise was attributable to the effects of forest clearance and changes in land use. Replanting the forest is a rapid way of fixing or sequestering carbon, but replacing it with agriculture adds further carbon to the atmosphere; 50 per cent of the carbon held in the organic matter of a soil is released under intensive cultivation. In addition, methane concentrations have increased by 140 per cent, primarily from rice paddies and from the various emissions associated with cattle rearing. Developed nations also burn colossal amounts of fuel to sustain their agricultural production, from the manufacture of fertilizers and pesticides to the machinery used in planting and harvesting (Figures 6.28, 6.29). Additional energy is then used to make the food more convenient—in food processing, packaging, transportation, and refrigeration.

We are also responsible for a range of new gases, some with very powerful 'greenhouse' properties. **Chlorofluorocarbons** (CFCs) and their relatives were unknown before 1930 but have since been were used extensively as coolants in refrigerators, as aerosol propellants, and in a variety of industrial processes. CFCs absorb infrared radiation in the one part of the spectrum where other greenhouse gases do not. They are also long-lived (Table 8.2) and contribute significantly to ozone depletion, but under an inter-national agreement, their large-scale use has now been successfully phased out (Box 8.5).

Some of our emissions help to ameliorate the greenhouse effect. Sulfate aerosols, themselves a product of fossil fuel combustion, can reduce short-term warming. They increase the reflectivity of the atmosphere, both by their presence and by inducing cloud formation. Sulfate aerosols readily wash out of the atmosphere, so their cooling effect tends to be highly localized, close to the source of emissions. The sulfate pollution in industrial areas may be the main reason why the northern hemisphere has warmed less rapidly in the last 200 years than the southern hemisphere. Persistent sulfate emissions from industrial areas are a major cause of acid rain (Box 7.3).

The role of ecological and biogeochemical processes in regulating the composition of the atmosphere/ocean system is still poorly understood, mainly because we lack information on the role of microbial and other communities, especially in deep-ocean cycling. But we can learn much from unexpected perturbations to the system. For example, the millions of tons of ash released by the eruption of Mount Pinatubo in the Philippines in June 1991 led to cooler weather, globally, the following year (Box 9.2). Although global mean temperatures were lowered by 0.25 °C, this short-term variation only interrupted the long-term trend of global warming.

 BOX 8.4   **The carbon balance and the missing sink**

Without anthropogenic inputs, carbon added to the atmosphere is balanced by that removed. We add an additional 7 gigatonnes (10^9 tonnes or 10^{12} kg) each year, yet the atmosphere is accumulating only 3.4 gigatonnes and the oceans an additional 2 gigatonnes annually, so where does the missing 1.6 gigatonnes go?

One possibility is that our estimates for the oceans have been too low, but since there is little evidence of major changes in phytoplankton productivity with the rise in atmospheric carbon over the last century this is, perhaps, unlikely. Most studies suggest that the global forests probably account for the deficit (Figure 8.32). About 31 per cent of the Earth's land area is under forest and that retains 86 per cent of total above-ground carbon. Around 73 per cent of global soil carbon is found in forest soils.

By virtue of their size and the rapid growth of trees at higher temperatures, the tropical forests seem to be the most likely long-term sink for the missing carbon, yet this biome is suffering the most extensive deforestation. Many argue that their scale of deforestation probably means the tropical latitudes are most likely to be net contributors to the carbon budget, though others suggest their effect is broadly neutral.

(continued overleaf)

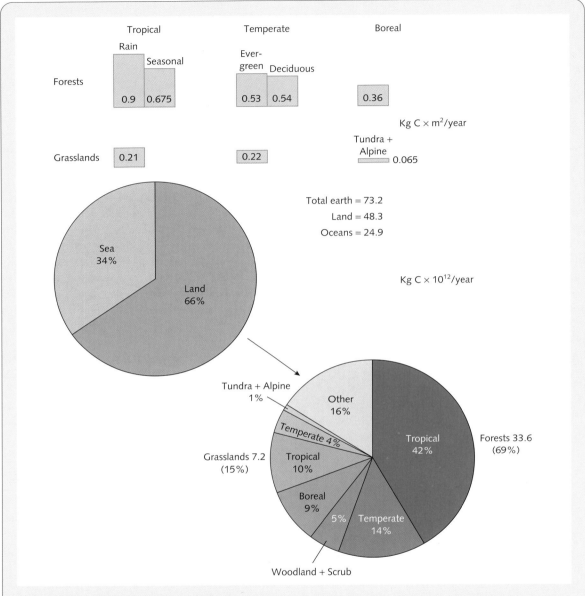

Figure 8.32 The productivity of the planet measured as the carbon fixed by different parts of the biosphere: (a) shows the differences in mean net primary productivity per square metre between forests and grasslands and between the three major latitudinal biomes. (b) Estimates of the contribution of land and oceans have been significantly revised with the advent of satellite data, and these suggest roughly equal amounts of carbon are fixed by each. Less carbon is fixed per square metre at sea (an average of 140 g compared to 426 g on land) because much of the photosynthetically active radiation is absorbed by water and its solutes, whereas leaves absorb about 31 per cent of the radiation falling on land where there is no permanent ice cover. However, turnover is much faster on the oceans (average 2–6 days, compared to 19 years on land). (c) The contribution of the different biomes to the annual net primary productivity on land.

In contrast, the temperate forests, especially those in the north, have been expanding. Commercial forestry and the abandonment of farmland have led to a spread in both the deciduous and boreal forests in the north, especially toward the end of the last century. In addition, the production of the northern forests may have been heightened because of the fertilizing effect of nitrogen pollution from regional industrial activity, as well as the widespread use of agricultural fertilizers. Sulfate deposition and the overall warming of the atmosphere would also have promoted the growth of established woodlands.

Roger Sedjo calculates that the increase in biomass in the northern forests, representing 0.7 gigatonnes each year of carbon, could account for a sizeable proportion of the missing carbon, and thus represent the single most important sink. Pointing to evidence that nitrogen may not always limit carbon sequestration, other ecologists suggest that a more integrated answer is needed, one which allows for soil processes as well as the photosynthesis and respiration of plants. These note that carbon and nitrogen ratios vary in both the growing plant and in the soil with heightened atmospheric carbon dioxide. Recent satellite-based studies suggest that global carbon fixation in terrestrial ecosystems increased by around 6 per cent in the final two decades of the last century, principally associated with tropical and northern temperate forests. These estimates amount to less than 0.2 gigatonnes per year, so we still fail to account for all the missing carbon.

This increase in net primary production in tropical latitudes is attributed almost entirely to an increase in solar radiation in these regions, as their cloud cover reduced over this period. This suggests significant changes in tropical rainfall patterns. Overall, it seems that there may not be one single major sink, but instead a more even distribution of carbon sequestration between temperate and tropical zones, and between forest and grassland.

BOX 8.5　Ozone depletion

Changes in the chemical composition of the atmosphere can all be related to the radiation balance of the atmosphere—what is being let in and what is not being allowed to escape. The effect of one of these constituents, ozone (O_3) depends on its concentrations at different altitudes.

In the lowest layer of the atmosphere, the troposphere, ozone levels have been increasingly locally, as pockets of pollution around cities, generated through photochemical reactions involving car exhaust gases. High local concentrations may be responsible for forest dieback around major industrial regions. However, low-level ozone is also a major source of hydroxyl radicals (OH^-), important for reacting with methane, carbon monoxide, sulfur, and nitrogen oxides, and helping to reduce their impact on the atmospheric energy budget.

In the next atmospheric layer, the stratosphere, ozone has been depleted. Here, the gas absorbs harmful incoming ultraviolet radiation (UV-B), protecting living systems from energy at wavelengths that are strongly absorbed by a range of organic chemicals including DNA. Increased UV-B has been blamed for the rapid rise in skin cancers amongst the peoples of the higher latitudes. It also depresses plant productivity and therefore reduces the fixation of atmospheric carbon.

The depletion of stratospheric ozone is not uniform over the globe. It is most pronounced over the polar regions in the spring when low temperatures help produce a high-altitude luminescent cloud within discrete circulation patterns. The ozone 'holes' recorded over Antarctica and the Arctic since 1977 follow from chemical reactions with sunlight and a number of gases, most especially CFCs. Ultraviolet radiation slowly breaks down these compounds to release chlorine monoxide, which goes on to react with ozone to produce molecular oxygen.

In addition, CFCs themselves account for perhaps 15 per cent of the greenhouse effect. Because of their longevity and impact on UV levels, action to phase them out has been relatively swift. According to the Montreal Protocol agreed in 1987 and revised in 1992, large-scale use of all CFCs and related gases has largely ended in most industrialized nations. The success of the protocol demonstrates that international agreements can work: forecasts made in 2002 suggested that the stratospheric ozone layer could be restored within 100 years but more recent measurements suggest recovery could come much sooner.

Predicting climate change

Of all the greenhouse gases, carbon dioxide is responsible for about half the greenhouse effect. The Intergovernmental Panel on Climate Change (IPCC), commissioned by the United Nations and the World Meteorological Organization, concluded in 2001 that carbon dioxide emissions would continue to rise for 100 years. Even if we reduced human inputs immediately, temperatures will continue to rise because heat transfer—to the seas and to the ice sheets—takes time. The high temperatures mean the biological machine will also run on. For example, when carbon dioxide concentrations reach 450 ppm, mean atmospheric temperatures will take several centuries to stabilize. We can expect to reach this level by 2030 and 500 ppm by 2 100: the longer we delay in reducing our emissions the longer it will take for stabilization and the higher the eventual temperature of the atmosphere at equilibrium. Sea levels will continue to rise for several centuries.

These predictions are based on the outputs from a number of **global circulation models (GCMs)**. These are immensely complex mathematical models of the air/water system of the planet, originally derived from the weather forecasting simulations used on more local scales. There are several such models, but all require very powerful computers running for long periods. The current UK model (the Hadley Centre CM3) requires four months to run on one of the fastest supercomputers. Even so, these models are not complete descriptors of the system, primarily because they do not include the ecological interactions mediated by the biotic components of the planet. Nevertheless, the first models used by the IPCC have been considerably refined and now take greater account of the effect of cloud cover, deep ocean currents and sulphate aerosols. The IPCC uses the predictions from a series of models and, allowing for a variety of scenarios, sets upper and lower limits based on their collective efforts. The consistency between models indicates a large measure of agreement and the trend has always been for a substantial rise—at least 1.5 °C by 2060 (Figure 8.32a).

Predictions from the models differ, primarily because of the different weights each attaches to the critical factors, especially deep ocean currents and cloud cover. Such differences and the associated uncertainty have been seized upon by their critics (Box 8.7). One way to test the predictive power of these models is to 'hindcast' past temperature movements, starting from a known set of conditions at some date in the past. The closer a GCM can match past observations from this date, the more confidence we can have that it captures the essential features of the system. Models are becoming more effective at this, so the latest output from the Hadley CM3 model catches both the scale of temperature change and the major fluctuations. It also demonstrates that the observed increase in the last 50 years is largely anthropogenic, a result of human activity (Figure 8.32a). Two recent attempts to measure the 'biological amplification' of carbon dioxide release with rising temperatures, using different data from recent and more long-term records, concur that physical-only models probably underestimate the scale of future warming, possibly by as much 78 per cent.

Complex computer models, particularly those associated with weather-forecasting, may not enjoy widespread public confidence. Nevertheless, the vast majority of scientists and the major national science academies now accept the reality of human-induced climate change. A small number continue to question the efficacy of the models or point to the natural climatic variability on both regional and global scales. But there are more obvious indicators the public themselves can use—the natural thermometers of long term trends the Earth conveniently provides across the globe.

Glaciers are retreating up their valleys, like mercury up a thermometer. Because large masses of ice are melting and because this is happening in all regions of the world, the indication of a persistent global trend is clear (Figures 8.34 and 8.35) and indisputable. The perennial sea ice in the Artic is now declining by almost 10 per cent per decade and the major ice shelves of the Antarctic are breaking up at unprecedented rates.

There are, of course variations and it is not always easy to explain local differences. A small number of glaciers are growing and, in some parts of the Antarctic, snow fall has increased. Taking the

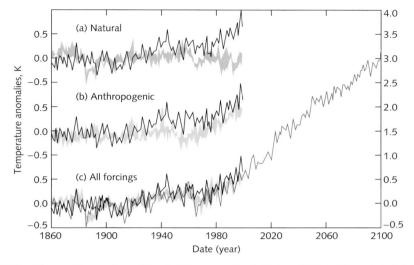

Figure 8.33   The mean temperature of the air 1.5 m above the Earth since 1860 and its projected rise over the next 100 years. This figure, derived from a Hadley Centre GCM uses a variety of starting positions and includes the effect of sulfate aerosols. These scenarios model the effect of (a) natural, (b) anthropogenic factors, and (c) their combined effect. The black lines shows the observed changes since 1860 in each case and the coloured area the range of predictions from different starting positions (each of which is based on an IPCC scenario that predicts mid-range changes in greenhouse gases, set to 0 for 1860). The match between the predicted and observed data is generally close. Using the left-hand scale, models of natural processes show little real effect, but human-induced warming (b) begins to be detected from 1940. Overall, and primarily due to human activity, global air temperatures are predicted to rise by 2010 (shown by the red line in c), a rate close to that observed over the last three decades.

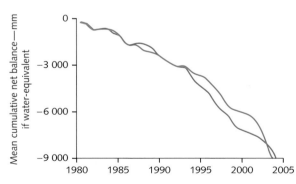

Figure 8.34   The average mass balance data for 30 glaciers (green line) and for nine regions (red line) in 2004, the latest data published by the World Glacier Monitoring Service. Glaciers are melting all over the world, demonstrated even over this relatively short sampling period since 1980.

temperature of the planet is not a simple process, even with sophisticated equipment. Studies on the thawing of the permafrost in Alaska show how even relatively simple systems are complicated by local factors. Matching soil temperatures with overhead air temperatures is not without its complications: Tom Osterkamp has reviewed the 30-year records for a transect running north–south through the permafrost of Alaska (Figure 8.36). The overall pattern is one of warming, which, at the regional scale, is obvious by the northward movement of the line demarcating the permanently frozen tundra. Each year, Alaska has a larger area of its upper soil thawed for some part of the year.

After peaking early in the 1980s, permafrost temperatures fell back again. This corresponded with a cooling of the air and reduced snow cover to insulate the permafrost. Warming resumed by 1986 though this is only detectable during the winter months. However, there are local differences—two sites have

Figure 8.35 The melting foot of the Grindelwald glacier in Switzerland.

(a)

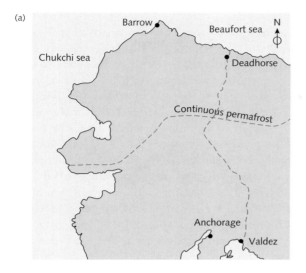

(b)

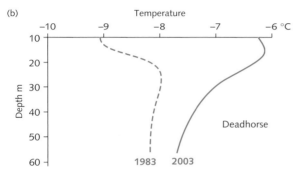

Figure 8.36 (a) Osterkamp's transect to record the temperature of permafrost across Alaska. The line of continuous permafrost has retreated since records were kept and although stations on the transect differ, the more northerly stations have experienced considerable warming in the last 20 years in their upper layers. (b) The change in the temperature profile of the Deadhorse monitoring station between 1983 and 2003. The gradual warming of this most northerly site is evident in the plots over several years.

cooled continuously and there were different rates of warming in different places. Surface soils have warmed most in the extreme north, on the Alaskan coastal plain. Nevertheless, the overall trend is clear and local variations do not disguise the important signal from this data—since the late 1970s annual mean temperatures have risen by 1–2 °C in the air and by about 2 °C at soil depths of 20 m. Over the region, this scale of change in such a large mass of

permafrost could only follow a persistent change, indicative of a massive increase in the energy balance of the atmosphere affecting Alaska.

Of course, that means a major increase in the energy balance of the global atmosphere. Like the glaciers, changes in the amount of permafrost support the predictions of the computer models. Here, then, is the weather forecast for the rest of this century: warming will be uneven over the globe but overall there will be more frequent heatwaves, intense storms, droughts, and floods. Temperatures will rise most rapidly in the northern hemisphere, perhaps more so as sulfate emissions are controlled in this region. Northern winters will be milder and the middle of these continents will have hotter and drier summers. The land will warm rather more than the oceans and the temperature range between day and night will become smaller. Continental margins will be wetter with higher precipitation at mid and high latitudes, though much less rainfall is forecast for the sub-tropical zones.

The Mediterranean Basin is likely to experience longer, hotter summers, promoting more arid conditions throughout the year. Similarly, much of the western edge of North America will have higher temperatures and a reduction in summer rainfall. Particular models of the Sacramento Basin suggest that runoff in the summer will decline by up to 50 per cent should average temperatures rise by 4 °C. In the winter, there may be too much water—the Sierra Nevada mountains are predicted to receive more winter precipitation and they will also experience more days with extreme rain and snow falls. The consequence could be more winter floods in coastal California. Critically, Amazonia is likely to become much drier and, if this leads to widespread forest fires, more rapid change in the climate of the planet could be triggered.

Sea levels will rise, primarily because of the thermal expansion of the water (2–4 cm per decade) and only partially because of melting ice (1.5 cm per decade). A number of countries face inundation during storms, including many low-lying Pacific islands, and large areas of Bangladesh and Northern Europe. By the end of 2006, at least one large island in the Sundarbans, the delta upon which much of

Bangladesh stands, had been abandoned to higher sea levels, and with it the livelihoods of tens of thousands of people.

More sea lanes will become available throughout the year in the Arctic. Sustained warming over the long term, with rapid melting of the polar ice sheets will almost certainly slow the thermohaline circulation (Figure 8.15). This 'oceanic conveyor belt' drives massive amounts of water and heat around the globe. Where it fails the oceans will become increasingly stratified, with little mixing between depths for long periods—a change known to have occurred with previous climatic oscillations. The significant melting of the Greenland ice sheet will introduce additional cold freshwater into the North Atlantic, which could block the thermohaline here. Ironically, without this source of heat, northwestern Europe is likely to cool.

Ecological changes likely with atmospheric warming

If some part of the thermohaline circulation ceases, the carbon cycle and the biogeochemistry of the planet will undergo profound changes. Then the living biosphere and its interactions with the water/atmosphere system become important. Because this system is so complex we should expect surprises, responses we do not anticipate, and rapid changes as thresholds are crossed.

With increased oceanic stratification, nutrients released from the stirring of marine sediments will decline, and this in turn will limit primary productivity in the surface waters. The most productive fisheries occur where deep upwellings bring nutrients to the surface and with fewer nutrients to support primary production, major fisheries will quickly collapse. Additionally, the higher atmospheric levels will cause more carbon dioxide to dissolve in the sea, increasing its acidity. This reduces the concentration of calcium available to many shelled animals and some types of phytoplankton that fix carbon in their shells or skeletons. It may also shift the trophic relations of some important plankton groups. One consequence of global warming could thus be dramatic changes to the communities in the upper waters of some oceans, with declines in both phytoplankton and zooplankton

biomass, and a reduction in this major carbon sink (Figure 8.30). Over time the larger animal community of these waters would also be reduced.

The distribution of terrestrial biomes will also shift as temperature and moisture regimes change. Current estimates suggest that the northern tree line will move 100 km northward for every 1 °C rise in mean temperature. The tundra will be warmer for longer and experience longer periods of biological activity. Some predictions see the tundra and taiga shrinking by about one-third as carbon dioxide levels double and temperate forests extend their range. Several of the Alaskan sites measured by Osterkamp have surface soil temperatures raised by 4°C, and the active layer of their permafrost extends deeper and remains thawed for longer, in turn allowing biological processes to operate faster for longer. We should expect biological amplification as the thawing of the permafrost continues, with increasing rates of decomposition opening up massive reserves of carbon for release as carbon dioxide or methane.

Some more southerly plant communities, most probably the grasses, will encroach polewards. Deciduous hardwoods and grasslands will follow. Temperate mixed forests are expected to lose their conifers and become dominated by broadleaf trees. Parts of Canada (Northern Alberta, Saskatchewan, and Manitoba) may lose their coniferous forests altogether. The extension of the monsoon rains towards the poles will increase the moist tropical forest cover in some areas, such as northern Australia. A higher frequency of natural forest fires in a drier Amazon basin will not only release significant amounts of carbon into the air, it will also reduce the pumping of water by transpiration, the largest single source on the planet. The ecology of the Amazon itself, the largest river on the planet, will, in turn, undergo dramatic change.

Patterns of cultivation will also shift. Growing seasons will be extended in northern latitudes, so that leaf and root vegetables could be grown in Central Alaska. Crop yields will increase in northern Europe (up one-third in Denmark) but decline where water becomes limiting (down one-third in Greece). More significantly, the drying of the prairies and the steppes in the southern part of their range will reduce cereal production. The overall impact will be increased food shortages in Russia and neighbouring countries. The Sahara will extend into the Sahel and increase its range in central Africa. With persistently drier conditions, changes in the ecology of Mediterranean catchments will become obvious much sooner. The rich flora of the Mediterranean Basin will suffer major species loss as the drier topsoil becomes more easily eroded. Agricultural productivity in these regions is also likely to fall.

Perhaps these changes are already detectable at the lowest levels in the hierarchy. Oscar Gordo and Juan José Sanz have used annual maximum temperature and rainfall data collected since 1943 at an observatory near Tortosa to review changes on the surrounding ecosystems, close to the Mediterranean coast of central Spain. This shows a consistent rise in temperatures since the 1970s and seems to correspond with an advancement of stages in the life cycle of most plants, especially leaf unfolding and flowering. However, later processes, such as fruit ripening or leaf fall did not change, despite the extended summers. Warmer spring temperatures were key—insect pollinators now become active before many of the early spring flowers have opened. This, along with changes in insect herbivore activity led Gordo and Sanz to suggest that the plant–insect interactions of these ecosystems are becoming 'decoupled'. These lower-level changes will eventually be transmitted to the higher levels of the hierarchy as the entire community adjusts to a new seasonal regime. The migrant birds may already be making such adjustments, with some staying longer and others arriving earlier and leaving sooner.

Such seasonal shifts may herald major change, but species and communities have always needed to adjust to the vagaries of the weather. This much is not new, but changes in the interactions between plants and their pollinators and pests can have serious consequences in some of the more fragile agricultural systems.

Changes in primary productivity

Can we expect an increase in photosynthesis to mop up the abundant carbon dioxide and so offset the warming? It would seem reasonable to expect photo-synthesis to increase as atmospheric levels rise, helping to remove some of the excess. Once again the devil is in the detail. The complication is that plants have to respond not only to higher carbon dioxide and higher temperatures, but also to local variations in the supply of water or nutrients. Higher rates of photosynthesis demand more water, nitrogen, and phosphates. These are needed to sustain both photosynthesis and growth, and without them additional carbon cannot be locked into new plant tissues.

An absence of nutrients might thus limit increases in photosynthesis in nutrient-poor ecosystems. Nitrogen is a key component in the chloroplast and one of the main enzymes responsible for photosynthesis, but different plants have different strategies for apportioning resources. For example, in the tundra, nitrogen and phosphates are limiting for most plants, yet an increase in phosphate availability (say by greater melting of the permafrost, or through pollution) often leads to greater seed production, rather than a more general increase in biomass. There is evidence that the tundra community accommodates the raised carbon dioxide and shows little increase in net primary productivity. The pattern for temperate grasslands is also not simple: these add large amounts of carbon to the soil each year, and the soil becomes a major site of carbon storage. The decaying plant material also fixes some of the available nitrogen, and retains it for as long as this organic matter persists. Ecosystem-level studies on calcium-rich grasslands also suggest that nitrogen becomes limiting as organic matter builds up in the soil, for the same reason. In each case, organically bound nitrogen checks plant growth and represents higher-level buffering, constraining carbon fixation by individual plants.

At the level of the leaf, elevated carbon dioxide levels cause partial closing of stomata, so reducing transpiration and making more efficient use of the plant's water (Section 6.1). For some species this is offset by the plant producing more leaves, so any shortage of water remains limiting. Experiments on a variety of crop plants have shown that photosynthesis may rise initially with carbon dioxide but then some species adjust to the higher concentration, and eventually show no overall increase in carbon fixation. One limit is the speed at which their photosynthetic enzymes can be regenerated, but others include a build-up of the end-products of photosynthesis—too many starch granules forming in the chloroplast impair its function. Together such effects represent a form of lower-level buffering.

But some plants do fix more carbon, at least when measured as the weight of carbon fixed per unit of nitrogen in the leaf. These plants, in particular some grasses, use their nitrogen more efficiently and appear to have no fixed ratio of carbon to nitrogen in their tissues. Also, many plants use the extra biomass to grow longer root systems, to forage further in the soil for the water and nutrients they need. Some grasses and sedges are likely to become more productive, even when nitrogen is in short supply, though as a consequence, plant communities can be expected to change with these groups becoming more dominant.

Changes in decomposition

Higher temperatures are likely to accelerate decomposition in the colder areas of the planet and increase respiration more generally. Alone this would add carbon to the atmosphere, though experiments in the Alaskan tundra show carbon storage actually increases with warmer temperatures. This is because the increased growth adds more carbon to the soil, but its low nutrient levels slow the decomposition of this litter, at least over the short term. In contrast, decomposition rates for older, deeper, and more nutrient-rich deposits will indeed increase as mean temperatures rise.

Other changes follow if the permafrost thaws to a greater depth: with a greater volume of waterlogged and oxygen-poor soil, there is a greater capacity to generate methane. Wetlands produce about five times as much methane as dry areas (whereas dry soils are an important sink for methane—particularly neutral woodland soils). Overall, microbial decomposition and respiration rates are expected to increase with high temperatures, releasing more carbon from thawed soils rich in organic matter.

Another significant source of methane are rice paddies. These have expanded considerably in the drive to produce more food—often at the expense of tropical forests. A large proportion of their methane production is oxidized to carbon dioxide in the upper layers of the soil, but significant amounts still enter the atmosphere. Methane is also generated in large quantities by animal husbandry (the wastes, gaseous and otherwise, of cattle), by landfill sites (domestic refuse), and by mining and oil extraction. Termites are a major natural source, especially in the savannas, through the significant role their symbiotic bacteria play in breaking down the vegetable matter the termites collect.

Other interactions between species—competitive, exploitative and cooperative—may change as the balance shifts favouring some physiologies and metabolisms over others. As the nutrient status of their diet improves, some insect herbivores may enjoy higher population growth rates. For others, longer summers prolong their active phases, perhaps to the detriment of species they exploit. For example, some ecologists have suggested that the heathlands of northwestern Europe may be threatened if a key herbivore, the heather beetle (*Lochmaea suturalis*), is able to achieve two generations each year. In this case, the heather will suffer sustained attack and may lose its competitive battle with invading grasses in these endangered habitats.

More ominous, perhaps, is the evidence that the composition of the topical forests of Amazonia is undergoing long-term change. Oliver Phillips and his co-workers have shown that the proportion of lianas in the rainforest of the western Amazon has been rising over the last 20 years, at the expense, it seems, of the trees. These vines appear to benefit more readily from the raised carbon dioxide and have a faster response than the trees from which they hang. Because lianas invest less in support tissues they fix less carbon in their stems and the total carbon storage in the forest declines as lianas increase.

Stabilizing the atmosphere: ecological methods

The main mechanism for controlling the rise in atmospheric carbon dioxide is to limit our emissions. Next would be to limit our deforestation programmes. Titus Bekkering suggests that together these two measures could reduce the *growth* in atmospheric carbon by 12 per cent by 2100. Yet even with a massive, perhaps unrealistic, programme of reforestation and regeneration extending forest cover by 865 million hectares, carbon levels in the atmosphere are still projected to be 54 per cent higher in a hundred years time. The problem for policy makers is that this requires planning for the long term, making changes (and sacrifices) now that will benefit only our grandchildren.

A variety of schemes for increasing carbon fixation have been suggested. One is based on a proposition that oceanic phytoplankton communities are limited by a shortage of iron. Adding iron as an artificial fertilizer may increase sequestration by the large oceans, hopefully leading to greater sedimentation to the ocean floor (Box 8.6). Given the popularity of carbon-offset schemes (where the carbon-producer pays somebody to fix an equivalent amount of carbon on their behalf), this may be commercially viable. However, it raises important ecological questions about the wisdom of inducing large-scale change in marine ecosystems.

 BOX 8.6   Can the oceans be engineered to reduce atmospheric carbon dioxide?

Walker Smith, Virginia Institute of Marine Science

Most ecologists and geochemists now accept that carbon dioxide concentrations in the atmosphere are increasing as a result of man's activities, leading to changes in the Earth's climate. We might moderate this increase by sequestering more carbon dioxide in the oceans and one suggestion is to fertilize the oceans with iron, the element limiting marine phytoplankton production in many areas.

This might be particularly effective in the Southern Ocean, surrounding Antarctica. The impact on atmospheric carbon dioxide is likely to be large here because of the extremely high nutrient concentrations present in the surface waters (a result of the massive convective overturn of water in this region). With iron enrichment this nutrient supply could support greater phytoplankton production, perhaps reducing the increase of atmospheric carbon dioxide by some 72 ppm over 100 years. In contrast, the same treatment applied to the nutrient-poor Pacific Ocean would lead to a reduction of just 3 ppm over the same period. However, attempts to model the impacts on the marine communities suggest that the implications are far from straightforward.

The principle is simple enough—the fixation of the nutrients by the phytoplankton would increase the rate at which organic matter, represented by its dead cells, sank to deeper water. According to early models this would produce a layer devoid of oxygen at depth, especially south of Africa. In this case, the entire ecosystem of the Southern Ocean would be changed, largely because a main herbivore in its food chain, Antarctic krill (*Euphausia superba*), would be destroyed. Krill migrate to depths of 1 000 m during their life cycle and would die under the anoxic conditions of the deeper waters.

More recent models suggest that such anoxia would be unlikely, but depleting nutrients in the Southern Ocean could nevertheless reduce the nutrient supply in waters upwelling in the equatorial region. In this case the productivity would be greatly reduced within 100 years, impacting tropical fisheries such as tuna.

Others have suggested that the phytoplankton community of the Southern Ocean would change under iron enrichment, perhaps favouring species that are not so readily grazed. One species *Phaeocystis antarctica*, a close relative of a species known to cause algal blooms in the North Sea, may well flourish because it is incompletely grazed, causing substantial changes in the local food web. This species is also important as a source of dimethyl sulfide, a key compound that in the atmosphere serves as cloud condensation nuclei. This would increase cloud cover, reduce surface ocean temperatures, and therefore make the iron enrichment counter-productive. Other scientists have suggested that fertilizing sections of the Antarctic would increase krill abundance and ultimately enhance the reproductive success of baleen whales, a species overexploited at the beginning of the twentieth century.

Since much of the Southern Ocean is covered with ice for long periods of the year, adding iron to its waters is not without its practical difficulties. Nevertheless, substantial interest remains, and some commercial organizations recognize that an industry which removes carbon dioxide can claim a potential 'energy credit' (a licence to emit carbon which can be sold to other industries unable to reduce their carbon dioxide emissions). It may well be that the international trade in carbon emission credits starts between companies rather than between nations, even though the enrichment would be of ecosystems to which no nation or company could lay claim.

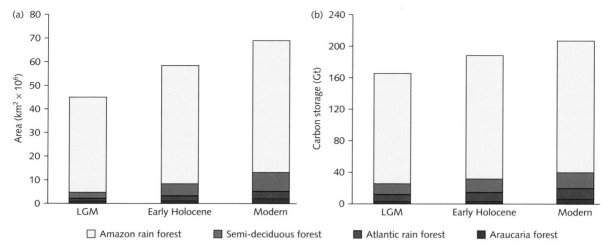

Figure 8.37 Behling's estimate of the total carbon fixed in the forests of Brazil since the last glacial maximum (LGM, 18 000 years BP) and the Early Holocene (7000 years BP), based on pollen deposits in sediments. Using samples from 45 locations, Behling mapped the spread of four forest types. The 'modern' value is an estimate of the maximum extent of each type in the recent past (a) and the amount of carbon fixed in mature forests of each type (b). The latter use estimates ranging from 260 tonnes C per hectare for semi-deciduous forest to 300 tonnes C per hectare for Amazonian tropical rainforest.

On land, forests are the principal means of fixing atmospheric carbon in terrestrial ecosystems, and the one component of these biomes over which we have the greatest control. Estimates in the IPCC report suggest that 370 million hectares planted at 10 million hectares per year over the next 40 years would eventually fix 80×10^{15} g carbon (80 gigatonnes), around 5–10 per cent of the carbon derived from fossil fuel combustion. After this time, when tree death eventually matches tree growth, the forest ceases to be a sink for carbon. By then, the total area planted would be equal to about half of the Amazon basin, but again, questions are asked about community change.

Can planned reforestation be combined with a conservation strategy that protects the biological diversity of the tropics? This may only be realistic in those countries currently able to feed themselves. Bekkering suggests this comprises 15 countries, which together have the potential to replant a total of 580 million hectares. The UN's Tropical Forest Action Plan (TFAP), begun in 1985, attempted to introduce forest management techniques into developing nations, bringing about an eight-fold increase in reforestation in participating countries. Unfortunately, the total area is small (1.9 million ha) and more carbon is released by these countries (282 million tonnes per year) than is actually fixed (12.4 million tonnes).

For a historical view of tropical carbon fixation, Hermann Behling estimated the increase in area of the four major types of Brazilian forest since their smallest extent at the last glacial maximum (18 000 years ago—Figure 8.37). The total forested area in Brazil has increased by 2.4 million km² (55 per cent) since then, mostly (1.1 million km²) in the last 7 000 years. This means that the trees of Brazil now hold nearly 43 gigatonnes (26 per cent) more carbon, about 6 per cent of the current mass in the atmosphere.

Such calculations allow us to estimate the impact of deforestation on carbon release and storage. Today in Brazil, 90 per cent of the tropical rainforest remains, 9 per cent of the Atlantic rainforest, and 19 per cent of the *Araucaria* forests. It is the latter, Behling suggests, that would offer the most significant increase in carbon sequestration, were they encouraged to regenerate.

Rates of deforestation have begun to decline in Amazonia, though elsewhere in the tropics they are increasing. Recent evidence indicates that the scale of deforestation may have been over-estimated, because of a poor understanding of long-term land-use patterns and a misinterpretation of the patterns on the ground. Too often, it seems, local farmers have been blamed for poor woodland management practices, whereas their activity helps to sustain the forests at their margins—as long as their population density stays low.

Tropical reforestation can be a realistic prospect only where the needs of local people are taken into account. Forests can be exploited without destroying the habitat. Besides their potential for timber and fuel production, they can also produce higher value goods (Case study 7). If exploitation is managed with a proper regard to the natural processes and community structure, there is every indication that this can be sustainable. Elsewhere, the blatant disregard for the impact of commercial agricultural methods is astonishing. In South East Asia, the deliberate burning of the tropical forest to create agricultural land, setting fire to the peat layers which are left to burn over extended periods and over large areas, emits immense amounts of carbon. Despite local peoples' attempts to patrol and extinguish these fires, the habitat destruction and threat to the resident species is an annual tragedy.

The real issue is not carbon fixation by forests but carbon release by fuel consumption. The solution depends on the will of industrialized and industrializing nations to curb their fossil fuel usage. Perhaps, the major user nations should pay tropical countries to maintain their carbon-fixing potential—a service they rely on, but for which they currently pay nothing. Costed on a *per capita* basis we, as nations or as individuals, might begin to recognize the true price of cheap energy (Box 8.7). As in any hierarchy, we should remember that our individual rates of consumption collectively determine our species' impact on the planet's atmosphere.

BOX 8.7   The writing on the wall

The uncertainty of predictions from the different GCMs, and a recognition that small-scale measures are unlikely to be effective, prompt many to argue against international proposals to cut carbon emissions. Further, the key role that cheap energy plays in industrial economies means that for some, emission targets equate to checks on economic growth. Better, they argue, to let market forces constrain fuel use through the costs of pollution and environmental degradation.

As Stephen Schneider and Janica Lane point out, the disconnection between social and economic scientists on the one hand and ecologists and meteorologists on the other is part of the problem in reaching a consensus on action. This long-standing cultural divide, with the two communities separated by the different languages of their disciplines as much as by their different assessments of the scale of the threat, constrains international agreements. At one extreme, some economists argue that even a rise of 1.5 °C would not put an undue strain on global economic systems, and poorer nations should be allowed the benefits of cheap energy to fuel their economic growth. No action, they argue, would give everybody time to find technological fixes to carbon abundance and fuel shortages. The majority of scientists do not see it in these terms and emphasize the shortcomings of using 'market-based metrics' to judge impact.

Trying to bridge this divide, Schneider and his co-workers have suggested five alternative measures to GDP (gross domestic product—the current measure of the wealth of nations). These new metrics embody some of the hidden costs of climate change, such as human lives lost per tonne of carbon released, refugees created per tonne or species lost per tonne. Crucially, they point out that all costs are relative and need to be expressed as a fraction of a nation's capacity to bear that cost.

Generally, scientists do not go in for polemic, and quite correctly, have been careful not to link any single event with climate change. Cautiously, climatologists predict more extreme weather events around the globe, even as some governments readily assert that a particular catastrophe cannot be linked to long-term climate change. Few climatologists would be so categorical.

(continued overleaf)

The reality is that higher energy levels in the atmosphere will drive larger movements of air and water around the planet, and the whole system will become much more fluid. We may argue over the details of local variations or singularities, but the overall pattern is now set. Carbon dioxide concentrations will continue to rise in the foreseeable future, with or without checks on human production, simply because of the time needed for atmospheric stabilization. The atmosphere is set to become much more energetic, the ice caps will continue to shrink, and ecological change is bound to follow.

Its detractors argue that the Kyoto protocol will not reduce carbon emissions by a sufficient amount quickly enough to have any impact. Others say that it puts an undue burden on the larger polluters—effectively those nations who are profligate in their use of carbon-based energy. Most of its proponents acknowledge that the agreement is primarily a gesture, symbolic of the sacrifice all peoples have to make.

Since there are few straightforward comparisons of its impact on different nations, perhaps the protocol will simply serve to demonstrate that we are all prepared to be good neighbours. Eventually, the targets on which we agree will depend on a range of factors, including the scale of environmental change we expect, the compromises we are prepared to make and the technologies we have available. Our strategies should not injure the economic prospects of many poor and malnourished peoples. And we should remember that our individual choices are the most important element in this hierarchy—choices that will impact on all peoples of future generations.

● SUMMARY

The global distribution of the major biomes reflects the climatic zones across the planet, most especially the mean temperature ranges and patterns of precipitation. Where water is abundant forests develop, where an extended dry season occurs grasslands are found. The tropics are characterized by forests with an immense diversity of plants and animals or highly productive savanna grasslands. Temperate regions have forest and grassland communities adapted to distinct seasonal cycles. The taiga and tundra are limited by short growing seasons and long winters.

A region can be divided into a series of landscapes. Within a landscape, ecosystems are subject to the same large-scale processes, such as climate or indeed, human interference. These ecosystems are usually defined by natural discontinuities and comprise a series of habitat patches that exchange energy, nutrients and species which each other. Landscape ecology describes these interactions and recognises that different species respond to different scales of resolution within a landscape.

Hierarchy theory suggests large-scale systems change slowly because they are big and complex. It sees biological systems, from an individual cell to a biome, as a series of levels, with the smaller functional units nested inside the higher levels of a system. Much of the organization of an ecosystem and its stability comes from the interactions between processes operating at these different levels. This becomes apparent when we look at the carbon balance of the atmosphere. A large-scale warming of the climate, primarily due to human fossil fuel combustion and forest clearance, is underway and our attempts to predict its consequences depend on understanding such interactions operating at various scales.

FURTHER READING

Archibold, O. W. 1995. *Ecology of World Vegetation*. Chapman & Hall, London. A comprehensive account of the major biomes.

Forman, R. T. T. 1995. *Land Mosaics. The Ecology of Landscapes and Regions*. Cambridge University Press, Cambridge, UK. An exhaustive account of the principles and practice of landscape ecology.

Houghton, J. 2004. *Global Warming. The Complete Briefing,* 3rd edn. Cambridge University Press, Cambridge. An excellent introduction to the science behind global warming from the Chairman of the first IPCC.

WEB PAGES

The IPCC page gives the latest data and reports:
http://www.ipcc.ch/

as does the Pew Center on Climate Change. This is an excellent and informed site, which is always up-to-date, and provides an accessible overview:
http://www.pewclimate.org/

The World Resources Institute offers a comprehensive and up-to-date summary on global climate change and data tables giving a country-by-country breakdown of emissions. The commentary in the tables provides qualifiers and sources of uncertainty associated with these data.
http://earthtrends.wri.org/index.cfm

The US EPA site provides a user-friendly and accessible summary of current data, especially as it relates to US emissions:
http://epa.gov/climatechange/index.html

This page of the UK Meteorological Office site offers a description of its global circulation model (including a slide set and downloadable images of the predictions from its last two models) and the associated datasets. There are links to several key documents and animations to illustrate its predictions:
http://www.metoffice.gov.uk/research/hadleycentre/index.html

The World Glacier Monitoring Service provides detailed records and mass balances from around the world:
http://www.geo.unizh.ch/wgms/

The Biology Browser is a good general-purpose directory with directions to details on the major biomes and landscape ecology:
http://www.biologybrowser.org

An a ccessible account of the decline of the Aral Sea with some stunning photography is found at:
http://visearth.ucsd.edu/VisE_Int/aralsea/

CASE STUDY 8

The world as rabbits find it

On the evidence of the speed with which rabbits achieved pest status there, Australia must have seemed a paradise to early rabbit colonists. However, a detailed analysis of their distribution and abundance shows that not everywhere is perfect. Indeed, with detailed analysis, mapping the landscape as rabbits perceive it could help us to control their numbers.

Rabbits (*Oryctolagus cuniculus*) form a metapopulation divided into burrows and home ranges dug in favoured areas, between which individuals move. Favoured areas have rich soils that offer nutritious grazing, where a population can grow. Other areas, with poor fodder, are used only as corridors and support little, if any, population growth. Hamilton *et al.* (2006) have used maps of soil conditions to partition a landscape according to the factors to which the rabbits themselves appear to respond. A simple landscape matrix, of habitat patches connected by corridors, is replaced in this study by a classification of patches according to their capacity to support rabbit populations: those which allow migration and those which present barriers to movement. The landscape patterns based on this classification reflect the gene flow between the rabbit's home ranges.

In the semi-arid areas of Australia, forests are unsuitable for burrows, and rabbits are also slow to cross them. On the other hand, shallow, slow-moving streams represent no barrier because rabbits are capable swimmers. Previous work has found they move freely over an open landscape up to distances of 1.5 km, but longer migrations take place as a series of steps. Not surprisingly, the number making the journey reduces with distance.

Hamilton *et al.* (2006) interpreted an area around Mitchell in Queensland in terms both of these environmental factors and the grain sizes to which the rabbits respond, creating a model of their metapopulation. Data from a soil survey was used to map three classes of habitat—poor, intermediate, and optimal—according to the densities they could support (relative to the optimum—these were 0.31, 0.61, and 1.0 respectively). Their model allowed each population to grow according to the availability of resources (determined by the relative nutrient status of the soil), using parameters from 19 years of data for rabbit populations in semi-arid Australia.

On this were superimposed the woodlands—unsuitable patches with few resources to support population growth

and barriers to rabbit migration. In other areas, a population can be established if the soil is sufficiently fertile, but if it is not, these places are used as routes between home ranges. The model allowed every population to connect with every other and for alternative routes between them. However, all migrations required a journey comprising several steps and these intervening habitats had different 'dispersal functions'—that is, the proportion of individuals surviving a journey across a patch depended on its length and its type. For example, 100 per cent were assumed to cross unsuitable patches of less than 1.5 km but only 10 per cent would pass through intermediate patches of 1.5—3 km. The number attempting to migrate depended on the size of the resident population, and the net movement between patches depended on the relative growth rates of the populations: one whose numbers were rising rapidly would donate more migrants to a patch where population growth was slow.

The output from the model gave the connectivity between populations, measured as the average number of individuals migrating in the two-way movement. This output was compared to their real-world connectivity, as indicated by the shared number of alleles at particular loci in their mitochondrial DNA. These **haplotypes** are known not to separate at meiosis (Figure 1.12) and so share the same sequence between maternally related individuals. In essence, female dispersal was being measured, and the model allowed for this by calculating the proportion of females leaving a population as its numbers grew. If the model was a fair reflection of these movements, we would expect some correspondence between the connectivity predicted by the model with that indicated by the rabbit's mitochondrial DNA.

In fact, this is exactly what was found. There was a highly significant negative correlation between pairwise connectivity and genetic distance: those predicted to be highly connected had fewer genetic differences, in line with simple population genetics (Section 3.6). The relationship remained even when two isolated populations, 'hidden' behind extensive woodland patches, were excluded from the analysis. These might have been entirely responsible for the result (they showed the most genetic distance from other populations); is so, excluding them would have caused the correlation to disappear. However, the inverse correlation was confirmed, suggesting that the model makes reasonable predictions about the frequency of exchanges amongst the remaining six populations.

In effect, the model captures features of the landscape as the rabbits perceive it—a heterogenous collection of habitats representing risks and opportunities. Forests reduce connectivity in arid areas and are more important than geographical distance, a critical observation when disease is used to control rabbit numbers. Indeed, Hamilton *et al.* (2006) show that the dispersal function is key to the model and, in Queensland at least, forested patches check the dispersal of rabbits. Along with their dependence on soil quality—a factor also known to be important in temperate areas—this suggests that such insights could have value in pest control strategies in semi-arid landscapes.

Interestingly, the model has other potential applications in the management of lagomorphs (hares and rabbits). In Britain, conservation efforts to promote population growth of the hare (*Lepus europeaus*), a protected species, are also advised to make use of its preference for heterogeneous landscapes, to create the most suitable habitats (Smith *et al.* 2004).

References

Hamilton, G. S., Mather, P. B., and Wilson, J. C. (2006) Habitat heterogeneity influences connectivity in a spatially structured pest population. *Journal of Applied Ecology 43*, 219–226.

Smith, R. K., Jenning, R. V., Robinson, A., and Harris, S. (2004). Conservation of European hares *Lepus europaeus* in Britain: is increasing habitat heterogeneity in farmland the answer? *Journal of Applied Ecology 41*: 1092–1102.

● EXERCISES

1. (a) Outside the polar regions, what single environmental factor is most important in determining the distribution of forests?
   (b) Explain the distribution of temperate grasslands in Figure 8.19.
   (c) What sequence of ecosystems (based on the types described here) would you expect to find as you ascended an alpine valley?

2. The hurricane season in North America lasts from 1 June 30 November. Suggest why hurricanes and intense storms are a feature of the Gulf of Mexico.

3. For each of the following biomes what factor determines a seasonal change in biological activity?

   Tundra
   Temperate forest
   Arctic polar region
   Savanna
   Sout East Asian tropical forest

4. Living and fossil marsupials are only found in America, Antarctica, and Australia. The oldest fossils are found in America. Suggest the pattern of dispersal of these mammals, and account for their absence from the other land masses once connected to Antarctica 65 million years ago.

5. Why should a persistent thawing of the permafrost lead to greater methane releases from this biome?

6. List the ways in which the abiotic ecology of the River Amazon may change if the Amazon Basin dries and experiences a higher frequency of forest fires.

7. Consider Figure 8.32. Suggest the main reason why turnover is so much faster in oceanic planktonic communities compared to terrestrial ecosystems.

Tutorial/seminar topic

8. Various mechanisms for trading carbon credits have been suggested to check fossil fuel usage by industry, and these are to be used inside the European Union. However, anthropogenic sources of atmospheric carbon include a significant component arriving from the waste of individual consumers. Should individuals face an equivalent carbon tax fully representing the 'carbon cost' of their activity? If so, how might it work?

9. Put together a proposed hierarchy for the decomposition processes of an ecosystem with which you are familiar. Choose one which is tractable (e.g. a pond, rather than the Pacific Ocean) and suggest how you might establish some of the factors that control methane generation.

Checks

'. . . the beauty of the cosmos derives not only from unity in variety, but also from variety in unity.'

Umberto Eco: *The Name of the Rose*

CHAPTER OUTLINE

- Predicting the number of species—species–area relationships.

- Island biogeography theory—the balance between colonization and extinction.

- The scale of species extinctions.

- Concepts of stability in ecological systems.

- Diversity and stability—the problems of sampling and of measuring biodiversity.

- Global patterns of species richness and the theories that attempt to explain them.

- Functional redundancy and stability in ecological communities.

- Soil processes, diversity, and stability in microbial and invertebrate soil communities.

- Human population growth and global patterns of soil productivity.

- Valuing biological resources and deciding priorities.

← Humankind walked this way. Rubbish left by walkers in the Himalaya.

The evolution of human beings, you will remember, is closely tied to that of the ice ages and the major climatic changes that began 7 or 8 million years ago. Back in Chapter 1 we should perhaps have asked what initiated the cooling and drying of the tropical African climate that prompted our evolution.

The short answer is we do not know. One suggestion is that there was a general cooling of the atmosphere caused by a shift in the Earth's orbit relative to the sun. Another key event was the creation of the Gulf of Mexico around 3 million years ago, when the Panama isthmus closed. This deflected a major flow of warm water, the Gulf Stream, north towards western Europe, and above it warm, moist air. Close to the pole this moisture fell as snow, allowing the Arctic ice cap to build. An expanding ice sheet would have reflected more solar radiation back into space,

so beginning the temperature oscillations that became the sequence of ice ages.

The Earth was perhaps more easily toppled into this instability by older and larger-scale continental movements. One theory suggests that global cooling followed the raising of the Tibetan plateau and the Himalaya as the Indian crustal plate pushed up into Asia, around 40 million years ago (Figure 9.1). This is thought to have induced vast amounts of precipitation as the atmospheric circulation was deflected over the new upland mass, depriving surrounding areas of moisture. In contrast to the drier conditions developing to the west (including in the Mediterranean Basin and East Africa—Section 1.1), massive rivers began draining and eroding the new upland, pouring minerals and nutrients into the seas. This fed the primary productivity of the oceans and

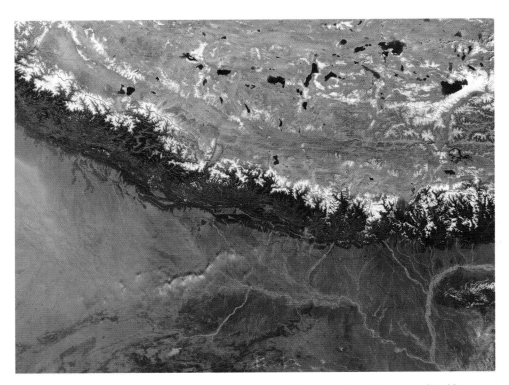

Figure 9.1 Was the raising of the Himalaya the prime cause of human evolution? Raymo and Ruddiman propose a sequence of events to link the collision of India with Asia about 50 million years ago with the climate changes in East Africa that gave rise to the first humans. This image shows the eastern Himalaya, below which are the two great catchments draining them, the Ganges and the Brahmaputra. Mount Everest is to the right of centre.

resulted in the explosive growth of their phyto-plankton. The increase in biological activity lowered atmospheric carbon dioxide levels, reducing the greenhouse effect, and allowed global temperatures to fall. Primed as it was in this way, it was easy for the cooler Earth to flip into the oscillations of the ice ages. If this was the ultimate cause of the climatic change in East Africa, then the birth of humanity was prompted by a collision of the continents. Literally, the Earth moved.

That new species arise following climatic change should come as no surprise to us. Nor that other species are lost. But it is spectacularly ironic that low atmospheric concentrations of carbon dioxide might be part of the explanation for our appearance on Earth, just as we learn that our enhancement of the greenhouse effect threatens to tip global ecosystems into new states. Ironic too that, under the pressure of our numbers and activity, the melting of the Arctic ice sheet could shut off the thermohaline in the North Atlantic and with it the Gulf Stream.

Rates of climate change have been particularly rapid over the last 100 years, primarily because of increased atmospheric pollution and large-scale deforestation (Section 8.3). As natural habitats disappear, so do species. Currently, extinctions are running at a rate perhaps 1 000–10 000 times higher than estimated background rates. We are at the beginning of a mass extinction event, comparable to those documented in the fossil record. The major extinctions of 65 or 250 million years ago have been attributed to massive volcanic eruptions or the impact of large meteors. Today, that same scale of loss has only one principal cause, *Homo sapiens sapiens*.

How many species are there and how many should there be? We start this chapter by looking at the number of species that ecological theory predicts for an area and the rate at which species are going extinct. We see how this might be related to the pattern of species diversity across the planet. The rapid decline in biodiversity, measured as the loss of the big and obvious, is now at the focus of many calls for environmental action.

We go on to consider which species can be lost without impairing ecological processes, especially those that provide essential services on which our lives and economic activity depend. In particular, we look at the significance of microbial and invertebrate diversity in maintaining soil fertility—the agricultural productivity of their soil is, along with the availability of water, critical for the quality of most people's lives. Yet, the species richness of soils goes largely uncounted and we have only a limited knowledge of the role these communities play in the stability of terrestrial ecosystems. It may be the activity of these small and inconspicuous species which determines how many humans the planet can support over the next century.

9.1 Predicting the number of species

Some patterns are easy to find but not always easy to explain. Here is a pattern of biodiversity you can find yourself: choose some way of sampling a large, uniform habitat using repeatable units—say the number of kinds of insect found on a leaf of a large bush or perhaps the number of plant species in a quarter square metre of lawn. Then double the size of your sample—two leaves or a half square metre and count again. Continue doubling your sample size until the count becomes impractical or your enthusiasm wanes (whichever comes later). Now plot the total number of species found against the area sampled. The chances are that you will produce a graph rather like Figure 9.2. As your sample size increases, so does the number of species recorded. This relationship is rarely linear—that is, S (the number of species) flattens out as the sample size gets larger. New species are easy to find early on but the rise in S does not keep pace with the increasing sample area. As Figure 9.2 suggests, the rate of addition of new species slows considerably until very large areas need to be surveyed to add the rare and the infrequent. If you are a

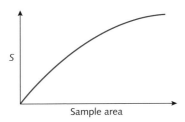

Figure 9.2 The species–area relationship. As the area sampled increases, more species are counted, at least in the initial stages. *Within* a habitat, greater sampling effort will eventually find few new species so S tails off with area sampled. This is a very important relationship—it provides some indication of the size of sample we need to sample an area adequately and also tells us how many species of a particular group to expect in a habitat of a given size.

birdwatcher you may have already experienced the difficulties of adding rarities to your tally and the frustrations this pattern can provoke.

This pattern is called the **species–area relationship** and is so common in ecology that we can make predictions based on its regularity. It can tell us the total number of species to expect in an area or how much sampling effort we need to detect the rarest species. It works best for a particular group—for example, herbivorous beetles or annual flowering plants—and shows that species number is some function of the available space, or some space-limited resource, such as food or water.

The relationship between species number and area can be summarized as:

$$S = cA^z$$

where S is the number of species (often termed species richness), A is the area, c is a constant representing the number of species in a unit area (or sample), and z is the rate of increase of S with A. Notice that z is a power term, which means that there is no simple arithmetic increase in species with area. The increase is logarithmic, so, for example, when $z = 0.3$ a 10-fold increase in the area will only double the species count:

For a 100 ha habitat with 10 species found in a unit sample:

$$S = 10 \times 100^{0.3} = \text{about } 40$$

and for 1 000 ha

$$S = 10 \times 1\ 000^{0.3} = \text{about } 80$$

Calculating z using data for different groups of plants and animals shows that it typically ranges from 0.15 to 0.35. z is high for species that do not travel readily between patches: birds or reptiles are likely to have different abilities to colonize isolated patches and will consequently have very different values of z.

When z is large the rise in S falls away sharply with increasing area, as usually happens in isolated habitats. In more connected patches, where S flattens slowly with area, z is low. This should make sense: well-connected habitats are easily reached so most patches will share most species with most other patches and the number of species does not change greatly between one patch and another. When colonization is rare, patches will tend to have different collections of species. Consequently, z is high for oceanic islands and low for habitat patches within a landscape mosaic.

Reaching a habitat is only part of the problem. A species only colonizes a habitat patch if it can establish a reproducing population. This may require a male and female of breeding age (which are reproductively compatible) to be present at the same time, and then to find each other. For many animals, colonization begins only when they cease to be passing strangers, but remain to raise a new generation.

An island habitat accumulating species from a nearby mainland or 'source community' can also lose them. Species may become locally extinct even after establishing a breeding colony. Small islands with a small carrying capacity can only support small populations, so the chances of going extinct here are high compared to a larger island (Section 3.7). However, if islands are close to the mainland, and new colonists arrive frequently, the local population may be sustained by new arrivals, as part of the movements within a metapopulation (Sections 3.7 and 8.1).

Continuous immigration and extinction means there is no fixed collection of species in a habitat, rather a turnover of species, the comings and goings of colonists and failed colonists. Over time, the total number of species within a group—say the birds or the reptiles—will settle down to some long-term value, even though the list of species actually present will change with every immigration or extinction event. An island is thus said to reach a dynamic equilibrium for *S*. A simple model of this was first set out by Robert MacArthur and Edward Wilson in their **island biogeography theory**.

If a new island is large and close to the mainland (that is, easily found and easily reached by the migrating group), it will rapidly acquire species from the source community. In its early days, habitats and resources on the island are unoccupied and new arrivals have little difficulty establishing themselves. Later, when most of the mainland species have colonized the island and the majority of its niches are occupied, there are fewer opportunities, and rates of immigration decline (Figure 9.3a).

As *S* increases, competition for the increasingly scarce resources rises and so does the extinction rate (Figure 9.3a). In a process comparable to succession (Section 5.3), individuals or species able to make best use of the limited resources are those most likely to colonize and persist. Since *S* is, in part, fixed by island size, area must be a summary of those important resources that place a check on both the number and types of species that can colonize.

With fewer migrants to supplement their numbers, populations on more distant islands suffer higher extinction rates than those close to the mainland, and these will be higher still if the island is small (Figure 9.3b). So equilibrium *S* differs between near and far, and large and small islands: colonization rates are higher on near islands and extinction rates higher on small ones. When colonization is balanced by extinction the total count of species for a group stays more or less the same. This is a dynamic equilibrium because there is a turnover of species and the species list is continually changing.

Although we commonly think in terms of oceanic islands, the theory can be applied to any habitat isolated by hostile surroundings, much as we described habitat patches within the landscape mosaics of Chapter 8. Indeed, the model has been instrumental in the development of landscape ecology and underpins much of what we had to say about habitat patches and metapopulations. Since its proposal in 1967, the theory has been tested against real data for different groups on different islands and found to be a consistent and effective predictor of equilibrium *S*. The best match to the real world occurs when we have data for well-defined groups, both functional and taxonomic—for example Jared Diamond found the limits to the number of species of large predatory bird on islands of different sizes through South East Asia (Box 5.3).

In its simplest form, island biogeography theory makes no predictions about which species are found on the island and says little about their interactions as they establish themselves. Nor does it distinguish between species' roles, even though some will be more important to a community than others: some may facilitate the arrival of later colonists, or preclude

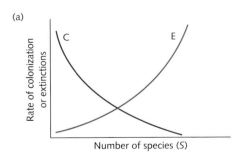
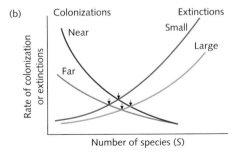

Figure 9.3 The MacArthur–Wilson theory of island biogeography. (a) Rates of colonization (*C*) decline as species are accumulated and rates of extinction (*E*) rise as niches are occupied and competition for resources is intensified. (b) Colonization rates will also vary with island proximity and extinction with island size. When *C* is balanced by *E* there is a dynamic equilibrium with a relatively fixed number but continuing turnover of species as local extinctions are replaced by new arrivals. For any particular group the value of *S* reflects the habitat size and the ease with which it is reached from the mainland.

subsequent colonization by competitors (Section 5.3). Additionally, over long timescales, new species are likely to evolve. Close to an equilibrium S, when local niches are occupied, there is intense competition for resources and the pressure to differentiate would be high (Section 2.5). Indeed, a relatively high proportion of endemic species is a feature of many islands, where isolated populations, undergoing genetic drift, have become adapted to the local conditions.

The original model did not consider speciation, but in a new variation upon it, Stephen Hubbell includes this process, at least in the source community. His 'unified theory' also seeks to incorporate the relative abundance of species—the rarity or otherwise of each species. Which species are common and which are not is some indication of the availability of key resources and numbers will change from community to community, reflecting differences in the way resources are partitioned between competing species (Box 9.1).

By allowing for speciation in the source community (assumed to have its own equilibrium S, where speciation on the mainland is balanced by

---

BOX 9.1   Measuring diversity

As with so many terms, the meaning of diversity depends very much on context. At different times diversity (or biodiversity) refers to species richness (the number of species, S), to genetic variability (the variety of genotypes), or to the variety of niches within a habitat. The term biodiversity may thus encompass everything from genes and nucleotides to biomes, and whilst it may be used collectively, we have to decide whether the prime focus of a discussion is species diversity, genetic diversity, or habitat diversity. Ecologically these are, of course, interlinked as high species richness is an indication of a large number of niches and therefore high habitat diversity.

Ecologists first used measures of diversity to compare communities. Much time and effort has been devoted to finding a universally acceptable measure or index, one that combines two components: **species richness** (the number of species) and **species equitability** (the proportion of individuals belonging to each species). Species richness alone cannot fully represent diversity—two communities may share the same number of species and the same total number of individuals, but the community with the more even distribution of individuals (the highest equitability) is the most diverse (Figure 9.4).

Species equitability is an indication of the resource space that each species occupies. A rare species with only a few individuals utilizes a small proportion of the resources available (Section 2.4). High equitability implies that resources are evenly shared between species and that the system is not dominated by just one or two abundant species. For example, surveys of the deep-sea fauna of the North Atlantic show that equitability is a significant factor in the latitudinal gradient in the diversity of bivalve molluscs towards the equator (i.e. there are few highly abundant species domination the samples).

Some indices of diversity assume an underlying species distribution, that is the relative abundance from the commonest to the rarest species, and so make assumptions about how resources are partitioned between them. That some of these patterns repeat themselves in very different ecosystems suggests similarities in the way their communities are organized (Section 5.2).

Others assume no underlying distribution and treat all species as equivalent. This may seem rather arbitrary, especially if the taxonomic distance between groups is not measured: two samples may have the same species richness, but one with four species of ant would be regarded by most ecologists as less diverse than a sample with a fly, a dragonfly, a beetle, and an ant. Because the higher levels of a phylogenetic hierarchy represent a larger range of adaptations, taxonomic diversity becomes a measure of the range of niches sampled, and therefore an indirect measure of habitat diversity (Figure 9.4c).

Sampling is critical in any attempt to properly quantify the diversity of a community, simply because of the change in S with sampling effort (Figure 9.3). Sampling a larger area will collect more species, and indices are thus sensitive to sample size. Selecting the appropriate sample size can be problematic if we are unsure about our sampling efficiency, or the

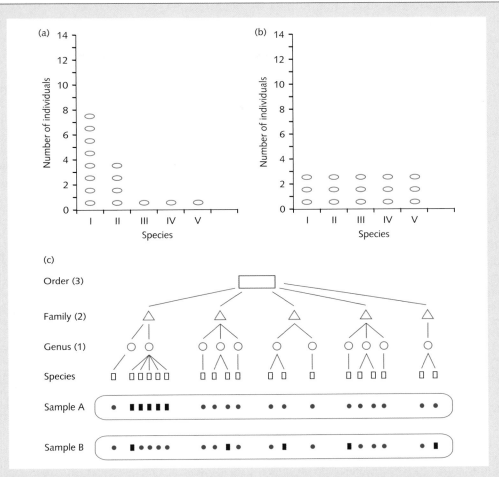

Figure 9.4 Measuring diversity. These samples from two different habitats contain the same number of species ($S = 5$) and the same total number of individuals ($N = 15$). However, we would not take them to be equally diverse because community (a) is dominated by just two species, whereas (b) shows equal numbers of all species. (b) has high equitability. This would not be indicated by a simple ratio of $S:N$ and so ecologists have sought more indicative measures of diversity. (c) Calculating taxonomic distance as the number of steps connecting individuals within a sample. Steps are weighted (shown in brackets) so that differences higher up the hierarchy score more—for example, with this simple weighting, two individuals from different families have a taxonomic distance of 3. Here, a sample of five individuals from 5 different families (sample B) has an average taxonomic distance three times that of one with five species from the same genus (sample A).

ecological processes that determine distributions and species richness at different scales within a landscape. Not surprisingly, species richness has been shown to increase most rapidly when sampling across several distinct landscape units, as a greater range of niches becomes part of the sample. This has important consequences for our attempts to conserve diversity—we are likely to protect more species by conserving a range of disparate ecosystems within a landscape.

Although no single measure of diversity has been shown to be effective under all circumstances, recent indices that incorporate taxonomic distance appear to be useful even when sampling efficiency is not known.

extinction), Hubbell derives the value of S for a receiving island. For each island, S is a function of its size, the migration rate, and the combined effect of the total number of individuals and the speciation rate in the source community. Colonization is more likely for relatively abundant species, since they have more individuals who could make the journey to the island. Compared to their source community, islands will thus have more common species and fewer rare species. Common species quickly dominate the resources on the island, whereas rare species, travelling less frequently, will tend to arrive later when resources are already occupied. Rare species on the mainland will thus be under-represented in the species inventory of an island.

The theory also provides an insight into the larger picture—the number and relative abundance of species in the metacommunity—across all the islands and the mainland, or across a landscape of habitat patches. Based on his model, Hubbell concludes that the large-scale regional stability in S seen in many landscapes is a consequence of species mobility, rather than local species interactions and competitive battles for niche space. In simple terms, the more mobile the colonists the more stable the local communities. If the model is accurate, the assembly rules of a community are less important for its stability than its frequency of colonization and extinction. On this basis, Hubbell has predicted that coral reefs will show higher stability in S than tropical rainforests simply because species move readily between reefs. The fossil record and evidence from living reefs suggest they recover more rapidly from a disturbance than tropical forests, and also that recovery is slower in isolated reefs (Case study 9).

Such models confirm the need for the frequent exchange of individuals and genetic material between patches for the long-term persistence of the populations within a metapopulation (Section 3.7). Refuges may be important for a species reduced to small numbers but on their own, small populations do not have long futures.

9.2   Extinction

Species come and go at different rates according to where you look in the fossil record. Based on what can be distinguished from the sequence in sedimentary rocks, we know that the Earth's species number has built steadily over millennia. Currently, the oceans have a variety of animals about twice the average for much of the fossil record and today the planet is at its most diverse. Robert May estimates that perhaps as many as 10 per cent of all multicellular species that ever lived are alive now.

This cranking up of S through geological time follows several episodes when there has been large-scale and rapid extinction of species. At least five mass animal extinction events feature in the fossil record—the 'Big Five'—the results of major upheavals in the global environment, when a sizeable proportion of multicellular species were lost. Yet, on both land and sea there have been impressive recoveries: for example, up to 95 per cent of oceanic species were lost in the Permian extinction 250 million years ago, but species richness recovered quickly thereafter.

Most extinctions occur outside these mass events —the fossil record shows that around 90 per cent of losses are part of the normal background rate of species turnover. This is the loss of species outside major upheavals or catastrophes, the extinctions resulting from species interactions or small-scale habitat loss. The Big Five, in contrast, signal global-scale change and record the sequence of momentous disruptions experienced by the Earth's biosphere over geological time. Not coincidentally, human beings have only ever known rapid rates of extinction, and we are currently in the midst of the latest mass event. Indeed, with some methods of counting, extinctions are happening faster today than at any time in the Earth's history. At background rates, the average life of a species in the fossil record is around 5–10 million years (though with great variation between groups and species), but Robert May estimates this average has now dropped to just 10 000 years.

Of course, our arrival and their departure are not unconnected. We can readily track human colonization in different continents by the disappearance of their larger species. Europe, Asia, and America lost many of their large mammals soon after humans arrived. Our spread through Australia 40 000 years ago marks the loss of its large vertebrates including nearly all the large marsupials. Much of the mammal and bird fauna of Madagascar was more recently devastated following large-scale human settlement. We can see the same process operating today across America, with the increasing impact of human activity threatening a variety of groups (Figure 9.5b). Extinctions are most obvious for the species we can count easily: Edward Wilson notes that one-fifth of all bird species have been lost in the last 2 000 years and a further 1 000 species are currently endangered. Beyond these are the unknown extinctions of the unclassified and uncounted.

One estimate suggests that 3 per cent of the entire flora of the United States is endangered by habitat loss. In contrast, the flora of the Mediterranean Basin is relatively resilient, probably because its habitat-sensitive, specialist species were lost long ago (Section 5.1). Just 0.15 per cent of its higher plants have been lost in the recent past, presumably because those which remain can accommodate change. A much greater proportion have disappeared from the other mediterranean climatic regions over the same period, especially from the mallee of Western Australia (0.66 per cent or 54 species), the most recently exploited region. Across these areas, the scale of the threat again shows some relation to the degree of human disturbance. While current losses are not yet significantly above the background rate, as aridity increases, all of these areas face much higher losses in the near future.

Globally, the World Conservation Union estimates that 13 per cent of all plant species are threatened. Others think this a gross underestimate, taking no proper account of the unrecorded species in the tropical latitudes. In the late 1980s and early 1990s, the area of tropical rainforest was declining at about 1.8 per cent each year. Using the normal range of z values (0.15–0.35), Edward Wilson calculated that this would amount to about 0.5 per cent of forest species going extinct each year. The Brazilian government subsequently reported a slowing in the rate of deforestation in Amazonia, though by the beginning of this century somewhere between 13 and 20 per cent of its naturally forested area had been lost.

Some extinctions are more important than others. In any inventory of diversity, losing the last member of a family must be a greater loss than a sub-species or an ecotype. Yet, within a species every individual has a value and adds to the variation in the gene pool, the variation upon which the adaptability of future generations depends. This genetic diversity, along with the variety of species (or higher taxa) and the diversity of habitats within an ecosystem are, collectively, what constitutes **biodiversity** (Box 9.1).

The variation lost with the extinction of a species or ecotype is a resource that we might one day have used: genes and alleles from wild types are continually used to invigorate our domesticated species (Section 2.6). There is also the value of a species' traits of which we have no knowledge: the range of potential drugs, foods, fibres, and other materials represented by this unknown biodiversity is the centre of intense research by the biotechnological industries and something over which nations are beginning to stake proprietory rights. As it stands, much of our current economic activity is based on the small number of species that we found easy to domesticate many years ago.

Some extinctions have greater ecological significance than others. Corals, for example, are keystone species upon which whole marine communities depend (Figure 9.6) and their loss causes the collapse of entire species assemblages. Yet, despite major fluctuations in sea level, there have been relatively few species extinctions within tropical reef communities over the last 2 million years. Instead, the animals associated with reefs have tracked the shift in the climatic regions, moving with them. Whilst most species move freely between reefs, the corals themselves are often highly localized and relatively immobile, and a coral species frequently disappears when its only population goes extinct.

Disturbance is an essential feature of coral reef ecology and the regularity with which reefs are

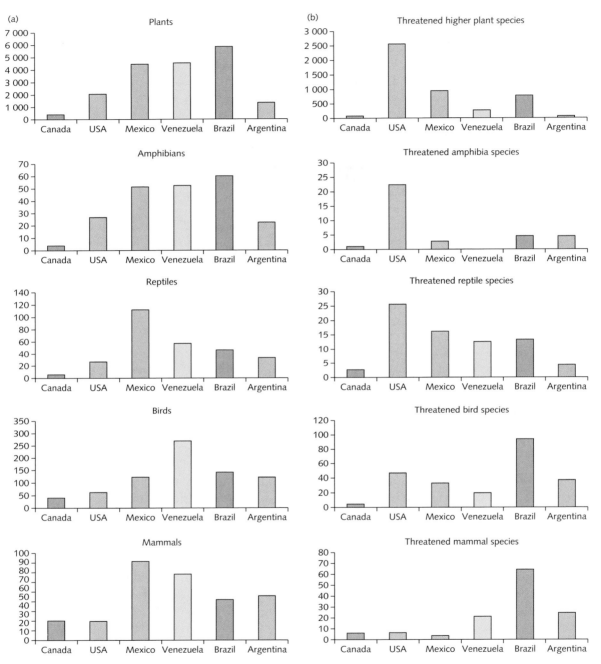

Figure 9.5 Latitudinal diversity through America. (a) This shows the number of species per 10 000 km², and thus standardizes the species count for countries of different sizes. Because some countries cover very large latitudinal ranges this will tend to obscure some of the latitudinal change in species richness, but nevertheless, it does pick out the higher diversity for each group towards the tropics (the Equator runs through Brazil, close to its border with Venezuela). (b) The total number of species classified as threatened in each country. This is not standardized to area and is thus a simple count of those thought to be endangered.

Figure 9.6 Tropical coral reef.

disrupted contributes to their high species richness. However, the current scale and frequency of the disturbances appear to be testing the elastic limit of reef communities everywhere. Mass-bleaching events, involving the rapid death of corals over entire oceans and sometimes simultaneously across hemispheres, are becoming increasingly common. Seas remaining at the upper end of their temperature range are one cause of coral bleaching; when this happens the coral polyps, the animals that collectively form a living skin of over the reef, evict their algal partners and lose their colour (Figure 9.7). Eventually, the polyps of most species will themselves die, leaving behind the white skeleton they secreted. Global-scale disruption of this kind, principally associated with persistently warmer oceanic temperatures, can now be observed with some regularity.

Why the symbiosis between animal and plant should become unworkable at higher temperatures is not understood, though one possibility is that the algae (termed zooxanthellae) begin to secrete toxic waste products. The loss of zooxanthellae from the polyp cells is the loss of an important nutrient source with the result that they cease to grow. These

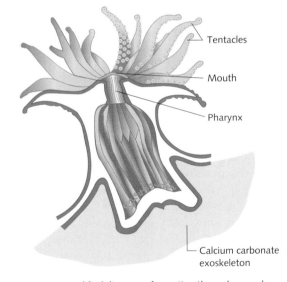

Tentacles

Mouth

Pharynx

Calcium carbonate exoskeleton

Figure 9.7 A simplified diagram of a section through a coral polyp. The zooxanthellae are held within a layer of cells called the gastrodermis (shown here in green). This forms the outer covering of the animal, which spreads over the calcareous exoskeleton and also lines parts of the central cavity. The zooxanthellae meet their nitrogen and phosphate requirements from the feeding activity of the polyp while the animal benefits from the sugars and oxygen released by the photosynthetic algae. The activity of zooxanthellae also helps the building of the exoskeleton.

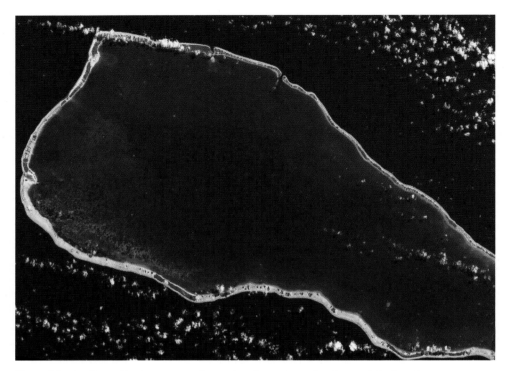

Figure 9.8   Satellites offer the prospect of surveying change in coral reefs, especially those in remote locations. Although only gross changes will be detected with this method, the frequency with which these reefs can be surveyed should allow for rapid quantification of regional- and global-scale changes. The Rangiroa atoll in French Polynesia suffered major damage in the global bleaching event of 1998 when its water temperatures were held 1 °C above normal for three months. This image was captured by the Landsat satellite in 1999.

cellular-level events can translate into the devastation of large areas of reef, at scales visible from Earth orbit (Figure 9.8).

In some areas, reef communities have shifted to entirely different configurations as a result of the bleaching and death of the coral. Over the last 30 years, as overfishing, eutrophication, disease, and hurricane frequency have increased, Caribbean reefs have been replaced by communities dominated by algae. Macro-algae—the larger seaweeds which effectively compete with the corals for space and light—have flourished on the abundant nutrients arriving in the islands' effluent. For many years algal growth was checked by the fish that grazed on them, but when the fish became the target of the fisherman's effort (because the larger predators had been fished out), they ceased to be effective herbivores. A

sea urchin, *Diadema antillarium*, took over this role and kept the seaweeds in check, but it succumbed to disease in the early 1980s. With no large herbivores, the seaweeds became dominant. Reducing the diversity of these ecosystems, here through the loss of a sequence of herbivores, reduced their stability and allowed the reefs to shift to a new state. Communities once structured by animals are now dominated by plants.

Mass bleaching is predicted to occur with increasing regularity, and perhaps annually in the Caribbean and South East Asia, in the next decades. The Great Barrier Reef, off the western coast of Australia, experienced large-scale bleaching events in 1998, 2002, and 2006 when water temperatures stayed close to 30 °C, the upper end of their normal range, for much of the summer. It may be that, with time, local adaptations to these higher sea temperatures will prompt new reef communities to evolve but this will neither happen quickly nor on timescales to which human beings respond. The immediate prospect in several areas of the tropics is one of massive decline in local fish and invertebrate diversity.

The fossil record suggests that coral reefs will eventually re-establish themselves under the new regimes. Reef communities have historically shown rapid diversification after each of the Big Five, requiring only 5–10 million years to recover their species richness. Indeed, a more rapid evolution rate and an early restoration of species number may be a feature of all tropical marine ecosystems. We find this elasticity, 2 million years ago, in the mollusc communities of the Caribbean. As sea temperatures fell during the glacial advances, mollusc extinction rates increased markedly, but this was more than offset by an increase in speciation rates. Indeed, over the last 500 years, the reduction of manatees, turtles, jewfish, and conches through hunting appears to have had little effect on Caribbean reef communities, possibly because other species, primarily invertebrates, have increased their abundance in response. Many of the ecological functions of these communities were

thus preserved, even though their species richness declined markedly. Inevitably, widespread loss of the coral itself, the keystone species, results in loss of habitat and loss of function, just as it does when we chop down the forests.

The low-lying islands of the Pacific are perhaps most at threat. Relying on reefs for their existence, in every sense of the word, several are under increasing threat from rising sea levels, storms, the regular flooding of homes, and the loss of fishing and tourism. Not surprisingly, given this economic and ecological reality, some peoples are seeking relocation to Australia.

Perhaps more of us will react when the important services of other ecosystems are under threat. Yet, despite the rates at which species are being lost, most people will not witness an ecological collapse on the scale of the Aral Sea (Section 8.1). Ecosystem functions seem fairly robust to small-scale species loss and we rarely see an ecosystem flipping to a new state. If species are coming and going all the time, what does it mean to say that an ecosystem shows stability? For most of us, the balance of nature seems pretty stable and fairly resistant to change. When would we know we had reached tipping point?

9.3 **Ecological stability**

Strangely, when communities change we expect them to remain the same. Even as we prepare for winter, we expect spring to return and the luxuriant forest to re-establish itself more or less as we saw it last year. Predictable change is incorporated into the biology and behaviour of plants and animals, and is even written into their genes. Squirrels hoard caches of food and plants swell their tubers to survive the times of shortage, banking on the times of plenty returning soon.

This continuity is what many people mean by a 'balance of nature'. It implies stability and the capacity of the system to restore itself. Beyond seasonal changes, natural ecosystems can withstand occasional shocks, so that even a major forest fire

only resets the system (Section 8.1). Stability implies checks and control mechanisms that keep things the same from one year to the next, or that follow a predictable path of recovery.

Much of what we have to say about stability can be applied to any level in the ecological hierarchy, from an individual to an ecosystem. To encompass these different levels we use the term 'system', meaning two or more interacting elements that remain together. In this way we can talk of cellular systems in the same terms as ecosystems. **Stability** is a property of systems that change little following a disturbance, or which return quickly to their previous condition. In fact, stability can be divided into several components, describing how a system responds to a disturbance.

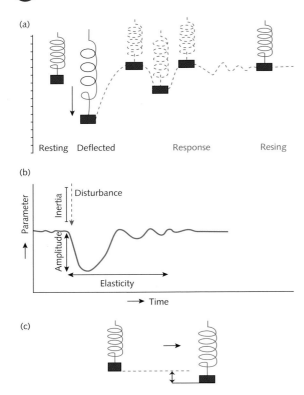

(a)

Resting   Deflected   Response   Resing

(b)

Parameter

Disturbance

Inertia

Amplitude

Elasticity

Time

(c)

Figure 9.9 A simple representation of stability in a system, shown by a weighted spring. In its resting position the spring and weight extend to a certain length (a and b). If we disturb the system, the amount of force we need to deflect the spring is termed its inertial stability (or **resistance**). The scale of any deflection (its amplitude or **resilience**) is the increase in length. The time taken for the spring to settle down to its original position is its **elasticity** (or adjustment stability). (c) If we exceed the capacity of the system to return to its original position (we stretch it), then it will return to a new (slightly lower) equilibrium position. In ecological systems, we might measure such dynamics as changes in population size, biomass production, species richness, or a range of other measures.

A good analogy for a stable system is the weighted spring (Figure 9.9). At rest, undisturbed, the spring has a particular length. Pull the weight gently (the disturbance) and it bounces around but eventually returns to its original position. The larger the disturbance the greater the amplitude of its fluctuations and the longer it takes to return. Exceed its elastic limit and the spring will be permanently stretched, coming to rest at a lower position.

Ecologists have used a variety of measures of ecosystem stability: for example, fluctuations in the population size of a key species, or coarse measures like decomposer biomass or rates of nutrient transfer. We concentrate here on populations and communities, where the quantified elements are individuals and species, respectively. A population or community might be judged stable if it resists a disturbance, changes little, or quickly resumes its previous pattern or state after being deflected. Many disturbances cause no permanent shift and some, like the seasons, are anticipated by the system. Such change is said to be incorporated, implying no lasting change and the capacity of the system to restore itself. A more sizeable or unpredictable disturbance may exceed this capacity and then individuals or even species are lost. A very severe winter can decimate squirrel populations if their food stocks become exhausted. We may already be detecting the signature of climate change as species assemblages in some well-documented communities have been seen to shift with changing seasons (Section 8.3).

For any organism to survive in a variable environment it must have mechanisms to maintain or restore itself after some sort of change. We understand how this has evolved in the cell or the whole organism: the internal environment has to be relatively constant to maintain metabolic efficiency, so that its complement of enzymes function close to their optimum. Homeostatic mechanisms have been refined by natural selection to maintain the efficiency of the individual and enhance its reproductive potential. We are familiar with these processes from observing our own biology —when we are hot or thirsty our behaviour and physiology respond to restore our poise and equilibrium. But how could such mechanisms have arisen in a community of species, when each is responding to different selective pressures? Communities, unlike individuals, do not have a heritable code that can be selected, so how does a collection of species come to operate together with any regularity and constancy?

In fact, any constancy we observe at this level will only be sustained from one generation to the next if traits imparting these properties, either directly or indirectly, are selected in the individuals that pass on their genes. It is the collective effect of natural

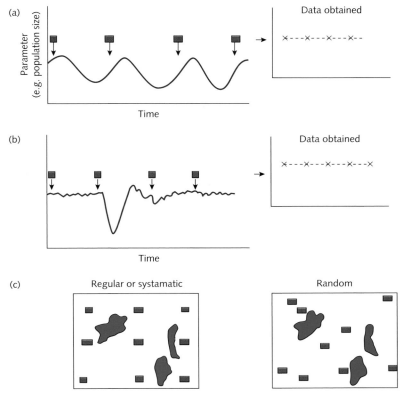

Figure 9.10 The effect of sampling frequency on detecting temporal and spatial patterns. Sampling at regular intervals (shown by the rectangles) (a, b) or randomly (c) in time and space can fail to detect the dynamics of a system, a disturbance or its patchiness. This makes detecting significant change and stability in variable ecological systems very difficult. We have to use preliminary surveys to match our sampling programmes to the pace of processes or the distribution of features in the real world.

selection operating on each individual that generates community stability: genes favoured in individuals must confer some advantage to their owners in their cooperative or competitive interactions, in their biotic environment. Host and prey, symbiont and symbiont, adjust to each other, and change in one is likely to induce change in the other.

We have seen such finely tuned cooperative interactions between flowers and their pollinators (Section 2.2), in resource partitioning as niches becomes separated (Section 2.5), and in the evolutionary arms race of predator and prey (Section 4.4). Stability at the community level emerges from the behaviour of its constituent individuals and species. A functioning ecosystem appears coherent in the same way that a

flock of birds or a shoal of fish forms living shapes, a composite that swoops and swerves as a whole when individuals move together. Each individual is following the same imperative, to protect itself or to reduce its energy costs, and collectively this gives the group its form. In the same way, a community's coherence flows from the selected traits of each individual. Moving too far out of line risks being at a selective disadvantage—no longer being adapted to the environment created by the community.

Some relations are very obviously constrained by selective forces—a parasite that kills its host before its eggs have been shed will not leave a long lineage and its genotype will soon be lost. Neither do most predators outstrip their prey—most shift to alternative species as these become relatively abundant and easier to find. Similarly, symbiotic relations only persist while the selective advantage of one partner does not override the interests of the other. Change the relation and the selective pressure shifts for one or both species. When the zooxanthellae become a liability their association with their coral hosts ends, so it is likely that selective pressures on them will be intense if long-term sea temperatures remain high.

Very often, the terms of the relationship are not decided simply by the adaptive capacity of the two interacting species. For example, a number of competing predators and their alternative prey species will collectively determine how closely the abundance of one particular predator follows that of one particular prey. In some cases, the fortunes of one species are determined by species several trophic steps away. This makes predicting the effect of species loss extremely difficult. Even in a system as relatively simple as that of the marine community in the Arctic seas, ecologists find it hard to predict the consequences of the loss of its top predator, the polar bear (Box 9.2)—we know too little about the species interactions, or the rules by which these communities are assembled (Section 5.1).

Comparative studies can then be useful. Some communities are more able to resist disturbance and some recover more rapidly than others. For example, forests with a mix of conifer and hardwood trees have a more diverse and resilient soil microbial community than pure conifer stands. Root exudates from their hardwood trees promote bacterial communities and this confers greater stability, and higher rates of decomposition. A greater diversity of soil decomposers seems to allow for **functional redundancy** in the system—that is, when several species carry out equivalent roles. A system with high redundancy can still perform a function when some of these species are lost.

Claire Kremen and her colleagues have shown how functional redundancy is reduced under intensive agriculture, compared to less intensive or 'organic' methods which conserve the diversity of natural habitats. The greater distances to wildlife refuges and the use of pesticides under intensive cultivation of watermelon in California requires farmers to import bee colonies at critical times to achieve the necessary levels of pollination for their crop. Compared to organic cultivation, the resident pollinator community contains fewer functional species. Kremen and her colleagues found that a diverse community, including native honey bee populations, allowed for variation between years, so that different insects fulfil the pollinating function when conditions change.

Finding common features between communities, and configurations that persist, can tell us much about the mechanisms which impart resilience to an ecosystem. Our understanding of the link between

---

BOX 9.2   Making the connections

As we indicated in First Words, one of the major problems in ecology is establishing cause and effect. Although we may think some things are linked, proving that one environmental phenomenon causes another is often very difficult. Today, we see patterns that appear to indicate climate change but we rarely have sufficient data to prove that local conditions reflect a larger picture of long-term global change. An example is given by the fortunes of the polar bears in western Hudson Bay (Figure 9.11).

Ian Stirling and his colleagues in the Canadian Wildlife Service have studied this population, again using radio collars to track the movements of 41 females. They asked whether a trend of reducing sea ice in this part of the bay over the last 50 years was reflected in the condition of the bears. Sea ice varies considerably in extent and duration from year to year but a trend is now fairly clear, with an annual decline in the Arctic of between 3 and 5 per cent, principally around Siberia. Because of the distribution of land and air movements, the eastern Hudson Bay has, in fact, cooled since 1950, but the western bay has warmed by 0.2–0.3 °C per decade. Since the bay is almost a closed system, largely unaffected by ocean currents, the decline in

Figure 9.11

sea ice in the western bay results almost entirely from the rise in mean atmospheric temperatures.

The critical season is the late spring when females have to grow fat reserves for themselves, for their unborn cub, and for their nursing thereafter. At this time there is still sufficient ice cover for the bears to hunt ringed seals (*Phoca hispida*), and the longer this lasts, the longer they have to feed. Since the bears tend to stay close to a particular part of the coast, the population in the western bay is well defined (Box 3.4). Stirling and his colleagues weighed sedated bears, and assessed their condition using an index of their weight divided by the square of their length. Heavy and well-fed bears have a higher index, but on average their values have declined since 1981. Between 1991 and 1998 the loss of condition of the females correlates with the date at which the ice breaks up in the western bay, suggesting that the bears fed less well in years of reduced cover. However, a correlation does not prove that the two variables are connected. This would be difficult to demonstrate without some form of controlled experiment that would rule out other possible factors.

In 1992, the ice lasted longer—three weeks beyond the average—and the bears came ashore in good condition. This extended period of pack ice was almost certainly because of the cooling effects of two global weather events—the eruption of Mt Pinatubo in the Philippines in 1991 and an El Niño in 1991/1992. Temperatures were reduced by around 2 °C globally after the eruption, as they were in the bay. It is a measure of the connectedness of this planet that a volcano erupting in the South Pacific may have meant, for one year, richer pickings for polar bears on the other side of the globe.

Thereafter, the condition of the bears resumed its decline. There was also a reduction in the proportion of yearlings (bears 1–2 years old) captured, and whilst the size of the population has remained more or less the same, its natality rate has fallen. Stirling and his colleagues suggest that the condition of the bears in the Western Hudson Bay is responding to long-term global warming—Pinatubo briefly halted the warming and allowed one or two years' respite, but the long-term trend has been resumed.

They point out that this population of polar bears is not under threat—compared to other populations they are in good condition, have a higher natality rate, and are a more stable population. But life is getting harder for them and other populations around the pole. The duration of the sea ice is predicted to continue to reduce, and the bears' means of feeding themselves is disappearing from beneath their feet.

species richness, community organization, and eco-system stability is crucial if we are to predict changes over long timescales and large spatial scales. Yet some ecologists doubt whether stability is really a property of ecological systems over the longer term or even whether most species associations are especially long-lasting. Instead they ask whether any natural ecosystem really does have an equilibrium configuration.

Stability and diversity

One of the most persistent ideas in ecology is that a diverse ecosystem is a stable ecosystem. A link between the number of species and stability was first suggested in the 1950s, based on comparisons of species-poor and species-rich communities. Pest outbreaks were thought to be more prevalent in species-poor agricultural systems and populations more variable in temperate and boreal biomes compared with the species-rich tropical forests. Indeed, it is possible to find experimental evidence for such effects (Figure 9.12).

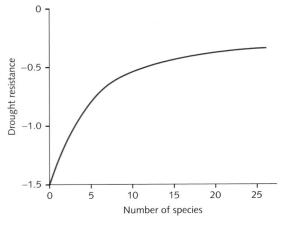

Figure 9.12 David Tilman and his co-workers have shown that temperate grassland plots with more species have a greater resistance to the effects of drought (a smaller change in total plant biomass between a drought year and a normal year). However, there was a limit—each additional plant species contributed less and less, so that the most diverse plots showed only marginal increases in resistance. The reason for the greater stability seems to be that species-rich plots are more likely to contain some drought-resistant plants. Beyond a certain level, however, new species are less likely to differ in this ability from established species.

The stability of species-rich communities was thought to come from their functional redundancy—more species allowed for more pathways through a food web. If one species was lost, others provided routes by which energy or nutrients could still flow, say when a predator switches to a different prey species. Earlier we saw how different herbivores became important in Caribbean reef communities as intense fishing pressure removed the principal fish grazing the large seaweeds. In the same way, populations in a diverse community show less variability than those in a species-poor community.

In the early 1970s, Robert May challenged this idea. He constructed mathematical models of food webs with different numbers of species and different numbers of connections between them, so varying the complexity of the system. A connection represented some interaction between two species (predation, competition, etc.) and was of variable strength (how much the population size of one species affected the size of the other). May was able to show that the link between diversity and stability was far from simple.

If everything else remained unchanged, adding more species would actually reduce the stability of the constituent populations. In May's models more species could be accommodated only if the number of connections or the strength of these interactions declined: stability in its populations followed when the system's complexity was reduced. The evidence from 40 published food webs seemed to support this—connectance within a web falls with species number (Figure 6.20). A more elaborate model by Kevin McCann and co-workers suggests that it is the strength of these interactions, rather than their number, that is critical for constancy in the web. Most of the complex webs seen in nature have a few strong links (where the abundance of one species has a major effect on another) but a large number of weak links. The McCann model shows these webs are stable because a hierarchy of relations, involving many species, dampens the oscillations of populations, even those that are partners in a strong interaction. Webs with many weak links serve to buffer population swings and thereby check movements in the larger system.

According to May, a highly diverse community could only persist if its interconnectedness was reduced and McCann and his colleagues conclude something similar. A food web divided into a series of **compartments** might achieve this. A compartment consists of a group of species sharing strong interactions with each other, but more tenuous links with the rest of the community. The abundance of one species can have major implications for the other members of its compartment but not necessarily for the rest of the community. John Moore and William Hunt found evidence of such compartments in their study of nitrogen transfers in the below-ground communities of North American grasslands. They describe compartments distinguished by their food source that only connect with each other further up the food web. For example, clusters of soil nematode species (roundworms) feeding solely on bacteria had no connection with other nematodes feeding on fungi. Top predators, in the form of predatory mites, eventually linked these compartments, two trophic steps away. Within each compartment, however, Moore and Hunt were able to show a close relationship between connectance and species richness.

Nigel Waltho and Jurek Kolasa also found evidence of compartments in their study of the fish populations in communities inhabiting Jamaican coral reefs. They describe a community structured according to habitat use, with specialist fish confined to narrow reef habitats and generalists occupying larger ranges. Not only were clusters of species (compartments) evident from the analysis, the population stability of each species also depended on its ecological range. Generalists, feeding high up in the hierarchy, were the least variable.

Robert May concluded that complex communities are found only where the environment was unchanging and large-scale disturbance was unlikely. Their complexity was possible because members of the community could have a long history of evolution in each other's company. With time, new species evolve and, through competition and other interactions, define their role precisely (Section 2.5). Long-lasting, unchanging environments would tend to accumulate more species in more well-defined niches, and only under these circumstances would species-rich communities and their balancing act develop. However, such finely engineered assemblages would also be most easily tilted from this position by a disturbance.

There is evidence that these are indeed the characteristics of complex communities. As we shall see below, this may help to explain global patterns of species richness.

9.4 The big questions: latitudinal gradients in diversity

There are large and small gradients of species richness across the planet. Diversity declines with altitude up a mountainside. Down the length of a river the variety of invertebrates and plants increases, from highland stream to lowland river. Many open-water (pelagic) marine animals show a maximum diversity some 1.5 km deep. Horizontally, across the oceans, species richness also reflects the gyral circulation of water in each of the major oceans (Figure 8.14).

At the global scale, the diversity of many groups increases from the poles to the equator so that the greatest number of species is found between the tropics. A range of benthic fish and invertebrate groups (especially molluscs and crustaceans) illustrate this pattern, at least in the North Atlantic (Figure 9.13). The gradient is less distinct in the South Atlantic and Indo-Pacific oceans, where regional factors dominate and there are 'hotspots' for some sea-floor invertebrates in the temperate latitudes. The greatest variety of marine isopods, for example, is found off the coast of Argentina, in temperate waters, possibly reflecting ancient ocean currents. The pattern is also confounded in shallow water communities, where local features, such as coastal configurations, are important.

Comparable processes create the patterns found in terrestrial ecosystems, though again regional contrasts are superimposed on the global gradient because of local climate or landform, or the history of the area

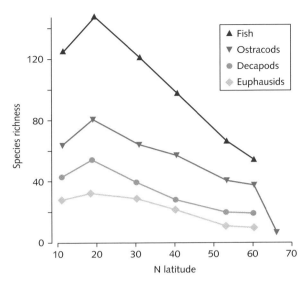

Figure 9.13 Four latitudinal gradients in the diversity of a variety of marine vertebrates and invertebrates (Euphausiids—planktonic krill, Decapods—shrimps, crabs, etc., and Ostracods) described by Martin Angel for the North Atlantic. Mean body size of benthic invertebrates also decreases toward the Equator. The length of food chains tends to be longer and the cycling of nutrients is faster closer to the Equator.

(Figure 8.12). An interesting example is the high diversity of mammals in the mountainous regions of North America, probably caused, in part, by the greater range of habitats in upland areas. However, despite such local hotspots, species richness of nearly all vertebrate groups rises toward the Equator in North America (Figure 9.14) and the pattern becomes more marked across America (Figure 9.5a).

Globally, the distinction between tropical and temperate biomes could not be more dramatic: 6 per cent of the Earth's land surface is covered with tropical forests but it is home to perhaps 70 per cent of all multicellular species. Table 9.1 provides a checklist of the factors that might explain this gradient. A brief review of the long list of theories follows, though many are interlinked.

Ecologists have looked for some prime abiotic factor to explain this increase in S towards the tropics, one which would apply in both marine and terrestrial communities, and to the variety of organisms showing the pattern. Temperature is the obvious candidate. Not only is it warmer near the equator, it remains

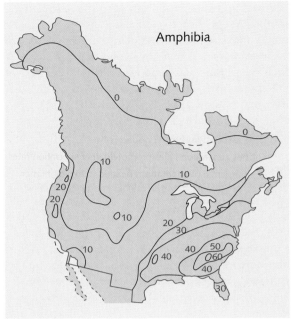

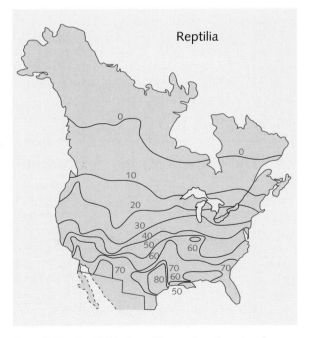

Figure 9.14 The distribution of the number of species of amphibians and reptiles in North America. David Currie attributes this primarily as a response to solar radiation and its impact on water availability and temperature, both of which change with latitude and continentality.

| **TABLE 9.1** | **Possible factors contributing to global patterns of species richness** |

Physical conditions

Climate generally more benign and less variable in tropics, harsh and highly seasonal in higher latitudes

Highly contrasting seasons at the higher latitudes limit close associations between species

Shorter generation times promote higher speciation rates

Greater range of adaptations needed to exploit higher latitudes—major groups originate in tropics and 'leak' to higher latitudes

Limited nutrient supply allows coexistence of many tropical plants in a small area

Larger area

Greatest area of the globe found in the tropics, allows for larger populations and, therefore, fewer extinctions

Fragmentation of populations allows for higher speciation

Productivity

Higher net primary productivity in tropics, more elaborate food webs

Disturbance

Moderate levels of disturbance may create a greater range of opportunities for colonists in the tropics, more distinct regeneration niche for trees in tropical forests

Constancy and long history

Longer periods and larger areas undisturbed by climate change

Longer period of co-evolution of species in tropics without major change in biotic or abiotic parameters

Species accumulated since extinction rates are low

Greater complexity

Complex plant communities provide more niche opportunities for animals in the tropics

Greater competition and predation

High levels of competition, herbivory, and predation prevent any species from becoming dominant

Sexual reproduction is favoured to promote variation, to escape especially parasites

Greater diversity

Promotes greater diversity

so throughout the year. Perhaps a consistently warm climate allows for more specialization and greater differentiation of niches (Section 2.5). So, for example, tropical forests have birds and mammals that feed exclusively on fruit, something that is possible because this resource is available all year round. There are no equivalent fruit-eaters in the seasonal forests of temperate regions and, similarly, most of their insectivorous birds have to migrate to avoid winter shortages.

In contrast, some argue that the marked seasonality of the temperate zones should promote greater niche differentiation. We know, for example, that changes in day length are used to cue plant growth and flowering. In the early spring of an oak forest, bluebells and a host of other flowers bloom before the tree canopy closes and the competition for sunlight becomes limiting. The seasonal clock sequences the availability of a resource, allowing species to specialize and avoid competing with their neighbours. Even so, the temperate forests cannot match the net primary productivity of the tropical forest, nor temperate grassland that of the savanna (Figure 8.32, Table 6.1), if only because their carbon fixation is turned off for several months when temperatures are too low for growth.

With the stark seasonality of the higher latitudes, all significant biological activity may cease for some part of the year. In this case, close associations that tie the fortunes of one partner to another can be a risky strategy when the association has to be re-established next year. Species in temperate areas also have to adapt to a wider range of conditions, favouring adaptable, most likely generalist species, rather than narrow niche specialists.

The adaptations required include the capacity to survive the seasonal cold—fur, feathers, or, in some cases, even anti-freeze in the body fluids. Often such adaptations are so extensive that they can only be the result of a long period of selection, by which a species gradually colonizes a habitat. The latitudinal leakage model suggests that the decline in S towards the poles is due to species only slowly invading the higher latitudes after the retreat of the ice sheets. Moving further from the Equator, a smaller 'leakage' of species has occurred, roughly corresponding to the scale of adaptation required.

Based on sunlight received we might expect primary production to show a simple latitudinal gradient. In the oceans this pattern is masked by local turbulence where turbid waters limit the radiation reaching the phytoplankton or where there are changes in the supply of nutrients to fund primary production. The most productive areas are often at continental margins where sediments are stirred and nutrients are periodically mixed with the surface waters. The seas of the higher latitudes are distinctly seasonal and at their most productive when nutrients and sunlight are abundant. Because tropical waters have little overturning of surface waters, marine primary production shows only localized maxima around the Equator.

Nevertheless, beyond the coral reefs, the greatest diversity of marine animals is found on the tropical sea bed (Figure 9.13) thriving on the year-long 'rain' of organic matter from above. This high diversity is most probably a consequence of the reliability of the food supply rather than the amount arriving. High productivity, therefore, does not guarantee more species, only more biomass (Section 6.3).

Many productive plant communities are dominated by one or two species occurring in very large numbers, often at the expense of others. Estuarine and wetland communities typically have only a small number of competitive, fast growing, and abundant plants (such as the cord grass *Spartina* or the rush *Typha*), whereas there is a prodigious variety of flowering plants in the nutrient-poor, semi-arid soils of the mediterranean biomes (Section 5.1). A shortage of the major nutrients is a feature also of tropical forests and several studies point to the effect of a latitudinal gradient of soil fertility on species richness. The distribution of the trees often reflects the presence of key minerals in the soil. David Tilman suggests that low nutrients in tropical soils allow many species to coexist, preventing any one species becoming abundant and dominant (Section 4.3).

Much of the plant and animal richness of the tropical forests is attributable to the diversity of the trees and the opportunities they themselves represent, much as the variety of corals and their superstructure provide niches for other species. Adding more species creates opportunities—each becomes a niche or a collection of niches available to other species. Thus,

to some extent, the diversity of tropical forests and reefs follows from the diversity of resources they represent. Species richness promotes species richness.

Some large areas of tropical forest have survived intact for millions of years, although much of it did change with the climatic shifts of the ice ages. Despite changing sea levels and habitats, most tropical areas were able to maintain large-scale forest refuges that today represent 'hotspots' of biodiversity. An unchanging environment causes fewer extinctions and, with time, greater specialization and niche differentiation. It may be that the tropics have simply had longer to accumulate species. In contrast, the higher latitudes have had a changeable history, subject as they have been to the migrations and extinctions associated with the glacial advances. The deep ocean floors are buffered from significant change and the benthos of a tropical sea typically has a low density of individuals but high species richness. Not surprisingly, these unchanging habitats harbour some of the oldest taxa.

High species richness may also be linked to the ease with which newly created habitats are reached by potential colonizers. This is one explanation for the rapid recovery and high diversity of most tropical coral reefs. The age of a reef determines how far the colonization process has proceeded and consequently its species richness Only tropical corals have zooxanthellae and only such corals will build substantial reefs providing a complex three-dimensional habitat attractive to other species. Large reefs take time to build. Although speciation is rapid in these ecosystems, especially amongst the corals themselves, this still operates over a long timescale. Once again, age is important.

The tropical biomes straddle the girth of the planetary sphere, where its land area is greatest. One possible effect of such large ranges is that populations may become discontinuous, divided into a metapopulation with local ecotypes that can go on to differentiate into new species (Section 2.6). This area has expanded (and contracted) in geological history, but as Michael Rosenzweig has pointed out, it nevertheless represents a vast region of relatively uniform climate that stretches back many millions of years. The size of this biome means many species can develop large populations over large areas, with consequently little risk of extinction. This is another reason for expecting taxa to be older in the tropics and there is supporting evidence not only for trees, but also for birds. The trees in tropical forests indicate a long evolutionary history because they are represented by different families and genera, with relatively few species belonging to the same genus. In contrast, the temperate zones have more species per genus, with relatively little taxonomic separation between them, presumably the result of recent speciation.

Animal species richness in tropical forests and coral reefs seems linked to the architectural complexity of these systems. Their three-dimensional complexity is in contrast to the limited vertical structure of a grassland or tundra. This distinction was found to be a good predictor of food web complexity by Frédéric Briand and Joel Cohen. Their survey of 113 webs from all types of ecosystem linked structural complexity and a relative constancy of the environment to longer food chains (Section 6.3). Without major disturbance, a complex community can assemble itself, and these conditions have pertained in tropical habitats far longer than in those of higher latitudes.

Disturbance too has its part to play in maintaining diversity. Fugitive species rely on gaps to maintain a population and can only sustain themselves if the gaps appear with sufficient regularity (Section 5.3). Again, the scale of the disturbance is critical, since diversity will only be maximized if the adjustment stability of the system is not exceeded (Figure 9.9). We saw in our survey of mediterranean-type communities how a certain frequency of minor fires is needed to maintain the scrubland mosaic (Figure 5.9), a scale of disturbance from which these communities could recover rapidly and upon which some species rely.

Intermediate levels of disturbance remove competitive species that would otherwise exclude new arrivals (Section 5.3). The intermediate disturbance hypothesis has been used to explain species richness in more variable environments, especially in coastal zones and temperate forests. For this to create a latitudinal gradient, however, disturbance rates would have to be close to optimum in tropical regions and minimized in the higher latitudes. In fact, turnover rates for trees in temperate and tropical forest biomes do not differ.

Robert Ricklefs argues that it is the nature of the gaps, not their quantity, which is key. He points out that an opening in the tropical forest creates a habitat (a regeneration niche—Section 4.1) that contrasts sharply with the forest floor under a closed canopy, with a major increase in light penetration. This opens up a much wider range of very different niches, allowing for very different species than would be found under the continuous canopy. In temperate forests the gaps remain in partial shade because of the sun's angle at these latitudes (Figure 9.15). Ricklefs believes this explains the astounding variety of trees in tropical forests despite the fact that their understorey vegetation is so surprisingly uniform.

Temperate forests

Tropical forests

Figure 9.15 Gaps in the canopy in forests are much more fully illuminated in the tropics because of the angle of the sun, producing a much greater contrast with the closed canopy. Robert Ricklefs suggests that this creates very different niche conditions favouring a wider variety of trees compared to temperate forests.

With its vast array of animal species, predation and herbivory are probably more severe in tropical forests and this has important implications for the organization of these communities—a species consumed is a species constrained. At a high intensity, the consumed species is prevented from becoming dominant. That pressure, operating throughout a food web, freeing up resources, promotes greater species diversity. Certainly, the intense seed predation observed around the parent tree would explain the large distances between individuals of the same species in tropical forests (Figure 9.16).

Finally, there seems to be more sex in the tropics. Sexual reproduction may be favoured here because it promotes variability, essential in the continual battle to escape the attentions of the large number of pathogens, parasites, and predators. At higher latitudes, the frequency of asexual reproduction increases in some plant and animal groups, presumably because the selective advantage of sexual reproduction is less when there are fewer natural enemies. Again, species richness promotes species richness.

Despite its simple pattern, it seems that the latitudinal gradient in diversity is probably the result of several ecological factors operating together. It may be that two or three abiotic factors set off a chain of biotic reactions that accentuate the differences between latitudes. Perhaps the greater energy influxes and primary productivity of the tropics provide the resources for a greater variety of herbivores, which, in turn, provide for more carnivores. As the community builds, it amplifies the opportunities for other species. In each case, the consumers serve to constrain species they feed on, preventing any one species becoming dominant. The contrast between latitudes then follows from this potential for positive feedback. Adding more species provides more opportunities, which, in the tropics at least, can be supported by their higher productivity.

But can we keep adding species? Highly diverse communities may be more fragile, yet through natural selection, evolution will try to add new species. Perhaps only the tropics provide the stable conditions necessary for complex communities to assemble, and here, because of their long history, many 'trials'— different combinations of species—have taken place.

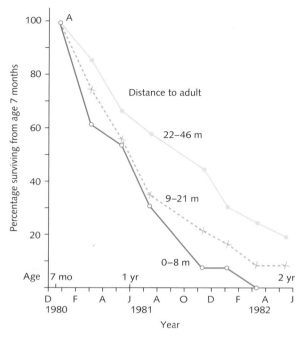

Figure 9.16 The high diversity of trees in tropical forests may be due to high mortality of juveniles around the parent tree. Janzen and Connell proposed that the pressure of herbivory on seedlings would mean that distance between the parent tree and its offspring would be very large. Clark and Clark were able to demonstrate this for one canopy tree, *Dipteryx panamensis*, by following the survival of its seedlings. Although 80 per cent of the seeds fall within 13 m of the tree none survived within this zone after two years. The nearest saplings from this cohort were 36 m and 42 m away from their parent. The graph shows how seedling survival decreases closer to the parent tree with time. This provides some evidence that biotic interactions within the community, in this case herbivory or attack from pathogens, help to promote high diversity. In the same way, predation or competitive pressures may prevent a species from becoming dominant.

Compartments may help to buffer the larger system, as a hierarchy of sub-sets that flex and absorb change, whilst maintaining a continuity of service to the larger community. Overall, the community we see today is the current result, the current balancing act.

But can we keep removing species? Diverse ecosystems may remain stable only if there are redundant species whose removal does not cause the community to topple. We have seen how the plant communities of the Mediterranean have, under human influence, lost species in the past and become relatively stable configurations (Section 5.1). As more are lost the role played by the remainder is increasingly critical to the stability of the whole system. Lose too many and collapse becomes inevitable—a threat facing mediterranean-type communities around the globe today.

On the one hand, the numbers lost in a mass extinction are dramatic, but, on the other, they are not dramatic enough: it is often the small species, especially bacteria and invertebrates, which are critical for ecological processes. Their diversity is not easily counted and is poorly recorded in the fossil record; living specimens are difficult to type and their contribution remains unmeasured in much research. We assume prokaryotes are resilient species, able to survive cataclysm, and presume that few have been lost with their habitats.

9.5 The big questions: stability and sustainability

The question of how many species are needed for an ecosystem to function is critical, both to ecological theory and to political practice. Perhaps we could afford to be unconcerned about species going extinct if their ecosystems continued to work in their absence. If ecosystem services were unaffected by their loss we might be able to live without some species and the economic argument for protecting large and talismanic species would be weaker.

We rely on natural ecosystems to purify our water, replenish our oxygen supply, and moderate our climate. It is important to know whether the loss of some of their species is likely to affect the delivery of these, or other, essential services. It is a cliché, but we do take these for granted and they are rarely part of any accounting procedure. When the costs of alternative or technological means of providing these services are included the figures can be astounding—the value of pollination by the honey bee *Apis mellifera* to US agriculture is estimated to be 5–14 billion dollars annually.

Rather more parochially, ecologists are interested in the principle—the resilience of ecosystem functions as species are lost. If all species in a community were essential to a process, we might expect its performance to decline progressively as species were lost (Figure 9.17). Alternatively, with a large measure of functional redundancy, species could be lost with little loss of performance or stability in the system. Then a major reduction in function would only occur when non-redundant species disappear.

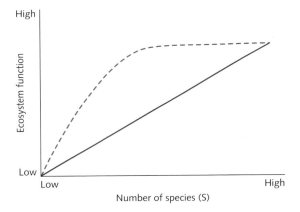

Figure 9.17 Functional redundancy in ecological communities. If every species is important in maintaining some ecological function then we would expect a linear relationship between S and the process (the solid line). If, instead, some species are more important—that is other species are to some degree redundant—the ecological function will not decline linearly as each species is lost (the broken line).

M. W. Schwartz and colleagues have looked at both the experimental and theoretical evidence for such patterns. Most observational and experimental work showed that ecosystem function did indeed fall as species number reduced, but only in three out of 19 cases was there a simple linear response—a decline in performance in direct proportion to the number of species.

A non-linear response means that the less common and easily lost species are not critical for the performance of the larger ecosystem. We saw this pattern before with the effects of drought on temperate grasslands (Figure 9.12), and indeed most studies point to a high degree of redundancy in other ecosystems. Simple theoretical models suggest why a linear response is rare and why many communities can remain stable even when losing species—again, it seems that a large number of weak interactions appear to be critical for dampening major fluctuations. Additionally, species often respond in different ways to an ecosystem change, with some species increasing their abundance as others decline.

It is often the little things that count. We saw that a large proportion of the productivity of the savanna (Section 8.2) is processed by termites, and their activity, more than that of any larger species, is critical to nutrient cycling in this biome. The same is true of soil invertebrates, bacteria, and fungi in all soils of the world. The functional redundancy in these communities appears to depend on the task: Lewis Deacon and colleagues found that nearly all fungi from an upland grassland in Scotland could degrade cellulose but only a small number could attack lignin, the major component of wood. Interestingly, the rare species of fungi in their samples could, in most cases, attack the greatest range of substrates and their absence would thus reduce the capacity of these soils to decompose less tractable organic matter. Because they can also use readily available carbon, these rarities may serve as a community in reserve, able to become dominant if the nutrient source or the abiotic conditions change. Significantly, the functional redundancy of this decomposer community is confined to easily degraded materials (Figure 6.14).

Experiments using laboratory soil microcosms have concentrated on the easily manipulated species—the obvious bacteria, the larger invertebrates, fungi, and plants. Its bacterial community is essential to nutrient fixation and turnover in a soil (Section 7.1). Bacteria are also a critical factor determining the above-ground plant community, and, through their interactions with the fungal and invertebrate communities, they shape the properties of a soil. However, there are formidable technical difficulties in sampling the microbial and micro-invertebrate diversity of a soil,

and relatively few studies come close to quantifying their species richness or relative abundance. Indeed, for practical purposes it may be impossible to measure the true diversity of the Bacteria and Archaea, though distinctions made on the basis of their ribosomal RNA allow us to glimpse organisms we have been unable to cultivate in the laboratory. Indeed, new phyla have recently revealed themselves from extreme environments (Box 6.1), an indication of the immense microbial diversity yet to be catalogued. Bess Ward points to these unsuspected communities beneath our feet, in soils we thought we knew relatively well, carrying out ecological functions of which we are largely unaware.

The way we treat the soil determines its microbial health, and intensive agriculture often simplifies soil communities. For example, dramatic change follows if the grazing of a pasture ceases, as the plant community shifts to a new configuration. For example, creeping buttercup (*Ranunculus repens*) replaces perennial rye grass (*Lolium perenne*) as the dominant plant when sheep are removed from upland pastures in Scotland. Roy Neilson and his co-workers found that the below-ground decomposer community also switches from one dominated by bacteria to a fungal-based economy. Grazing induces the production of root exudates by the grasses and these promote bacterial growth near the roots. Without this factor, fungi dominate the soil and a different community of soil invertebrates prevails—principally fungivorous nematodes and springtails. Springtails are a major consumer of soil fungi in grasslands, but some species become functionally redundant if one of the larger species—*Folsomia*—is present. On its own, *Folsomia* can set rates of litter decomposition and soil respiration, and other springtail species have little influence on these processes when it is present.

It seems that much of the immense diversity of bacteria and protozoa in soils is redundant. Even the coarsest of manipulations of bacterial and fungal communities (or the food chains above them) appear to have little effect on soil functioning. Mira Liiri and co-workers in Finland found no significant change in biomass or in drought resistance of coniferous woodland soils when a sizeable part of the invertebrate community was removed from field plots. Even when the community was disturbed by a pulse of nutrients (wood ash), the capacity of the soil to maintain its functional stability (measured as the loss of nitrogen from soil microcosms) was unaltered.

At the microscopic level, a soil is not one microbial community, but several—a series of habitat patches —with bacteria or fungi flourishing where opportunities present themselves. Perhaps this degree of compartmentation helps to buffer the response of the larger system. Many high-latitude and temperate soils appear to show such resilience, in part because they have incorporated major seasonal changes, including pulses of nutrients, into their annual cycle. These are also young soils, where many key plant nutrients, such as phosphates, are relatively abundant. Elsewhere, nutrients are not so readily available, and soils are more easily pushed beyond their capacity to restore themselves.

Gradients of sustainability

The global gradient of diversity also tells us something about the agricultural capacity of the different biomes. Close to the poles, successive glaciations have scoured and relaid deposits, young soils with relatively high phosphate levels. Close to the Equator soils are typically old and infertile (Section 8.2). Their communities are adapted for rapid decomposition, quickly assimilating what becomes available when organic matter arrives at the soil surface. Much of the nutrient capital of these ecosystems is locked in the trees, and easily lost when the trees are removed. The small amount of organic matter that glues the soil together also disappears with deforestation. Michael Huston argues that this alone should give us every incentive to stop the wholesale destruction of the tropical forests.

These forests are in regions where agriculture is still the principal means of wealth creation. Huston describes the global gradient of declining GNP (Gross National Product, a measure of the economic wealth of a nation) towards the Equator, in countries where human population growth is often rapid, but which rely on soils and ecosystems unsuited to

large-scale disruption (Section 7.2). With small populations, the indigenous practice of slash and burn, creating small clearances which are cultivated for 2–5 years, is a sustainable method of exploiting the short-lived fertility of these soils. At these scales, the resilience of the forest ecosystem allows the system to restore itself. Indeed, the disturbance may be important for the gaps it creates, promoting diversity through regeneration. Whilst it would be hard to demonstrate that its peoples were in any sort of balance with their forest, their numbers, the area over which they range, and the size and frequency of each disturbance may be the reason why native agriculture appears well-adapted to the forest's ecology.

In contrast, the nature and scale of commercial logging operations exceed the elastic limit of the ecosystem with the result that in much of Amazonia and South East Asia forest recovery is non-existent. The system switches to a new state, typically a semi-arid grassland. Even where there has been commercial reforestation (for rubber or paper pulp production) the poor soil means yields are low. The problem is the scale of the devastation. Up to 20 per cent of the Amazonian forest (600 000 km^2) has been lost through human activity in recent decades and overall it now has about half of its pre-human forest cover.

Much of the developing world relies upon wood as its principal fuel and their growing populations have stripped the land of its trees, especially in sub-Saharan Africa, India, and the Himalaya. The loss of tropical forest leads to local climatic change, usually with a major reduction in rainfall. As the trees disappear, so does the soil. Without the forest, steep slopes are more easily eroded by flash floods: between 100 and 200 tonnes of soil per hectare are lost annually from cultivated slopes in tropical areas. Drought and periodic famine are written into the history of many tropical and semi-arid areas, but the scale of habitat degradation today and, alongside it, the rate of species loss, indicate more permanent change. Henry Kendall and David Pimental describe how, in the latter half of the twentieth century, most nations ceased to be self-sufficient in food. Despite the advances in agricultural technology—the 'green

BOX 9.3   **The growth of the human population**

As far as we can tell, the human population has been relatively stable for much of its recorded history. Various methods have been used to measure the size of past populations in different parts of the world and the overall picture is one of a slow rise for much of its history, with one or two noticeable declines due to disease.

However, the global rate of population growth showed a marked increase in the last century and this will continue for most of this century. Most nations began a phase of rapid growth some time during the last 100 years as they adopted modern standards of hygiene and medical techniques, and as nutrition improved. Together these factors have led to major reductions in death rates.

Today (2006) the human population stands at 6.5 billion, an addition of more than 500 million people since the first edition of this book (1997), an increase of 8 per cent in less than ten years. At its current rate of growth (around 1.4 per cent per year), the global population will double in the next 50 years (Figure 9.18). Its age structure is characteristic of rapid growth (Figure 3.6), set by the youthful populations of the southern hemisphere, where population growth is largely concentrated. The United Nations gives annual population growth of 0.1 per cent for the economically developed nations, compared to 1.5 per cent in poorer countries. The highest rates continue to occur in Africa, where there are few economic, medical, and agricultural resources to support these increases. Fertility rates in many states of sub-Saharan Africa remain at around seven births per female. In contrast, population growth has all but ceased in developed countries, where fertility rates have fallen from 4.7 births per female in 1975–1980 to around 1.6 today.

These rates are critical in determining the future size of the global population. If the global average stays close to the current level (2.5–2.7 births per female), there will be 10 billion people by 2050 and 16 billion by 2150, when growth is expected to halt (mortality rates will then match natality rates). A rapid fall in fertility rates will bring this stabilization much sooner—at 1.6 births per female, global population

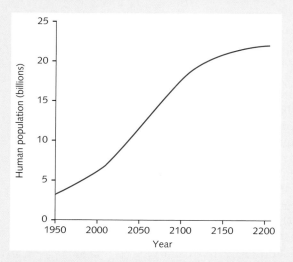

Figure 9.18   The growth in world population, modelled according to the logistic model described in Section 3.2. Henry Tuckwell and James Koziol found that this simple model gave the best fit to the data between 1950 and 1985, and that it accurately predicted the population in 1992, of 5.48 billion. If it continues to be accurate, a carrying capacity of around 24 billion, four times the current size, will be reached in the year 2200.

growth will flatten in 2050, at 7 billion people, and decline to 5.3 billion a hundred years later.

Predictions about the consequences of large human populations have been around for centuries (and perhaps longer). The rates of increase in the 1950s and 1960s provided momentum for the early environmental movement, and their premature forecasts of environmental collapse. But today we are seeing large-scale environmental degradation in many parts of Africa, where numbers do outstrip resources and natural services. Globally, water shortages are expected to affect two out of three people by 2025, and the struggle for limited regional supplies may provide, in the words of Kofi Annan, 'the seeds of violent conflict'.

revolution' based on new crop strains, artificial fertil-izers, and pesticides (Section 6.5)—per capita food production has slowed down. Food production in 70 per cent of developing nations can no longer match their population growth.

The usual response to food shortages is to bring more land into production, causing more natural habitats to be lost. This land is inevitably marginal and cannot sustain agriculture over the long term. Sixty years ago much of the central highlands of Ethiopia were covered in relatively lush forest. The once nomadic peoples of sub-Saharan Africa have today settled around centres where medicine or water is available and, in the process, destroyed the forests. In 1984, Ethiopia had its driest year since records began and the combination of civil war, a large settled population, and a depleted ecosystem based on unprotected soils conspired to produce one of the worst famines of the twentieth century. A series of droughts have hit the Horn of Africa ever since, which have been matched more recently by droughts further south, in the normally verdant areas of East Africa.

The consensus today is that famine relief does not provide a long-term solution, that a simple reliance on food aid undermines local agriculture. Whatever political and economic changes are needed in Africa, large areas will need to be reforested and the popula-tion helped to achieve densities that are compatible with a functional soil and a vegetation cover that protects it. Carl Jordan argues that tropical agricul-ture needs to avoid western methods with their dependence on fossil fuels. Better, he argues, to use practices that harness the natural processes in the local environment, with more imaginative cultivation practices. This was recognized by most world leaders at the Rio Summit of 1992 and the commitment of Agenda 21 to find sustainable strategies for all economic development. The cur-rent initiatives in debt-cancellation following the Gleneagles Agreement in 2005 perhaps go some way towards restoring an economic balance between Africa and the rest of the World, but consistent efforts to address its environmental degradation, and its political and health problems have to be part of the strategy.

So we end up where we started—in sub-Saharan Africa with a receding forest (Section 1.1). Today, most people living south of the great desert suffer some form of malnutrition. What was the cradle of humanity has become the place where large-scale human death is most prevalent. Much of the world population is hungry and perhaps one-fifth suffer from malnutrition. Yet, despite the fact that our numbers have grown, a smaller proportion suffer from a lack of food today than at the start of the twentieth century. That encourages some to argue that human ingenuity could eventually solve short-ages of food and water. We will inevitably look for a technological fix. Perhaps genetic manipulation might improve the disease resistance of key crops, or equip cereals with the capacity to fix their own nitrogen.

But, of course, there are no free lunches. Even assuming that this technology is freely shared (or that genetically modified organisms pose no environ-mental risks), we will still need sustainable agricul-tural practices to preserve the complex ecological interactions that makes food production possible. Functional redundancy may seem an abstract idea, of interest only to some ecologists, but it assumes immense practical significance as we try to restore the productivity of degraded agricultural soils. In the end, feeding the people will have to be based on an ecological solution.

Perhaps the extinctions that humankind have set in train are just the 'froth', the functionally redund-ant species the Earth has accumulated over the last 65 million years, and which it can afford to lose. Comforting as this might seem, it fails to recognize that current extinction rates are too rapid and too wholesale for the selectivity that this implies.

It is too simplistic to suggest that all environmental degradation can be traced back to the explosion of the human population. Instead, it is more the number of individuals multiplied by the amount of resources we each demand, set against the fragility of the environments we exploit. What we each consume and demand of our environment varies greatly across the globe and is shaped by our culture and tradition, and the political constraints on our activity. We have used our technology and ingenuity to bring us this far. Now we need the wisdom to give ourselves a future.

BOX 9.4   Final thoughts

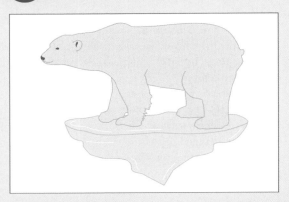

In 2006 the polar bear was officially declared an endangered species by the World Conservation Union. In the previous three or four years, the reality of human-induced global warming was accepted by most nations and their scientific academies. The connection between these two developments may not be obvious but whilst the loss of their hunting habitat and the threat to polar bears may be accepted by many ecologists, there is no proven link to climate change. For this, and for other reasons, the majority of political, economic, and social commentators are unable or unwilling to join the dots.

Here, in simple terms, are the likely connections between human-induced atmospheric warming and the threat to the polar bear:

- Human beings have been releasing fossilized carbon into the atmosphere.

- Human numbers and activity have dramatically increased atmospheric carbon levels in the last 250 years.

- High concentrations of carbon in the atmosphere increase its greenhouse effect and its average temperature.

- Higher global temperatures lead to melting of the Arctic ice sheets.

- Reduced ice cover over the Arctic seas reduces the feeding time for polar bears.

- Polar bear reproduction and condition decline and their population sizes decrease.

It now becomes a question of which of the above statements is false, or which implied connection is not demonstrated . . . and then what it would take to prove each statement or demonstrate the connection, either to a scientist or to a decision maker.

Perhaps polar bears will become part of the price of our economic progress, though only a small part. Thin polar bears on thinner ice are symbolic of the danger facing the living biota of our planet, emblematic of the shrinking space we allow other species. The bears speak for us, and the demands our numbers place on the capacity of the Earth.

We should not expect rapid changes or wait for obvious shifts before we take decisions. Like the melting ice, the shifts in the climate will be gradual. It can only get hotter. The fire lit by *Homo erectus* 3 million years ago is still burning. Who would have thought that one species, a clever ape, could threaten the existence of so many others? Who indeed?

● SUMMARY

The number of species increases with area and allows us to predict how many species a habitat or island might contain when colonization and extinction rates are balanced. At this point there is a dynamic equilibrium and the total species count (or S) stays more or less the same. Quantifying this diversity becomes problematic when our measures are sensitive to sample size, and more recent efforts now include some scoring of their taxonomic diversity.

Currently, the loss of habitats, largely through human activity, is the principal cause of a mass extinction event, possibly the most rapid in the history of the Earth. The loss of some species is more important than the loss of others, and ecologists are seeking to describe the proportion that can be seen as functionally redundant; lost without affecting the stability of the system. Much of the uncounted biodiversity at this level—of micro-organisms and small invertebrates—is critical for agricultural productivity.

An ecosystem is stable if it resumes its original state after disturbance. Stability can be divided into inertial stability (the capacity to resist deflection) and adjustment stability (the capacity to return to the undisturbed state). The association between the diversity of a community and its stability is not simple; often, the most diverse communities are the most fragile, though coral reefs appear to recover rapidly from small-scale disturbance. However, the oldest, least disturbed biomes—often nutrient-poor ecosystems—have the highest diversity and these are found in the tropics. A range of other factors, including a benign climate and high productivity, may be important, and also the positive feedback of high species richness that creates further niche opportunities.

The capacity of some of these ecosystems to withstand major disturbance is now being tested—especially in tropical forests and coral reefs—and this has major economic and political consequences. As habitats and species are lost, the ecological functions necessary for sustaining economic activity are being constrained, and this threatens the way of life of some local peoples.

● FURTHER READING

Jordan, C. F. 1995. *Conservation*. John Wiley & Sons, New York. Provides an excellent overview of the ecological, social, and economic issues surrounding conservation, especially in tropical regions.

May, R. M. and Lawton, J. H. (eds). 1994. *Extinction Rates*. Oxford University Press, Oxford. A useful collection of essays providing our current understanding of extinction rates in different ecosystems.

Wilson, E. O. 2002. *The Future of Life*. Abacus, London. A highly readable digest of the consequences of our individual choices, from the ecological to the economic.

● WEB PAGES

Details of the effort to map coral reefs and the beautiful Landsat 7 images from space can be found at:
http://earthobservatory.nasa.gov/Study/Coral/

and similarly, the scale of tropical deforestation and forest fires from Earth orbit can be seen at:
http://earthobservatory.nasa.gov

More especially for images of human impact on the Earth:
http://images.jsc.nasa.gov/

The latest Red List of threatened species is available at the IUCN site:
http://www.iucn.org

and also at:
http://www.redlist.org/

Attempts at measuring global biodiversity are described at:
http://www.all-species.org/

Useful and up-to-date data on various global environmental parameters, including the current estimates of world population trends from the UN, can be found at the World Resources Institute:
http://earthtrends.wri.org/index.cfm

For a detailed commentary on and insight into the important factors and issues associated with human population growth, visit the Population Reference Bureau:
http://www.prb.org

Grief

Local ecological change can have massive social and economic consequences. For many, climate change may seem a distant and nebulous threat, but for others it is their everyday reality. From the poles to the Equator, there are signs of long-term change—the collapse of native American villages with the thawing of the Alaskan permafrost, or the inundation of tropical islands because their coral reefs no longer provide protection against high seas.

Coral reefs are ecosystems constructed by animals, and the elaborate scaffold they create in warm, shallow, and well-lit waters supports a remarkable diversity of species. However, reefs across the globe are under threat as large areas have become 'bleached'—a loss of colour as pollution or high water temperatures induce the polyps to eject their algal partners (Figure 9.7). Coral bleaching on a global scale occurred in 1998 associated with an 'El Niño', a periodic shift in the major currents of the Pacific that cause warm waters to dominate the southern ocean. At this time, the Indian Ocean was particularly affected, especially the isolated reefs around the Seychelles (Figure 9.19). These lost more than 90 per cent of their live coral. Work by Graham and colleagues (Graham *et al.* 2006) has measured scale of change on the Seychelle reefs and the impact on their fish communities.

This study compared reef structures and fish diversity in 2005, seven years after the bleaching episode with data collected from a survey conducted four years prior to the El Niño. Although the reefs supported several important elements of the local economy (such as fishing), these pressures had been relatively constant for some time, and there had been no significant decline in reef condition prior to 1998. Consequently, the changes seen by Graham's team were almost certainly due to the persistently warm waters of the El Niño event.

On each occasion more than 50 000 m² of reef was sampled over 16 locations across 21 sites, and three distinct habitat types (fringing reefs, corals growing on a granite substrate and patch reef on sand, rubble or rock). The percentage cover of different types of coral (or its remains)

Figure 9.19 The Seychelles. These islands lie between Madagascar and India, more than a 1000 km from the coastal fringe of any sizeable land mass. The reefs can be seen marking the island outline in the offshore shallows.

and the structural complexity of the reef were scored on a scale from 1 to 6. Additionally, the abundance of 134 reef fish species was recorded in each sampled area.

The diversity of the fish was assessed using a derivative of an index of taxonomic diversity to represent fully the phylogenetic range within each community (Box 9.1). The taxonomic distance between all possible pairs of species (the number of taxonomic steps that separate them) was used to calculate the **average taxonomic distinctness** (AvTD). This index quantifies species equitability as the average taxonomic distance for any two individuals chosen at random (Clarke and Warwick, 1998). Put simply, the most abundant species are sampled most frequently and thus dominate the calculation of the average taxonomic distance for any sample of the fish communities. AvTD expresses this average taxonomic distance as a proportion of its minimum possible value (if all species come from the same genus). The higher the value, the greater the taxonomic spread within a sample, an index that is largely insensitive to sample size.

Prior to 1998, reefs around the Seychelles were 'coral-dominated' with a high proportion of live coral and high structural complexity. However, the 2005 survey found these largely replaced by seaweed, rubble, or dead standing coral, with live coral cover of just 7.5 per cent. The remaining living corals are massive encrusting forms (rather than soft or branching forms), providing little structural complexity and a reduced range of habitats for reef-dwelling animals.

In fact, this loss of structural complexity explains about half of the reduction in fish species richness, and is a more important factor than the loss of the living coral itself (Figure 9.20). Overall there was a significant decline in AvTD for the fish communities associated with all reef types, though some (those growing on a granite substrate) were more resilient than the others. Because it measures the scale of adaptive differences between taxa—groups further apart in the phylogenetic hierarchy have greater differences in their adaptive traits—AvTD is an indication of the range of habitats (and niches) represented in the sample. Its decline is therefore likely to be a good indicator of habitat loss on the reefs.

Reef fish are highly specialized and adapted to finely divided niches. The loss of habitat leads to increased competition for the remaining space and higher rates of predation of the smaller fish. Unfortunately, it may also allow the remaining fish to be caught more readily by fishermen. Additionally, there is a loss of species which feed directly on

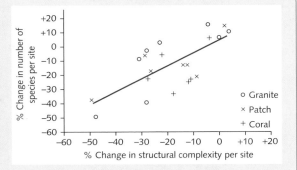

Figure 9.20 The decline in fish species richness correlates with the decline in the structural complexity of the Seychelles reefs between 1994 and 2005. The analysis of Graham *et al.* (2006) combined data from three types of reef (based on their principal substrate) and showed that the structural features of the reef were more important for fish diversity that the presence of living coral. Reefs formed on granite were the most resilient.

the coral or the plankton associated with a reef. Perhaps four species of fish could become locally extinct around the Seychelles but six other species have critically low populations. Interestingly, while the species richness amongst herbivorous fish has gone down, and the remaining populations are reduced, the overall biomass of this group has not changed. Consequently the reef is now dominated by older and larger herbivores, fish that were alive before the 1998 El Niño.

A variety of functional groups of fish have declined on the reef and this is thought to have lowered the stability and resilience of the reef community. Disturbance is part of the ecology of all coral reefs and probably contributes to their diversity (Nyström *et al.* 2000), at least under normal regimes of change. Recovery can be very rapid when the disturbance is moderate and when there are nearby sources for recolonization. The changes seen in the Seychelles repeat a pattern seen in Caribbean reefs, where some key functions, such as grazing the macro-algae (seaweeds), are now performed by fewer species. The reduced diversity within functional groups, especially amongst the herbivores, increases the fragility of an ecosystem, making it less stable (Nyström *et al.* 2000).

The Graham team asks why the Seychelles have failed to recover since 1998, when reefs elsewhere have restored

(continued overleaf)

themselves to a much greater extent. The poor recruitment of coral larvae and other key species is almost certainly due to the isolation of these islands, but other factors are known to be important. The lack of structural complexity and the inability of the herbivores—fish and others—to control the growth of the seaweed are thought to be critical. Fishing had already reduced the herbivore population, and the loss of habitat complexity is now threatening species from other functional groups. Intense fishing pressures have also shifted the balance between different trophic groups in the tropical and non-tropical reefs of Brazil, again with changes in their fish population structure (Floeter *et al.* 2006). Fish numbers and size increase here when there is protection but endemic species are lost when large areas of reef are destroyed.

The Graham team points out that reefs around other isolated islands may well succumb to the effects of climate change just as readily, but have yet to be surveyed properly. As the reefs lose their structural complexity they lose their resilience. They also lose their capacity to provide the important services upon which the ecology and economics of the islands depend, from food and coastal protection to income from tourism. Although coral reefs may have been able to reconstruct themselves quickly in the past (Nyström *et al.* 2000), persistent warming of the oceans may now be exceeding their elasticity.

References

Clarke, K. R. and Warwick, R. M. 1998. A taxonomic distinctness index and its statistical properties. *Journal of Applied Ecology 35*, 523–531.

Floeter, S. R., Halpern, B. S., and Ferreira, C. E. L. 2006. Effects of fishing and protection on Brazilian reef fishes. *Biologial Conservation 128*, 391–402.

Graham, N. A. J., Wilson, S. K., Jennings, S., Polunin, N. V. C., Bijoux, J. P., and Robinson, J. 2006. Dynamic fragility of oceanic coral reef ecosystems. *Proceedings of the National Academy of Sciences USA. 103*, 8425–8429.

Nyström, M., Folke, C., and Moberg, F. 2000. Coral reef disturbance and resilience in a human dominated environment. *Trends in Ecology and Evolution 15*, 413–417.

● EXERCISES

1. (a) Look at the two diagrams below and decide which sample shows the greatest taxonomic diversity and give your reason. The individuals collected in each sample are shown by the filled boxes—assume that in each case only a single individual was found for each species.

   (b) Calculate the average taxonomic distance for each sample, given that the phylogenetic links are weighted as follows:
   two individuals from the same species = 0
   two species belong to the same genus = 1
   two species belong to different genera = 2
   two species belong to different family = 3

   Remember that the average is derived from the taxonomic distance between all possible combinations of the individuals collected.

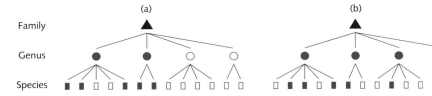

2. Why should the sea floors of deep oceans have
   (a) some of the oldest taxa;
   (b) high species richness;
   (c) few dominant (abundant) species?

3. Which of the following statements best explains how hierarchy theory (Section 8.1) might explain the buffering role of compartments in complex communities?
   (a) Energy stays within a compartment so disruption is not passed to the larger community.
   (b) There are strong links between compartments that hold the community together.
   (c) There are strong links within compartments so that the loss of one species only causes that compartment to collapse.
   (d) There are weak links between species in different compartments and the loss of one species in a compartment is not transmitted to the larger community.

4. Coral reefs are thought to recover from disturbance more readily than tropical forests because (select the best answer):
   (a) coral reefs are long lived and readily colonized from surrounding reefs.
   (b) coral reefs have 'incorporated' disturbance and are readily colonized from surrounding reefs.
   (c) corals are often endemic to an area and speciate rapidly.
   (d) corals use zooxanthellae as symbionts and that allows then to build the coral superstructure rapidly.

5. The graph below shows the change in nitrogen fixation in three different soil communities with bacterial species richness. Which has the greatest functional redundancy and which the least? Write a sentence justifying your choices.

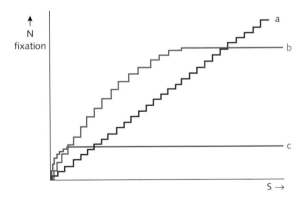

6. Which island is likely to have (i) the highest equilibrium S; (ii) the highest rate of colonization; (iii) the highest rate of extinction?
   (a) near and small
   (b) near and large
   (c) far and small
   (d) far and large

Tutorial/seminar topics

7. Review the data in Figure 9.4. To what extent does the number of threatened species match their species richness across America? You could prepare for this tutorial by

   (a) calculating (roughly) the proportion of endangered species in each country, for each group of organisms;

   (b) choosing one group and looking at the particular threats in two or three countries from this range;

   (c) offering reasons why simple comparisons between countries are not strictly possible.

   Suggest ways in which you could improve the comparability of these data.

8. The population of the polar bears in the western Hudson Bay has a declining natality rate but a reasonably stable population. What does this suggest about the age structure of its population and its capacity for population growth in future years?

9. Can individuals, companies, or nations own the rights to the genetic code of a species? Do we, as individuals, own our genotype? What are the moral implications of your conclusions?

GLOSSARY

Adaptation The collection of characters that enable an organism to survive. An organism may adapt in the short term, within its lifetime, by adjustments of its behaviour, physiology, or development. A species may adapt over generations by changes to its genetic code, through natural selection (*qv.*).

Admixture The flow of genes between two populations.

Age structures The proportion of individuals in each age class of a population. A stable age distribution is dominated by the younger age classes and is characteristic of a rapidly growing population. Closer to the carrying capacity a stationary age distribution is approached when the distribution is far more even.

Allele The various forms of a gene found at a particular position (locus) on a chromosome, and which thereby code for a particular character.

Allelopathy The inhibition of one organism by another using chemical means. Some plants may inhibit the growth of others (especially those of the same species) by the chemical nature of their litter or by special secretions.

Allopatric speciation See speciation.

Amino acid sequencing The chemical analysis of proteins by determining the sequence of individual amino acids along the polypeptide chain. This method can be used to measure the degree of relatedness between two species by comparing the similarity between the sequences.

Anaerobic (anoxic) An environment that is devoid of or has very low concentrations of oxygen; aerobic metabolism uses oxygen, anaerobic metabolism does not.

Annual A species that completes its life cycle within a year.

Antibody A protein produced by some animals in response to a foreign substance (antigen) as part of their immune system.

Apomict (micro-species) A species that reproduces without fertilization.

Aquifer A natural reservoir of groundwater held within porous rocks

Asexual reproduction Reproduction in which there is no swapping of genetic material between individuals.

Archaea A primitive form of bacteria typically from extreme environments often with a distinctive metabolism. A group forming one of three domains (*qv.*).

Assimilation efficiency A measure of the efficiency by which an organism assimilates energy from the food it consumes (this is given by the amount of energy assimilated divided by the amount of energy consumed).

Assortative mating The tendency of individuals to mate with those with a similar genotype. This form of mate recognition reduces variation within a population and favours certain traits.

Autecology The study of the relationship between a single species and its environment.

Autotroph An organism able to obtain its energy either from sunlight (*cf.* photoautotroph) or from chemicals (*cf.* chemoautotroph).

Authority The person who first describes and names a particular species.

Batesian mimicry A form of mimicry in which a harmless species resembles a harmful or poisonous species to deter potential predators.

Benthic communities Those found on the bed of an aquatic ecosystem (qv.). Pelagic communities are found in the open water.

Biennial A plant that lives for two years, with growth and establishment in the first followed by maturation fruiting and death in the second.

Binomial system The system used to formally name species, using a generi name followed by the specific name (*e.g. Canis rufus*).

Bioaccumulation An increase in the concentration of a pollutant from water and diet by a living organism.

Biochemical oxygen demand (BOD) A measure of the amount of organic matter in water.

Bioconcentration An increase in the concentration of a pollutant from water by a living organism (*qv.* bioaccumulation, biomagnification).

Biodiversity Often taken to be simply the number of species (cf. species richness), but also used to encompass biological variety at all levels, from the genetic diversity between the individuals of a species to the total number of species on the planet.

Biological control Pest control that makes use of a species' natural enemies.

Biological species concept Species defined as a group of individuals able to breed together and produce viable fertile offspring.

Biomagnification An increase in the concentration in the tissues of a living organism either from the soil (in plants) or from the diet (animals) (*qv.* bioaccumulation).

Biomanipulation Changing the structure of an ecological community to bring about long-term changes in ecosystem (qv.) structure and processes. This technique is used especially with eutrophic freshwaters (*qv.*).

Biomass The weight of living material.

Biome The characteristic vegetational community associated with a particular latitude or biogeographical region, such as tropical forests or temperate grasslands, and primarily defined by climate.

Biosphere The part of the Earth that supports living organisms; the global ecosystem (qv.).

Biotic indices The use of the species assemblage of a community as a measure of its habitat quality (*qv.* indicator species).

Brood parasite A parasitic organism that lays its eggs in a nest of another species or individual that then rears the young.

Bipedalism Walking on two legs.

Carrying capacity (K) The maximum number of individuals that can be supported by a habitat (*qv.*), the maximum population size for that habitat. As this number is reached, environmental resistance, the effects of competition for space and other resources, and the build-up of wastes, slows the growth of a population.

Catchable stock Individuals sufficiently large to be harvested, such as fish being caught in nets with a certain mesh size.

Cation-exchange capacity (CEC) The total number of cations (positively-charged atoms or molecules) that can be bound by a soil, a measure of its capacity to hold essential elements and buffer its acidity. CEC is high in organic and clay-rich soils, and one indication of their fertility.

Chamaephyte Low-growing woody shrub, typical of maquis- or chaparral-type habitats (*qv.*).

Character displacement The divergence of a trait over several generations and as a result of genetic change between two or more species competing for the same resource.

Chemoautotroph Micro-organisms that derive their energy from oxidizing inorganic compounds and use carbon dioxide as their carbon source. These include the nitrifying bacteria.

Chimera An organism that consists of two genetically distinct cell types.

Chromosome The main site of the genetic material in living cells, composed of DNA (*qv.*) and protein.

CITES Convention on International Trade in Endangered Species of Flora and Fauna.

Climax community A community of species found towards the end of a successional process, when species turnover is minimized and when the plant community assumes a persistent structural form.

Cohort A group of individuals of the same age in a population.

Commensalism An association between two species of benefit to one partner only, but of no detriment to the other.

Community A collection of plants and animals that live in the same habitat (*qv.*) and interact with each other.

Compartmentation The extent to which a community, or more specifically a food web, is divided into discrete units each with strong internal interactions between its members, but weaker links to others.

Compensation point Condition in plants when energy fixed in photosynthesis is balanced by energy used in respiration. The extinction of light down a water column means this is represented by a certain depth in aquatic ecosystems (*qv.*).

Competition The fight for a limited resource between individuals of the same species (intraspecific competition) or between different species (interspecific competition). Organisms can compete for food, water, space, or other resources.

Competitive exclusion principle Two species cannot co-exist in the same niche at the same time in the same place. Of two species with identical resource requirements, one will eventually be ousted.

Concentration factor The ratio of the pollutant concentration in an organism to that of its diet or the soil in which it grows.

Connectance The number of interactions between different species within a food web expressed as a proportion of the total number of possible interactions. This is a measure of the complexity of the ecosystem (qv).

Connectivity In landscape ecology, the extent to which habitat patches are connected to each other.

Consumer A species that feeds on another species or organic matter—that is, herbivores, carnivores, parasites, and detritivores.

Convergent evolution The evolution of two or more distinct species towards a similar adaptive solution under equivalent selective pressures, often resulting in similar morphologies or appearances.

Corridor In landscape ecology, a linear habitat (*qv.*) that facilitates the movement of organisms across an inhospitable matrix (*qv.*), e.g. a roadside verge or a stream.

Cultivar A distinct form of a cultivated species that has been produced as a result of artificial selection.

Cytochrome An important protein involved in the conversion of energy within chloroplasts and mitochondria.

Decomposer An organism that feeds on dead organic matter. Detritivores feed on detritus, fragmented organic matter.

Decomposer food chain A food chain made up of decomposer (*qv.*) organisms that feed on dead organic matter.

Definitive host See Parasite.

Denitrification The conversion of nitrate to nitrite and nitrite to molecular nitrogen by a range of anaerobic bacteria.

Density-dependence Populations that grow in a density-dependent way increase their numbers in relation to the density of the existing population; usually the rate of population increase falls as the habitat (*qv.*) becomes more and more crowded. Other organisms have density-independent growth.

Diploid The $2n$ chromosome state, having received one chromosome from each parent at fertilization. Polyploidy (*qv.*) may occur when more than one of each chromosome is passed to the offspring. For example tetraploids are $4n$, having received $2n$ from each parent.

Diversity Species diversity simply refers to the number of different species in a habitat, or species richness (*qv.*). Other measurements of diversity include the relative proportions of different species, or equitability. Genetic diversity refers to the totality of the variation in the gene pool (*qv.*).

Detritivore See Decomposer.

Detritus Dead and decaying organic matter.

DNA Deoxyribose nucleic acid, the molecule that carries the genetic code.

DNA sequencing A technique used to determine the sequence of bases along a piece of DNA (*qv.*).

Domain The major taxonomic division of organisms of which there are three—Archaea (*qv.*), Bacteria, and Eukaryota (*qv.*).

Ecology The science that studies the relationship between living things and their environment.

Ecological niche The totality of adaptations of a species to the biotic and abiotic factors in its environment; a species role in its community. Niche breadth measures the range of a resource exploited by a species. Niche overlap is the range of a resource used by two species occurring together. Fundamental niche is the niche breadth that a species would use (occupy) in the absence of competitors or predators. Realized niche is the niche breadth a species is restricted to in the presence of competing species.

Ecological species concept Used within a collection of very similar species to define species according to their adaptation and functional role within their habitat (*qv.*) (niche, *qv.*).

Ecoengineering The use of vegetation in the construction and stabilization of structures.

Ecosystem A community of living organisms and its physical environment.

Ecotone The boundary between two adjoining communities.

Ecotype A variety with sufficient genetic differences from other members of its species to be recognized as a separate race.

Ectotherms Animals which, when active, attempt to regulate their body temperature by gaining heat from, or losing heat to, the environment, primarily by their behaviour (*cf.* endotherms).

Effective population size The size of the breeding population at a particular time and location, taking into account the breeding behaviour of the species.

El Niño or the *El Niño Southern Oscilliation (ENSO)* The periodic influx of warm water shifting the circulation pattern within the southern Pacific Ocean. One major effect is the halting of the cold current off Ecuador and Peru, which causes drastic changes to the marine communities that depend on the nutrients this current delivers. An El Niño also causes significant changes in regional climates, most especially increased hurricane activity.

Eluviation The movement of soluble nutrients and fine particulates around a soil profile by water.

Emergent property A characteristic of a population or community that could not be detected by looking at individuals, which only emerges from seeing the assemblage operating together.

Endemic species Native to and only found in an area.

Endosymbiont theory The theory that suggests that a number of organelles within the eukaryotic cell have their origins in primitive prokaryotes (*qv.*) that evolved a symbiotic relationship with the host cell.

Endotherms Animals that regulate their body temperature by generating heat internally through their metabolism (*cf.* ectotherms).

Energy subsidy The supplemental energy used in agriculture to aid cultivation (in the form of fuel, fertilizers, and pesticides).

Epiphyte A plant that grows entirely on another plant. Typical epiphytes include the Bromeliads of tropical forests.

Eukaryotes Organisms whose cells have a discrete nucleus within a membrane. Eukaryotes include all algae, fungi, and all multicellular plants and animals.

Eutrophication Changes in the structure of an ecological community as a result of nutrient enrichment, most often phosphate and nitrogen enrichment.

Evolution A gradual, directional change in the characters of an organism. The process by which one species might arise from another (cf. natural selection).

Extinction vortex A descriptive title given to the three factors (genetic, population, and habitat, *qv.*) that can act in combination to drive a population extinct.

Evapotranspiration The total loss of water from the Earth's surface, from soil and water bodies, and from plants via their transpiration stream.

Evolutionary Significant Unit (ESU) A population considered to be on a separate evolutionary trajectory than other populations of that species and which, if left, may well give rise to a new species. (*cf.* cohesion species concept).

Extremophile An organism adapted to extreme conditions (frequently applied to microbes).

Facilitation An effect where the activity or presence of one species allows other species to colonize a habitat (*qv.*), as part of the process of succession (*qv.*).

Fishing Mortality (F) The loss (in terms of numbers or biomass) from a population to a fishery; usually the proportion caught.

Flux rate The rate of turnover of an element as it passes from inorganic reservoirs through living systems and returns to the reservoirs.

Food web The trophic or feeding relations betwen different members of a community. A food chain consists of a series of trophic levels—primary producer, herbivore, carnivore.

Fossil record The entire catalogue of fossilized plants and animals that have been dated in geological time.

Founder effect The genetic differentiation of a small population isolated from a larger (parent) population. In a small gene pool, some alleles may have different frequencies compared to the main population and these differences become more pronounced as the isolated population interbreeds.

Fugitive species Species that maintain a population by colonizing ephemeral gaps. Usually applied to plant species that persist in habitats with a predictable frequency of disturbance.

Functional redundancy The presence of a number of species within a community whose activity has no significance for its ecological processes. Such species can be lost without any major impact on the nature of the community.

Functional response A response of a predator to prey numbers in which it consumes more prey as they become available.

Gaia hypothesis The theory that the atmosphere of the Earth is maintained in its present form by the biological activity in the biosphere.

Gametes The sex cells, eggs (ova), or sperm, which are haploid, having a single set of chromosomes. The fertilization of an ovum by a sperm produces a diploid zygote with two sets of chromosomes.

GCM (Global Circulation Models) Computer models used to describe and predict the behaviour of the climate under the influence of the land, the oceans, and the chemistry of the atmosphere.

Gel electrophoresis A technique that separates proteins and other chemicals using an electrical current passing through agar gel. The molecules separate out according to their electrical properties and size.

Gene A sequence of nucleotides on the DNA (*qv.*) molecule that is inherited as a unit and which codes for a specific RNA (*qv.*) molecule or a polypeptide for which that RNA molecule is responsible.

Gene pool/Genome The totality of genetic material of a species.

Generalist species A species that is *r*-selected and has an opportunistic life-history strategy (*qv.*).

Generation The individuals belonging to one age class within a population.

Generic name See Binomial system.

Gene sequencing See DNA sequencing.

Genet An organism that grows from a fertilized egg and is therefore genetically distinct. In contrast a ramet is an individual that has arisen by asexual reproduction.

Genetic drift The genetic differentiation of a small, isolated population through random changes (and independent of any selective pressure).

Genetic fixation The tendency for individuals within a population to share the same allele (*qv.*), as variation is lost with time. This happens most rapidly in small populations, so that all individuals quickly become homozygous (*qv.*) at that locus.

Genotype The genetic code of an individual, or sometimes, rather confusingly, all individuals that share that code (cf. phenotype).

Genotypic variation Variation between individuals that is wholly attributable to differences in their genetic code. Genetic diversity refers to the totality of the variation in the gene pool (*qv.*).

Genus A category between species and family in which a number of closely related species are grouped.

Geomorphology The study of landforms. Topography is the study of surface features of an area.

Good gene hypothesis Genes that confer traits which signal the fitness of the individual to potential mates—see sexual selection.

Grain In landscape ecology, the scale of resolution by which an organism resolves significant features of its habitat (*qv.*).

Grazing food chain A food chain involving plant material as the primary producer and herbivores as the primary consumers.

Green revolution A term used to describe high-tech agricultural innovation in plant breeding and cultivation.

Greenhouse effect The capacity of the Earth's atmosphere to absorb long-wave (infra-red) radiation, while being largely transparent to the incoming short-wave radiation.

Gross primary production (GPP) The total amount of energy fixed by plants within a given area.

Growth efficiency A measure of efficiency by which an organism converts energy assimilated into new tissues (the amount of energy fixed in tissues divided by the amount of energy assimilated).

Guild A group of organisms sharing a similar set of adaptations and carrying out a similar role or exploiting the same resource, within a habitat (*qv.*).

Habitat The place where an organism lives; sometimes used in the general sense to refer to the type of place in which it lives.

Haploid Having a single set of chromosomes—the 1*n* state, usually associated with a gamete (egg or sperm).

Haplotype The type specimen for a species, whose characters define the species description.

Herbivore An animal that feeds on plants.

Heterotroph An organism that obtains its energy from other species; consumers.

Heterozygote An individual that, for a particular character, has different genes on each of its paired chromosomes. A homozygote individual has the same alleles (*qv.*) on each chromosome (*qv.*).

Hierarchy theory The proposal that large and complex systems are self-organizing by virtue of the constraints and limitations imposed by the interactions between different levels. A level is defined by the spatial and temporal scale over which key processes operate. Levels are nested, or formed into a hierarchy, in which a process occurs fastest at the smallest scales; change is slowest at the largest scales.

Holoparasite A plant parasite that is totally dependent on its host and cannot photosynthesize (see also parasite).

Homoplasy A structural resemblance arising out of the convergent evolution of two different species or groups.

Homozygous With the same allele (*qv.*) on each chromosome (*qv.*) (cf. heterozygote).

Humus The highly degraded organic matter within a soil.

Hunter-gatherer A non-settled form of agricultural activity in which people harvest animals and plants from the wild.

Hybrid A cross-bred individual derived from gametes (*qv.*) from different populations, species, or genera.

Hybrid breakdown A series of mechanisms that prevent hybrids (*qv.*) from breeding.

Hydrological cycle The movement of water from oceans onto land and back again.

Immunological methods Techniques that use antibodies to detect the similarity or difference between proteins and that operate on the basis on the specificity of the reaction between antibodies and antigen.

Inbreeding depression The reduced reproductive potential of a population in which individuals share much of the same genetic code. Conversely, outbreeding depression occurs when the genetic differences between individuals are too large to produce viable offspring.

Independent assortment The random allocation of alleles (*qv.*) to the gametes (*qv.*) at meiosis (*qv.*).

Indicator species Species that are characteristic of certain environments (soil, climates, etc.) or can be used to measure habitat quality. Sentinel species are used as measures of pollution levels in a habitat).

Infield–outfield A form of agricultural activity in which grazing and cultivation takes place in enclosed areas around a settlement whilst the surrounding areas are used for grazing.

Inhibition The capacity of one species to inhibit the colonization of another, especially in a successional sequence.

Illuviation The leaching and deposition of material from one part of a soil profile to another.

In-situ conservation Conservation effort to protect an endangered species in its natural habitat. *Ex-situ* conservation refers to captive breeding or culturing of an endangered species in a botanic or zoological garden.

Integrated pest management A method of pest control that includes the use of biological, chemical, and cultural techniques.

Intermediate disturbance hypothesis The suggestion that maximum diversity occurs in communities subject to disturbance that is frequent enough to prevent competitive species becoming dominant but which is not so frequent enough to depress colonization rates by immigrant species.

Interspecific competition See competition.

Intraspecific competition See competition.

In-vitro fertilization Fertilization of an egg in the laboratory.

Island-biogeography theory A model predicting a dynamic equilibrium of species number in a habitat (*qv.*), as a result of the opposing processes of colonization and extinction. This is allied to the species–area relationship (*qv.*).

Joule The SI unit of energy, the amount of work needed to move 1 kg through 1 m.

K The carrying capacity (*qv.*) for a population in a limited environment.

K-selection Applied to an organism whose reproductive and life history strategies are primarily adapted to life in an unchanging and limited environment where there is intense competition for resources (cf. *r*-selection).

Keystone species An organism whose abundance or activity is central to maintaining the nature of a habitat (*qv.*).

Kinetic energy Energy of a body due to its motion.

Kingdom Five broad taxonomic categories that are divided up between the Superkingdoms—Prokaryota (*qv.*) and Eukaryota (*qv.*).

Landscape ecology The study of ecological processes across several ecosystems (*qv.*) united by a shared landform, climate, and regime of disturbance. Adjacent landscapes under an equivalent climate or topographical area may be collected together as a region.

Laterite A hard red-brown clay, devoid of many soluble minerals and nutrients, and which forms in tropical areas dominated by ancient soils.

Leaching The loss of material when it is washed out of a soil.

Life history strategy The allocation of time and resources that an organism makes between different stages of its life cycle so as to maximize its reproductive potential.

Life table A summary of the rates of mortality and survivorship of different age groups in a population.

Loam A well-balanced mixture of sand, silt, and clay with abundant organic matter producing a well-drained soil with a good crumb structure (cf. soil crumb structure).

Locus The site of an individual gene (*qv.*) on a chromosome (*qv.*).

Macrophyte An aquatic plant that is either rooted or attached to a submerged substrate (e.g. sand, mud, or rock) and with a stem that holds the leaves close to or above the water surface.

Matrix In landscape ecology (*qv.*) the dominant component of the landscape—e.g. agricultural fields. Often an inhospitable space that the organism of interest has to cross to reach habitat (*qv.*) patches.

Maximum sustainable yield (MSY) The largest number of individuals or mass that can be harvested from a population year on year without damaging its reproductive potential. Detailed models use the age structure of the population to estimate the fishing intensity that will provide the maximum yield in terms of biomass or numbers.

Meiosis The division of the paired chromosomes (*qv.*) necessary to form a gamete (*qv.*). In contrast, mitosis divides the nucleus but not the chromosomes, as a preclude to cell division.

Metapopulation A population of populations. A series of populations that may swap individuals with each other, but are usually divided into discrete patches.

Microcosm An ecosystem (*qv.*) or part of an ecosystem isolated in the laboratory for experimental purposes, for example, an aquarium. A mesocosm is part of a real ecosystem partitioned for the same reason, sometimes called enclosures.

Mitochondria Organelles, the site of cellular respiration in eukaryotes (*qv.*).

Mitochondrial DNA Non-nuclear DNA (*qv.*) that is contained within the mitochondria essential for maintaining their structure, function and replication.

Monera The taxonomic category comprising prokaryotic (*qv.*) single-celled organisms which lack a true nucleus and membrane-bound organelles.

Mor and Mull soil types Mor are acidic forest soils, typical of cold, wet soils under coniferous forest where decomposition is slow. Mull soils have their organic matter incorporated into the mineral soil, are relatively dry and neutral to alkaline.

Mortality rate (m) The death rate.

Mullerian mimicry Warning signals (patterns and coloration) adopted by potential prey that signal danger to predators (e.g. the yellow and black stripes of the wasp).

Mutation A change in the genetic code that may be inherited.

Mutualism A form of symbiotic relationship between two species to their mutual benefit.

Mycorrhiza A symbiotic (*qv.*) association between plant and fungus, where the plant root acts as a host to a network of fungal threads. The plant benefits by having an enlarged 'root' system to collect more phosphates and water, and the fungus benefits from the supply of carbon from the plant.

Natality rate (b) The birth rate.

Natural selection The reproductive success of different individuals under the constraints placed upon them by their environment. Less fit individuals fail to reproduce and their genes are lost from the gene pool (*qv.*). The persistence of particular adaptive traits through the generations produces the evolutionary change that may produce new species.

Net primary production (NPP) The amount of energy fixed within plant tissues after metabolic and photosynthetic energy costs have been met.

Neutral variation Variation within a population or species that confers no selective advantage.

Niche, Niche breadth, Niche overlap See ecological niche.

Nitrification A process carried out by a range of soil microbes by which nitrite is oxidized to nitrate.

Nomadic-pastoralist A semi-settled form of agriculture involving the management of grazing herds by setting up camp and then moving along with them.

Nucleotides The basic unit of DNA (*qv.*) and RNA (*qv.*) comprising a phosphate group, a sugar group, and a nitrogenous base.

Nucleus In eukaryotic cells, the organelle in which chromosomes are held.

Numerical response The increase in numbers of predators as a response to the increased abundance of its prey.

Oligotrophic Nutrient-poor; oligotrophic ecosystems (*qv.*) have low primary productivity (*qv.*). Eutrophic (*qv.*) ecosystems are nutrient rich and have high primary productivity.

Ombrotrophs Plants that obtain their nutrients from precipitation.

Omnivore An animal that feeds on plants and animals.

Optimal yield The largest number of individuals that can be harvested from a population. In simple population models, this is at half the carrying capacity (*qv.*), when the fastest rate of population growth is achieved. Optimal yield takes no account of the age structure of the population.

Parapatric speciation See speciation.

Parasite An organism that is metabolically dependent on another, at the expense of the host. A definitive host is one in which the parasite matures (and may reproduce sexually); an intermediate host serves as a vector in transmitting immature stages between definitive hosts.

Parasitoid Insect parasites (primarily Hymenoptera and Diptera) that lay an egg within a host insect, which eventually leads to its death. Often seen as a specialized form of parasitism and predation.

Parthenogenesis The development of an egg without fertilization. Where the egg is diploid the offspring represents a clone of the parent.

Patch Element(s) of an organisms' landscape that it can utilize, e.g. a pond or woodland.

Pathogen An organism or virus that can cause disease.

Perennial A plant that lives for more than two years.

Permafrost The soil layer found at depth beneath tundra vegetation that remains frozen throughout the year.

Pest A species whose presence causes a nuisance and results in some economic cost.

Pheromone A chemical signal used by one individual to alter the behaviour of another, usually of the same species, perhaps to attract a mate.

Phenotype The characteristics of an individual, the product of the expression of its genetic code and its interaction with the organism's environment.

Phenotypic variation Variation between individuals that is wholly attributable to physiological or developmental (i.e. non-genetic) differences.

Photoautotroph An organism that obtains its energy from sunlight through the process of photosynthesis.

Phylogeny The evolutionary relationships and history of an organism.

Physiognomy The characteristic features of a plant community, reflecting their adaptation to life in that environment.

Plagioclimax A succession that has been deflected from its normal climax state (*qv.*), often a result of human activity, and with a different plant community as a result.

Polymorphism Variation within a species where individuals may take different forms.

Polyploidy Organisms that possess multiple copies of the entire genome.

Population Individuals of the same species living in a defined area at a defined time.

Predator An animal that kills another animal to feed.

Pre-zygotic barriers A range of mechanisms that prevent the formation of a hybrid zygote from gametes (*qv.*) from two different species or varieties.

Primary producer An autotroph (*qv.*). A secondary producer derives its energy from a primary producer.

Primary production The synthesis of complex organic molecules from simple inorganic ones by autotrophs (*qv.*), using either sunlight (photoautotrophs, *qv.*) or chemical energy (chemoautotrophs, *qv.*).

Principal component analysis (PCA) A statistical technique for summarizing a large dataset, where entities (e.g. skulls or ecosystems, *qv.*) can be defined by a collection of measurements. PCA reduces these to one or two 'principal components' which, in this combination, account for most of the differences between the entities.

Prokaryotes Single-celled organisms without a well-defined nucleus; the Bacteria and the Archaea.

Protista The taxonomic group that includes protozoa, primitive algae, and fungi.

Production efficiency A measure of efficiency by which an organism converts the energy from its food into tissues (the ratio of the amount of energy fixed in its tissues to the amount consumed).

Productivity The energy fixed in the biomass of an individual, population, species, or trophic level.

Post-zygotic barrier Some failure in the subsequent development of the fertilized egg or the individual, which makes it non-viable, typically in hybrids.

Potential energy Energy that is stored (e.g. chemically as starch and oils in plants, or fat and glycogen in animals).

r (r_m) The rate of change in a population per individual. r_m is the maximum possible rate of population growth per individual under ideal conditions.

r-selection Organisms that are r-selected are primarily adapted to life in changeable, short-lived habitats, where rapid population growth is favoured (*cf.* K-selection).

Recombination The swapping of fragments of genetic code between chromosomes (*qv.*) during meiosis (*qv.*) (sometimes called crossing over) and which may therefore produce a new genotype (*qv.*).

Recruitment The addition of new individuals to the catchable stock (*qv.*) from growth or immigration.

Reproductive rate (R_o) The average number of offspring per individual in a given time. The net reproductive rate (R_N) is the average number alive per individual in the next generation, allowing for births, deaths, and survivors.

Resource spectrum The range of a resource that is available to organisms in a habitat (*qv.*). Different species may use different parts of the spectrum—say food items of different sizes.

Realized niche See ecological niche.

Reclamation The process of restoring a degraded area to a state which is more ecologically sustainable and aesthetically pleasing.

Rehabilitation The restoration of an ecosystem (*qv.*) part way towards an intended endpoint, which allows the site to continue developing naturally through succession.

Replacement An alternative endpoint to a restoration programme, selected either because the original ecosystem (*qv.*) is impossible to recreate or alternatively had little ecological value.

Ribosome Site of protein synthesis in the cell.

RNA Ribonucleic acid. A polynucloetide molecule that exists in several forms in the cell and carries out various functions, especially in protein synthesis.

Ruderal Plant species characteristic of disturbed and temporary habitats (*qv.*).

Rules of assembly The suggestion that there are certain ways in which a community can be configured, particularly regarding the structure of food webs. If there are rules for assembling species into a long-lasting community, we expect to find equivalent niches, allocations of resources and interactions between species under equivalent abiotic conditions.

Saprotrophs Organisms that derive their nutrients from dead and decaying organisms.

Sclerophyllous Used to describe individual plants or a plant community with small, tough, and leathery leaves, adapted for drought. Typical of mediterranean-type communities.

Secondary consumer An organism that is dependent on primary consumers (herbivores) for its energy needs.

Secondary plant metabolites Biologically active compounds produced by plants that also have a defensive role against herbivores.

Segregation The separation of homologous chromosomes (*qv.*) at meiosis (*qv.*) into separate cells.

Sexual dimorphism Phenotypic differences between the sexes that have adaptive significance, e.g. cryptic coloration in female birds and bright display plumage in males.

Sexual reproduction Reproduction that requires the fusion of two haploid (*qv.*) gametes (*qv.*) to form a diploid (*qv.*) cell in a process of fertilization.

Sexual selection The selection of individuals according to particular traits by a potential mate. This often includes courtship behaviour and is commonly taken to be a means of selecting 'fit' partners.

Soil crumb The crumb structure of a soil is its capacity to form small cohering lumps. Soils with a good crumb will not waterlog but will have a high water-holding capacity.

Soil horizon A distinct layer (usually distinguished by colour or texture) in a soil profile that demarcates a zone of biological or chemical activity.

Specialist species A *K*-selected (*qv.*) species is closely adapted to its habitat (*qv.*) and a particular part of a resource spectrum. Invariably highly competitive species.

Speciation The formation of new species (*qv.*) as a result of evolution by natural selection (*qv.*). Speciation can be allopatric in which species become genetically distinct because they are geographically isolated, or sympatric, when genetic isolation occurs between populations with overlapping distributions. Parapatric speciation is where adjacent populations diverge, say along some environmental gradient or by exploiting different resources.

Species A collection of individuals able to breed with each other and produce viable (fertile) offspring. Note that this is a functional definition: the simple concept of a species as the basic unit of biological classification is now regarded as untenable, and is used largely as a convenient unit in taxonomy.

Species-area relationship A highly consistent pattern of increase in species number (or species richness) with larger area or sample size.

Species richness, species diversity The number of species (*qv.*) in a habitat (*qv.*) or S. Species equitability is the proportion of individuals in each species, which is used in combination with S in some indices to measure diversity (*qv.*).

Specific name See Binomial system.

Stability The capacity of a system to resist change (inertial stability) or to return to its original position or original rate of change following a disturbance (elasticity or adjustment stability).

Stabilizing selection The favouring of the common form by natural selection (*qv.*), working against extreme phenotypes.

Stable isotope analysis A technique using naturally occurring isotopes that do not decay to trace the trophic transfers through an ecosystem (*qv.*).

Stock recruitment potential (SRP) The capacity of a population to produce viable eggs and larvae.

Stress tolerators A category of plants that have a life history strategy (*qv.*) adapted to life in hostile conditions.

Succession A directional sequence of changes in the species composition of a community. A primary succession is where there has been no previous life on a site; a secondary succession re-establishes a community where some nutrients or organic matter remain from its previous occupants.

Superkingdom The broadest taxonomic groups. Two Superkingdoms encompass the five Kingdoms.

Surplus yield The production (either in terms of numbers or biomass) of a population in excess of that needed to maintain its reproductive capacity. This is the amount of fish that can be harvested without detriment to the reproductive rate.

Sustainability In the simplest terms, maintaining the capacity of an ecological system over the long term to carry out a particular function or to provide a given level of productivity.

Symbiosis The association of two species living together. A variety of associations are possible. Mutualism (*qv.*) describes where each species gains an advantage from the association.

Sympatric speciation See Speciation.

Systematics Classifying organisms according to their evolutionary relationships.

Taxonomic distance The number of taxonomic steps separating two individuals, or the average for an entire sample of a community. E.g. two species belonging to the same genus would be separated by a taxonomic distance of 1, two sharing the same family a distance of 2. Used in some measures of biodiversity within a community.

Taxonomy The description, naming, and classification of organisms.

Thermohaline The global circulation of water of different salinities in massive deep ocean currents. The thermohaline is a major disperser of heat across the planet.

Tissue culture Growing cells and tissues from higher plants and animals in the laboratory.

Tolerance (i) The ability of a species to survive adverse or toxic conditions; (ii) in a succession, the capacity of many late successional plants to persist in poor growing conditions, typically low nutrient levels.

Totipotent The capacity of some cells to develop and differentiate into a variety of specialized cell types.

Transcription The creation of a molecule of messenger RNA (*qv.*) using part of a DNA (*qv.*) molecule as a template. See also translation.

Transhumance A form of agriculture involving the movement of grazing herds between summer and winter pastures.

Translation The conversion of the code of the messenger RNA (*qv.*) to create a chain of amino acids.

Translocation (genetics) The movement of part of a chromosome (*qv.*) to a different part of the same chromosome or to become attached to another chromosome, through an error in the natural copying and replication processes of the cell.

Translocation (habitat) The transfer of a habitat (*qv.*) from a donor to a receptor site. This is an extreme form of restoration that is undertaken to rescue a site whose destruction is otherwise inevitable.

Transpiration The evaporation of water vapour from the aerial parts of a plant.

Trophic level Position along a food chain, from primary producer (*qv.*) through a sequence of consumers.

Variation Differences between individuals of the same species. Phenotypic variation is acquired, through growth, development, or physiological change, but is not passed on in the genes. Genotypic variation is differences coded in the genes, which may therefore be passed on to the next generation.

Variety A distinct form within a species that occurs naturally.

Vector An organism that carries the parasite from one host to another (e.g. malarial mosquitoes of the genus *Anopheles*).

Vegetative propagation Producing new individuals without sexual reproduction; in horticulture, producing individuals by promoting asexual reproduction.

Xerophyte A plant adapted to live in an arid environment.

REFERENCES

First words

Becker, S. and Berhane, K. 1997. A meta-analysis of 61 sperm count studies revisited. *Fertility and Sterility* 67, 1103–1108.

Bhatt, R. V. 2000. Environmental influence on reproductive health. *International Journal of Gynecology and Obstetrics 70*, 69–75.

Carlsen, E., Giwercman, A., Keiding, N., and Skakkebæk, N. E. 1992. Evidence for decreasing quality of semen during past 50 years. *British Medical Journal 305*, 609–613.

Cobb, G. P., Houlis, P. D., and Bargar, T. A. 2002. Polychlorinated biphenyl occurrence in American alligators (*Alligator mississippiensis*) from Louisiana and South Carolina. *Environmental Pollution 118*, 1–4.

Haave, M., Ropstad, E., Derocher, A. E., Lie, E., Dahl, E., Wiig, O., Skaare, J. U., and Jenssen, B. M. 2003. Polychlorinated biphenyls and reproductive hormones in female polar bears at Svalbard. *Environmental Health Perspectives 111*, 431–437.

Kucklick, J. R., Struntz, W. D. J., Becker, P. R., York, G. W., O'Hara, T. M., and Bohonowych, J. E. 2002. Persistent organochlorine pollutants in ringed seals and polar bears collected from northern Alaska. *Science of the Total Environment 287*, 45–59.

Pajarinen, J., Penttila, A., Laippala, P., and Karhunen, P. J. 1997. Incidence of disorders of spermatogenesis in middle aged Finnish men 1981–91: two necropsy series. *British Medical Journal 314*, 13–18.

Pant, N., Kumar, R., Mathur, N., Srivastava, S. P., Saxena, D. K., and Gujrati, V. R. 2007. Chlorinated pesticides concentration in semen of fertile and infertile men and correlation with sperm quality. *Environmental Toxicology and Pharmacology 23*, 135–139.

Polischuk, S. C., Norstrom, R. J., and Ramsay, M. A. 2002. Body burdens and tissue concentrations of organochlorines in polar bears (*Ursus maritimus*) vary during seasonal fasts. *Environmental Pollution 118*, 29–39.

Rauschenberger, R. H., Wiebe, J. J., Buckland, J. E., Smith, J. T., Sepúlveda, M. E., and Gross, T. E. 2004. Achieving environmentally relevant organochlorine pesticide concentrations in eggs through maternal exposure in *Alligator mississippiensis*. *Marine Environmental Research 58*, 851–856.

Sacco, J. C. and James, M. 2004. Glucuronidation in the polar bear (*Ursus maritimus*). *Marine Environmental Research 58*, 475–479.

Sonne, C., Leifsson, P. S., Dietz, R., Born, E. W., Letcher, R. J., Kiekegaard, M., Muir, D. C. G., Andersen, L. W., Riget, F. F., and Hyldstrup, L. 2005. Enlarged clitoris in wild polar bears (*Ursus maritimus*) can be misdiagnosed as pseudohermaphroditism. *Science of the Total Environment 337*, 45–58.

Suominen, J. and Vierula, M. 1993. Semen quality of Finnish men. *British Medical Journal 306*, 1579.

Swan, S. H., Brazil, C., Drobnis, E. Z., Liu, F., Kruse, R. L., Hatch, M., Redmon, J. B., Wang, C., and Overstreet, J. H. 2003. Geographic differences in semen quality of fertile U.S. males. *Environmental Health Perspectives 111*, 414–420.

Swan, S. H., Kruse, R. L., Liu, F., Barr, D. B., Drobnis, E. Z., Redmon, J. B., Wang, C. Brazil, C., and Overstreet, J. H. 2003. Semen quality in relation to biomarkers of pesticide exposure. *Environmental Health Perspectives 111*, 1478–1484.

Wiig, O., Derocher, A. E., Cronin, M. M., and Skaare, J. U. 1998. Female pseudohermaphrodite polar bears at Svalbard. *Journal of Wildlife Diseases 34*, 792–796.

1 Origins

Section 1.1

Brumm, A., Aziz, F., van den Bergh, G. D., Morwood, M. J., Moore, M. W., Kurniawan, I, Hobbs, D. R. and Fullagar, R. 2006. Early stone technology on Flores and its implications for *Homo floresiensis*. *Nature 441*, 624–628.

Darwin, C. 1872. *The Origin of Species*, 6th edn. John Murray, London.

Isaac, G. 1978. The food-sharing behaviour of protohuman hominids. *Scientific American 238*, 90–108.

Leakey, R. 1995. *The Origin of Humankind*. Weidenfeld & Nicholson, London.

Leakey, R. and Lewin, R. 1992. *Origins Reconsidered*. Abacus, London.

Lewin, R. 1984. *Human Evolution*. Blackwell, Oxford.

Parker, S. 1992. *The Dawn of Man*. Crescent, New York.

Pough, F. H., Heiser, J. B., and Mc Farland, W. N. 1996. *Vertebrate Life*, 4th edn. Prentice Hall, New Jersey.

Trinkaus, E. and Howells, W. W. 1979. The Neanderthals. *Scientific American 241*, 118–133.

Walker, A. and Leakey, R. E. F. 1978. The hominids of east Turkana. *Scientific American 239*, 54–66.

Wood, B. 1992. Origin and evolution of the genus *Homo*. *Nature 355*, 783–790.

Sections 1.2–1.5

Campbell, N. A. and Reece, J. B. 2002. *Biology*, 6th edn. Benjamin/Cummings, Redwood, CA.

Dawkins, R. 1982. *The Extended Phenotype*. Oxford University Press, Oxford.

Dawkins, R. 1995. *River Out of Eden*. Weidenfeld & Nicholson, London.

Dennett, D. C. 1996. *Kinds of Minds*. Weidenfeld & Nicholson. London.

Lincoln, R. J., Boxshall, G. A., and Clark, P. F. 1982. *A Dictionary of Ecology, Evolution and Systematics*. Cambridge University Press, Cambridge.

Mace, R. 1995. Why do we do what we do? *Trends in Ecology and Evolution 10*, 4–5.

Purves, W. K., Sadava, D., Orians, G. H., and Heller, H. C. 2001. *Life: The Science of Biology*, 6th edn. Sinauer, Sunderland, MA.

2 Species

Section 2.1

Camp, W. H. 1951. Biosystematy. *Brittonia 7*, 113–127.

Ford, E. B. 1940. Polymorphism and taxonomy. In J. Huxley (ed.), *The New Systematics*. Clarendon Press, Oxford.

Jeffrey, C. 1989. *Biological Nomenclature*, 3rd edn. Edward Arnold, London.

Linnaeus, C. 1753. *Species Plantarum*. Laur Salvius, Holmiae.

Linnaeus, C. 1758. *Systema Naturae*. Laur Salvius, Holmiae.

Mayer, E. 1942. *Systematics and the Origin of Species from the Viewpoint of a Zoologist*. Columbia University Press, New York.

Minnelli, A. 1993. *Biological Systematics: The State of the Art*. Chapman & Hall, London.

Stace, C. A. 1989. *Plant Taxonomy and Biosystematics*, 2nd edn. Edward Arnold, London.

Wagner, W. H. 1970. Biosystematics and evolutionary noise. *Taxon 19*, 146–151.

Section 2.2

Margulis, L. 1992. Biodiversity: molecular biological domains, symbiosis and Kingdom origins. *BioSystems 27*, 39–51.

Margulis, L. and Schwartz, K. V. 1982. *Five Kingdoms*. W.H. Freeman, San Francisco, CA.

Whittaker, R. H. and Margulis, L. 1978. Protist classification and the kingdoms of organisms. *BioSystematics 10*, 3–18.

Woese, C., Kandler, O., and Wheelis, M. 1990. Towards a natural system of organisms: proposal for the domains Archaea, Bacteria and Eucarya. *Proceedings of the National Academy of Sciences USA 87*, 4576–4579.

Section 2.3

Altschul, S. F. and Lipman, D. J. 1990. Equal animals. *Nature 348*, 493–494.

Carr, S. M., Ballinger, S. W., Derr, J. N., Blankenship, L. H., and Bickham, J. W. 1986. Mitochondrial DNA analysis of hybridisation between sympatric whitetailed and mule deer in West Texas. *Proceedings of the National Academy of Science of the United States of America 83*, 9576–9580.

Crandell, K. E., Bininda-Emonds, O. R. P., Mace, G. M., and Wayne, R. K. 2000. Considering evolutionary processes in conservation biology. *Trends in Ecology and Evolution 15*, 290–295.

Dowling, T. E., Moritz, C., and Palmer, J. D. 1990. Nucleic acids II. Restriction site analysis. In D. M. Hillis and C. Moritz (eds), *Molecular Systematics*. Sinauer Associates, Sunderland, MA.

Grant, V. 1985. *Plant Speciation*, 2nd edn. Columbia University Press, New York.

Gregg, K. B. 1983. Variation in floral fragrances and morphology: incipient speciation in *Cycnoches? Botanical Gazette 144*, 566–576.

Hewitt, G. M. 1988. Hybrid zones—natural laboratories for evolutionary studies. *Trends in Ecology and Evolution 3*, 138–147.

Hutchinson, G. E. 1957. Concluding remarks. *Cold Spring Harbour Symposium on Quantitative Biology 22*, 415–427.

Hutchinson, G. E. 1959. Homage to Santa Rosalia, or why there are so many kinds of animals? *American Naturalist 93*, 145–159.

Jansen, R. K. and Palmer, J. D. 1988. Phylogenetic implications of chloroplast DNA restriction site variation in the Mutisieae (Asteraceae). *American Journal of Botany 75*, 753–766.

Janzen, D. H. 1977. What are dandelions and aphids? *American Naturalist 111*, 586–589.

Janzen, D. H. 1979. Reply. *American Naturalist 114*, 156–157.

Janzen, D. H. 1979. How to be a fig. *Annual Review of Ecology and Systematics 10*, 13–51.

Knobloch, I. W. 1972. Intergeneric hybridization in flowering plants. *Taxon 21*, 97–103.

Lewis, K. B. and John, B. 1968. The chromosomal basis of sex determination. *International Review of Cytology 23*, 277–379.

Mallet, J. 1995. A species definition for the modern synthesis. *Trends in Ecology and Evolution 10*, 294–299.

Moritz, C. 1994. Defining 'evolutionary significant units' for conservation. *Trends in Ecology and Evolution 9*, 373–375.

Moritz, C., Dowling, T. E., and Brown, W. M. 1987. Evolution of animal mitochondrial DNA: relevance for population biology and systematics. *Annual Review of Ecology and Systematics 18*, 269–292.

Moyle, P. B. and Davis, L. H. 2000. A list of freshwater, anadromous, and euryhaline fishes of *California. California Fish & Game 86*, 244–258.

Nylander, W. 1866. Circa novum in studio lichenum criterium chimicum. *Flora 49*, 98.

Poulton, E. B. 1904. Presidential address. *Proceedings of the Entomological Society of London*, 73–116.

Ryder, O. A. 1986. Species conservation and systematics: the dilemma of subspecies. *Trends in Ecology and Evolution 1*, 9–10.

Sarich, V. M., Schmid, C. W., and Marks, J. 1989. DNA hybridisation as a guide to phylogenies. A critical analysis. *Cladistics 5*, 3–32.

Sterck, A. A., Groenhart, M. C., and Mooren, J. F. A. 1983. Aspects of the ecology of some microspecies of *Taraxacum* in the Netherlands. *Acta Botanica Neerlandica 32*, 385–415.

Wagner, D. F., Furnier, G. R., Saghai-Maroof, M. A., Williams, S. M., and Allard, R. W. 1987. Chloroplast DNA polymorphism in lodgepole and jack pines and their hybrids. *Proceedings of the National Academy of Science of the United States of America 84*, 2097–2100.

Walters, S. M. 1986. The name of the rose: a review of ideas on the European bias in angiosperm classification. *New Phytologist 104*, 527–540.

Wayne, R. K. and Jenks, S. M. 1991. Mitochondrial DNA analysis implying extensive hybridzation of the endangered red wolf *Canis rufus. Science 351*, 565–568.

Young, J. P. W. 1988. The construction of protein and nucleic acid homologies. In D. L. Hawksworth (ed.), *Prospects in Systematics*. Clarendon Press, Oxford.

Zuckerkandl, E. 1987. On the molecular evolutionary clock. *Journal of Molecular Evolution 26*, 34–46.

Section 2.4

Begon, M., Harper, J. L., and Townsend, C. R. 1990. *Ecology: Individuals, Populations, and Communities*, 2nd edn. Blackwell, Oxford.

Inhorn, M. C. and Brown, P. J. 1990. The anthropology of disease. *Annual Review of Anthropology 19*, 89–117.

Thornton, R. 1997. Aboriginal North American population and rates of decline, ca AD 1500–1900. *Current Anthropology 38*, 310–315.

Wirsing, R. L. 1985. The health of traditional societies and the effects of acculturation. *Current Anthropology* 26, 303–322.

Section 2.5

Grubb, P. 1977. The maintenance of species richness in plant communities: the importance of the regeneration niche. *Biological Reviews* 52, 107–145.

Pianka, E. R. 1976. Competition and niche theory. In R. M. May (ed.), *Theoretical Ecology Principles and Applications,* Blackwell, Oxford.

Pianka, E. R. 1988. *Evolutionary Ecology,* 4th edn. Harper and Row, New York.

Schoener, T. W. 1989. The ecological niche. In R. M. May (ed.), *Theoretical Ecology Principles and Applications.* Blackwell, Oxford.

Vandermeer, J. H. 1972. Niche theory. *Annual Review of Ecology and Systematics* 3, 107–132.

Section 2.6

Ammerman, A. and Cavalli-Sforza, L. L. 1971. Measuring the rate of spread of early farming in Europe. *Man* 6, 674–688.

Antonovics, J., Bradshaw, A. D., and Turner, R. G. 1971. Heavy metal nutrient tolerance in plants. *Advances in Ecological Research* 7, 1–85.

Arnqvist, G., Edvardsson, M., Friberg, U., and Nilsson, T. 2000. Sexual conflict promotes speciation in insects. *Proceedings of the National Academy of Sciences USA* 97, 10460–10464.

Baker, A. J. M. 1987. Metal tolerance. *New Phytologist* 106 (Suppl.), 93–111.

Beringer, J. E. 2000. Releasing genetically modified organisms. *Journal of Applied Ecology* 37, 207–214.

Chapman, G. P. (ed.). 1992. *Grass Evolution and its Domestication.* Cambridge University Press, Cambridge.

Ehrendorfer, F. 1970. Evolutionary patterns and strategies in seed plants. *Taxon* 19, 185–195.

Elias, D.O., Hebets, E.A., and Hoy, R. R. 2006. Female preference for complex/novel signals in a spider. *Behavioural Ecology* 17, 765–771.

Harlan, J. R. 1992. Origins and processes of domestication. In G. P. Chapman (ed.), *Grass Evolution and its Domestication.* Cambridge University Press, Cambridge.

Hellmich, R. L., Siegfried, B. D., Sears, M. K., Stanley-Horn, D. E., Daniels, M. J., Mattila, H. R., Spencer, T., Bidne, K. G., and Lewis, L. C. 2001. Monarch larvae sensitivity to *Bacillus thuringiensis*— purified proteins and pollen. *Proceedings of the National Academy of Sciences USA* 98, 11925–11930.

Hull, D. L. 1976. Are species really individuals? *Systematics and Zoology* 25, 174–191.

Hunt, P. F. 1986. The nomenclature and registration of orchid hybrids at specific and generic levels. In B. T. Sykes (ed.), *Infraspecific Classification of Wild and Cultivated Plants.* The systematics association special volume, Clarendon Press, Oxford.

Lai, J. and Messing, J. 2002. Increasing maize seed methionine by mRNA stability. *The Plant Journal* 30, 395–403.

Larson, A. 1989. The relationship between speciation and morphological evolution. In D. Otte and J. A. Endler (eds), *Speciation and its Consequences.* Sinauer Associates, Sunderland, MA.

Letourneau, D. K. and Burrows, B. E. 2002. *Genetically Engineered Organisms: Assessing Environmental and Human Health Effects.* CRC Press, Boca Raton, FL.

Losey, J. E., Rayor, L. S., and Carter, M. E. 1999. Transgenic pollen harms monarch larvae. *Nature* 399, 214.

Lupton, F. G. H. 1987. *Wheat Breeding: Its Scientific Basis.* Chapman & Hall, London.

Macnair, M. R. and Cumbes, Q. 1987. Evidence that arsenic tolerance in *Holcus lanatus* L. is caused by an altered phosphate uptake system. *The New Phytologist* 107, 387–394.

McNeilly, T. and Antonovics, J. 1968. Evolution of closely adjacent plant populations. IV. Barriers to gene flow. *Heredity* 23, 205–218.

Majerus, M., Amos, W., and Hurst, G. 1996. *Evolution: The Four Billion Year War.* Longman, Harlow.

Richards, A. J. 1986. *Plant Breeding Systems.* Allen and Unwin, London.

Salamini, F., Ozkan, H., Brandolini, A., Schafer-Pregl, R., and Martin, W. 2002. Genetics and geography of wild cereal domestication in the near east. *Nature Reviews Genetics* 3, 429–441.

Schilewen, U. K., Tautz, D., and Paabo, S. 1994. Sympatric speciation by monophyly of crater lake cichlids. *Science 368*, 629–632.

Schluter, D. 1994. Experimental evidence that competition promotes divergence in adaptive radiation. *Science 266*, 798–801.

Scriber, J. M. 2001. *Bt* or not *Bt*: Is that the question? *Proceedings of the National Academy of Sciences USA 98*, 12328–12330.

Sears, M. K., Hellmich, R. L., Stanley-Horn, D. E., Oberhauser, K. S., Pleasants, J. M., Mattila, H. R., Seigfried, B. D., and Dively, G. P. 2001. Impact of *Bt* corn pollen on monarch butterfly populations: a risk assessment. *Proceedings of the National Academy of Sciences USA 98*, 11937–11942.

Traverse, A. 1988. Plant evolution dances to a different beat. Plant and animal evolutionary mechanisms compared. *Historical Biology 1*, 277–301.

White, M. J. D. 1978. *Modes of Speciation*. Freeman, San Francisco, CA.

Wilkinson, M. J., Davenport, I. J., Charters, Y. M., Jones, A. E., Allainguillaume, J., Butler, H. T., Mason D. C., and Raybould, A. F. 2000. A direct regional scale estimate of transgene movement from genetically modified oilseed rape to its wild progenitors. *Molecular Ecology 9*, 983–991.

Wilkinson, M. J., Elliott, L. J., Allainguillaume, J., Shaw, M. W., Norris, C., Welters, R., Alexander., M., Sweet, J., and Mason, D. C. 2003. *Science 302*, 457–459.

Zangeri, A. R., McKenna, D., Wraight, C. L., Carroll, M., Ficarello, P., Warner, R., Berenbaum, M. R. 2001. Effects of exposure to event 176 *Bacillus thuringiensis* corn pollen on monarch and black swallowtail caterpillars under field conditions. *Proceedings of the National Academy of Sciences USA 98*, 11908–11912.

3 Populations

Section 3.3

Cushing, D. H. 1981. *Fisheries Biology*. University of Wisconsin Press, Madison, WI.

Section 3.4

Holmes, R. 1994. Biologists sort the lessons of fisheries collapse. *Science 264*, 1252–1253.

Jedrzejewska, B., Okarma, H., Jedrzejewska, W., and Milkowski, L. 1994. Effects of exploitation and protection on forest structure, ungulate density and wolf predation in Bialowieza Primeval Forest, Poland. *Journal of Applied Ecology 31*, 664–676.

Section 3.5

Lampert, K. P. and Linsenmair, K. E. 2002. Alternative life cycle strategies in the West African reed frog *Hyperolius nitidulus*: the answer to an unpredictable environment? *Oecologia 130*, 364–372.

Section 3.6

Land, D. and Taylor, S. K. 1998. Florida Panther Genetic Restoration and Management. *Annual Performance Report 1997–1998*. Bureau of Wildife Diversity Conservation. Naples, Florida.

Mauritzen, M., Derocher, A. E., Wiig, O., Belikov, S. E., Boltunov, A. N., Hansen, E., and Garner, G. W. 2002. Using satellite telemetry to define spatial population structure in polar bears in the Norwegian and western Russian Arctic. *Journal of Applied Ecology 39*, 79–90.

Section 3.7

Ashley, M. V., Melnick, D. J., and Western, D. 1990. Conservation genetics of the Black Rhinocerus (*Diceros bicornis*). 1. Evidence from the mitochondrial DNA of three populations. *Conservation Biology 4*, 71–77.

Caughley, G., Dublin, H. T., and Parker, I. 1990. Projected decline of the African elephant. *Biological Conservation 54*, 157–164.

Cohn, J. P. 1990. Elephants: remarkable and endangered. *BioScience 40*, 10–14.

Dinerstein, E. and McCracken, G. F. 1990. Endangered greater one-horned rhinoceros carry high levels of genetic variation. *Conservation Biology 4*, 417–422.

Douglas-Hamilton, I. 1987. African elephants: population trends and their causes. *Oryx 21*, 11–24.

Dublin, H. T., Sinclair, A. R. E., and McGlade, J. 1990. Elephants and fire as causes of multiple stable states in the Serengeti-Mara woodlands. *Journal of Animal Ecology 59*, 147–164.

Haynes, G. 1991. *Mammoths, Mastodonts and Elephants: Biology, Behaviour and the Fossil Record*. Cambridge University Press, Cambridge.

Hoare, R. 2000. African elephants and humans in conflict: the outlook for coexistence. *Oryx 34*, 34–38.

IUCN Status list; CITES Appendix 1, 2001. available at www.worldwildlife.org

Lombard, A. T., Johnson, C. F., Cowling, R. M., and Pressey, R. L. 2001. Protecting plants from elephants: botanical reserve scenarios with Addo National Park, South Africa. *Biological Conservation 102*, 191–203.

4 Interactions

Bronowski, J. 1973. *The Ascent of Man*. BBC Publications, London.

Section 4.1

Krebs, C. J. 2001. *Ecology: Experimental Analysis of Distribution and Abundance*, 5th edn. Benjamin Cummings, London.

Section 4.2

Addicott, J. F. 1986. Variation in the costs and benefits of mutualism: the interaction between yuccas and yucca moths. *Oecologia 70*, 486–494.

Beattie, A. J., Turnbull, C., Knox, R. B., and Williams, E. G. 1984. Ant inhibition of pollen function: a possible reason why ant pollination is rare. *American Journal of Botany 71*, 421–426.

Bond, W. and Slingsby, P. 1984. Collapse of an ant–plant mutualism: the Argentinian ant (*Iridomyrmex humulis*) and myrmecochorous proteaceae. *Ecology 65*, 1031–1037.

Giraud, T., Pedersen, J. S., and Keller, L. 2002. Evolution of supercolonies: the Argentine ants of southern Europe. *Proceedings of the National Academy of Sciences USA 99*, 6075–6079.

Goguen, C. B. and Mathews, N. E. 2001. Brown-headed cowbird behaviour and movements in relation to livestock grazing. *Ecological Applications 11*, 1533–1544.

Janzen, D. H. 1966. Coevolution of mutualism between ants and acacias in Central America. *Evolution 20*, 249–275.

Munro, J. 1967. The exploitation and conservation of resources by populations of insects. *Journal of Animal Ecology 36*, 531–547.

Wigglesworth, V. B. 1964. *The Life of Insects*. Weidenfeld & Nicholson, London.

Section 4.3

Brown, V. K. and Southwood, T. R. E. 1987. Secondary succession: patterns and strategies. In A. J. Gray, M. J. Crawley, and R. J. Edwards (eds), *Colonisation, Succession and Stability*. Blackwell, Oxford.

Case, T. J., Bolger, D. T., and Richman, A. D. 1992. Reptilian extinctions: the last ten thousand years. In P. L. Fiedler and S. K. Jain (eds), *Conservation Biology*. Chapman & Hall, New York.

Crombie, A. C. 1946. Further experiments on insect competition. *Proceedings of the Royal Society of London, Series B 133*, 76–109.

Grime, J. P. 1979. *Plant Strategies and Vegetation Processes*, John Wiley, Chichester.

Grime, J. P., Hodgson, J., and Hunt, R. 1988. *Comparative Plant Ecology: A Functional Approach to Common British Species*. Unwin Hyman, London.

Law, R. and Watkinson, A. R. 1989. Competition. In J. M. Cherrett (ed.), *Ecological Concepts*. Blackwell, Oxford.

Muller, C. H. 1966. The role of chemical inhibition (allelopathy) in vegetational composition. *Bulletin of the Torrey Botanical Club 93*, 332–351.

Southwood, T. R. E. 1977. Habitat, the templet for ecological strategies. *Journal of Animal Ecology 46*, 337–365.

Southwood, T. R. E. 1988. Tactics, strategies and templates. *Oikos 52*, 2–18.

Tilman, D., Mattson, M., and Langer, S. 1981. Competition and nutrient conditions along a temperature gradient: an experimental test of a mechanistic approach to niche theory. *Limnology and Oceanography 26*, 1020–1033.

Section 4.4

Akhter, Y., Ahmed, I., Devi, S.M., and Ahmed, N. (2007) The co-evolved *Helicobacter pylori* and gastric cancer: trinity of bacterial virulence, host susceptibility and lifestyle. *Infections Agents and Cancer 2*, 2–7.

Beeby, A. and Richmond, L. 2001. Intraspecific competition in populations of *Helix aspersa* with different histories of exposure to lead. *Environmental Pollution 114*, 337–344.

Beeby, A. and Richmond, L. 2002. Lead reduces shell mass in juvenile garden snails (*Helix aspersa*). *Environmental Pollution 120*, 283–288.

Beeby, A. and Richmond, L. 2007. Differential growth rates and calcium allocation strategies in the garden snail *Cantareus aspersus*. *Journal of Molluscan Studies* (in press).

Belsky, A. J. 1986. Does herbivory benefit plants? A review of the evidence. *American Naturalist 127*, 870–892.

Blum, M. S., River, L., and Plowman, T. 1981. Fate of cocaine in the lymantriid *Elonia noyesii* a predator of *Erythroxylum coca*. *Phytochemistry 20*, 2499–2500.

Cox, F. E. G. (ed.). 1982. *Modern Parasitism*. Blackwell, Oxford.

Dawkins, R. and Krebs, J. R. 1979. Arms race between and within species. *Proceedings of the Royal Society of London, Series B 205*, 489–511.

De Gregorio, E. and Lemaitre, B. 2002. The mosquito genome: the post-genomic era opens. *Nature 419*, 496–497.

Eisener, T. and Meinwald, J. 1966. Defensive secretions of arthropods. *Science 153*, 1341–1350.

Eisener, T., Alsop, D., Hicks, K., and Meinwald, J. 1978. Defensive secretions of millipedes. In S. Bettini (ed.), *Arthropod Venoms, Handbook of Experimental Pharmacology 48*. Springer, Berlin.

Feeny, P. P. 1970. Seasonal changes in oak leaf tannins and nutrients as a cause of spring feeding by winter moth caterpillars. *Ecology 51*, 565–581.

Fitzgerald, T. D., Jeffers, P. M., and Mantella, D. 2002. Depletion of host derived cyanide in the gut of the eastern tent caterpillar, *Malacosoma americanum*. *Journal of Chemical Ecology 28*, 257–268.

Gardener, M. J., Hall, N., Fung, E., White, O., Berriman, M., Hyman, R. W., Carlton, J. M., Pain, A., Nelson, K., Bowman, S., Paulsen, I. T., James, K., Eisen, J. A., Rutherford, K., Salzberg, A. C., Kyes, S., Chan, M.-S., Nene, V., Shallom, S. J., Suh, B., Peterson, J., Angiuoli, S., Pertea, M., Allen, J., Selengut, J., Haft, D., Mather, M. W., Vaidya, A. B., Martin, M. A., Fairlamb, A. H., Fraunholz, M. J., Roos, D. R., Ralph, S. A., McFadden, G. I., Cummings, L. M., Subramanian, G. M., Mungall, C., Venter, J. C., Carucci, D. J., Hoffman, S. L., Newbold, C., Davis, R. W., Fraser, C. M., and Barrell, B. 2002. Genome sequence of the human malaria parasite *Plasmodium falciparum*. *Nature 419*, 498–511.

Hammill, M. O. and Smith, T. G. 1991. The role of ecology in the predation of the ringed seal in Barrow Strait, Northwest Territories. *Canadian Marine Mammal Society 7*, 123–135.

Harborne, J. B. 1982. *Introduction to Ecological Biochemistry*, 2nd edn. Academic Press, London.

Hassell, M. P. and Anderson, R. M. 1989 Predator–prey and host–pathogen interactions. In J. M. Cherrett (ed.), *Ecological Concepts*. Blackwell, Oxford.

Jones, D. A. 1973. Co-evolution and cyanogenesis. In V. H. Heywood (ed.), *Taxonomy and Ecology*. Academic Press, New York.

Krebs, J. R. and Davies, N. B. 1987. *An Introduction to Behavioural Ecology*. Blackwell, Oxford.

LaPage, G. 1963. *Animals Parasitic in Man*. Dover, New York.

Linz, B., Balloux, F., Moodley, Y., Manica, A., Liu, H., Roumagnac, P., Falush, D., Stamer, C., Prugnolle, F., van der Merwe, S. W., Yamaoka, Y., Grahham, D. Y., Perez-Trallero, E., Wastrom, T., Suerbaum, S., and Achtman, M. 2007. An African origin for the intimate association between humans and *Helicobacter pylori*. *Nature 445*, 915–918.

Owen, D. F. 1980. *Camouflage and Mimicry*. Oxford University Press, Oxford.

Payne, C. C. 1977. The ecology of brood parasitism in birds. *Annual Review of Ecology and Systematics 8*, 1–28.

Peterson, S. C., Johnson, N. D., and LeGuyader, J. L. 1987. Defensive regurgitation of alleochemicals derived from cyanogenesis by eastern tent caterpillars. *Ecology 68*, 1268–1272.

Pierkarski, G. 1962. *Medical Parasitology*. Bayer, Leverkusen.

Pimental, D. and Stone, F. A. 1968. Evolution and population ecology of parasite-host systems. *Canadian Entomologist 100*, 655–662.

Potts, G. R. and Aebischer, N. J. 1989. Control of population size in birds: the grey partridge as a case study. In P. J. Whittaker and J. B. Grubb (eds), *Towards a More Exact Ecology*. Blackwell, Oxford.

Pundir, Y. P. S. 1981. A note on the biological control of *Scurrula cordifolia* (Wall.) G. Don. by another mistletoe in the Sivalik Hills (India). *Weed Research 21*, 233–234.

Rosenthal, G. A. and Janzen, D. H. (eds). 1979. *Herbivores: Their Interaction with Secondary Plant Metabolites.* Academic Press, New York.

Silvertown, J. W. 1980. The evolutionary ecology of mast seeding in trees. *Biological Journal of the Linnean Society of London 14,* 235–250.

Stirling, I., Øritsland, N. A. 1995. Relationships between estimates of ringed seal (*Phoca hispida*) and polar bear (*Ursus maritimus*) populations in the Canadian Arctic. *Canadian Journal of Fish and Aquatic Science 52,* 2594–2612.

Section 4.5

Boudouresque, C. F. and Verlaque, V. 2002. Biological pollution in the Mediterranean Sea: invasive versus introduced macrophytes. *Marine Pollution Bulletin 44,* 32–38.

Bram, R. A., George, J. E., Reichard, R. E., and Tabachnick, W. J. 2002. Threat of foreign arthropod-borne pathogens to livestock in the United States. *Journal of Medical Entomology 39,* 405–416.

Christian, C. E. 2001. Consequences of a biological invasion reveal the importance of mutualism for plant communities. *Nature 413,* 635–639.

Furukawa, A., Shibata, C., and Mori, K. 2002. Synthesis of four methyl-branched secondary acetates and a methyl-branched ketone as possible candidates for the female pheromone of the screwworm fly, *Cochliomyia hominivorax. Bioscience, Biotechnology & Biochemistry 66,* 1164–1169.

Mack, R. N. and Lonsdale, W. M. 2001. Humans as global plants dispersers: getting more than we bargained for. *BioScience 51,* 95–102.

Meinesz, A., Belsher, T., Thibaut, T., Antolic, B., Mustapha, K. B., Boudouresque, C.-F., Chiaverini, D., Cinelli, F., Cottalorda, J.-M., Djellouli, A., Abed, A. E., Orestoano, C., Grau, A. M., Ivesa, L., Jaklin, A., Langar, H., Massuti-Pascual, E., Peirano, A., Leonardo, T., de Vauelas, J., Zavodnik, N., and Zuljevic, A. 2001. The introduced green alga *Caulerpa taxifolia* continues to spread in the Mediterranean. *Biological Invasions 3,* 201–210.

Mooney, H. A. and Cleland, E. E. 2001. The evolutionary impact of invasive species. *Proceedings of the National Academy of Sciences USA 98,* 5446–5451.

Novak, S. J. and Mack, R. N. 2001. Tracing plant introduction and spread: genetic evidence from *Bromus tectorum* (Cheatgrass). *BioScience 51,* 114–122.

Thomas, J. A., Knapp, J. J., Akino, T., Gerty, S., Wakamura, S., Simcox, D. J., Wardlaw, J. C., and Elmes, G. W. 2002. Parasitoid secretions provoke ant warfare. *Nature 417,* 505.

Williamson, M. and Fitter, A. 1996. The varying success of invaders. *Ecology 77,* 1661–1666.

5 Communities

Section 5.1

Archibold, O. W. 1995. *Ecology of World Vegetation.* Chapman & Hall, London.

Begon, M., Harper, J. L., and Townsend, C. R. 1990. *Ecology: Individuals, Populations, and Communities,* 2nd edn. Blackwell, Oxford.

Bottema, S., Entjes-Nieborg, G., and Van Zeist, W. (eds). 1990. *Man's Role in the Shaping of the Eastern Mediterranean Landscape.* Balkema, Rotterdam.

Boucher, C. and Moll, E. J. 1981. South African Mediterranean Shrublands. In F. Di Castri *et al.* (eds), *Mediterranean-type Shrublands.* Elsevier, Amsterdam, pp. 233–248.

Cody, M. L. and Mooney, H. A. 1978. Convergence versus nonconvergence in Mediterranean-climate ecosystems. *Annual Review of Ecology and Systematics 9,* 265–351.

Cowling, R. M., Rundel, P. W., Lamont, B. B., Arroyo, A. K., and Arianoutsou, M. 1996. Plant diversity in Mediterranean-climate regions. *Trends in Ecology and Evolution 11,* 362–366.

Di Castri, F. 1981. Mediterranean-type shrublands of the world. In F. Di Castri *et al.* (eds), *Mediterranean-type Shrublands.* Elsevier, Amsterdam, pp. 1–52.

Di Castri, F. and Mooney, H. A. (eds). 1973. *Mediterranean Type Ecosystems: Origin and Structure.* Springer, New York.

Di Castri, F. and Vitali-Di Castri, V. 1981. Soil fauna of mediterranean-climate regions. In F. Di Castri *et al.* (eds), *Mediterranean-type Shrublands.* Elsevier, Amsterdam, pp. 1–52.

Di Castri, F., Goodall, D. W., and Specht, R. L. (eds). 1981. Ecosystems of the World 11. *Mediterranean-type Shrublands.* Elsevier, Amsterdam.

Greuter, W. 1995. Extinctions in mediterranean areas. In J. H. Lawton and R. M. May (eds), *Extinction Rates*. Oxford University Press, Oxford.

Hanes, T. L. 1981. California chaparral. In F. Di Castri *et al.* (eds), *Mediterranean-type Shrublands*. Elsevier, Amsterdam, pp. 1339–1373.

Le Houreau, H. N. 1981. Impact of man and his animals on Mediterranean vegetation. In F. Di Castri *et al.* (eds), *Mediterranean-type Shrublands*. Elsevier, Amsterdam, pp. 479–522.

Mills, J. M. 1983. Herbivory and seedling establishment in post-fire southern California chaparral. *Oecologia 60*, 267–270.

Naveh, Z. 1990. Ancient man's impact on the Mediterranean landscape in Israel–ecological and evolutionary perspectives. In S. Botteman *et al.* (eds), *Man's Role in the Shaping of the Eastern Mediterranean Landscape*. Balkema, Rotterdam.

Polunin, O. and Smythies, B. E. 1973. *Flowers of South-West Europe. A Field Guide*. Oxford University Press, Oxford.

Quezel, P. 1981. Floristic composition and phytosociological structure of sclerophyllous matorral around the Mediterranean. In F. Di Castri *et al.* (eds), *Mediterranean-type Shrublands*. Elsevier, Amsterdam, pp. 107–122.

Rundel, P. W. 1981. The matorral zone of central Chile. In F. Di Castri *et al.* (eds), *Mediterranean-type Shrublands*. Elsevier, Amsterdam, pp. 175–201.

Specht, R. L. 1981. Mallee ecosystems in Southern Australia. In F. Di Castri *et al.* (eds), *Mediterranean-type Shrublands*. Elsevier, Amsterdam, pp. 203–231.

Tomaselli, R. 1981. Main physiognomic types and geographic distribution of shrub systems related to mediterranean climates. In F. Di Castri *et al.* (eds), *Mediterranean-type Shrublands*. Elsevier, Amsterdam, pp. 123–130.

Trabaud, L. 1981. Man and fire: impacts on Mediterranean vegetation. In F. Di Castri *et al.* (eds), *Mediterranean-type Shrublands*. Elsevier, Amsterdam.

Van Andel, T. H. and Zanger, E. 1990. Landscape stability and destabilisation in the prehistory of Greece. In S. Botteman *et al.* (eds), *Man's Role in the Shaping of the Eastern Mediterranean Landscape*. Balkema, Rotterdam.

Weiher, E. and Keddy, P. (eds). 1999. *Ecological Assembly Rules. Perspectives, Advances, Retreats*. Cambridge University Press, Cambridge.

Section 5.2

Bach, C. E. 2001. Long-term effects of insect herbivory and sand accretion on plant succession on sand dunes. *Ecology 85*, 1401–1416.

Clements, F. E. 1936. Nature and structure of the climax. *Journal of Ecology 24*, 252–284.

Connell, J. H. 1978. Diversity in tropical rainforests and coral reefs. *Science, 199*, 1302–1310.

Connell, J. H. and Slatyer, R. O. 1977. Mechanisms of succession in natural communities and their role in community stability and organisation. *American Naturalist 111*, 1119–1144.

Connell, J. H., Noble, I. R., and Slatyer, R. O. 1987. On the mechanisms of producing successional change. *Oikos 50*, 136–137.

Diamond, J. and Case, T. J. (eds). 1986. *Community Ecology*. Harper and Row, New York.

Duckworth, J. C., Kent, M., and Ramsay, P. M. 2000. Plant functional types: an alternative to taxonomic plant community descriptions in biogeography? *Progress in Physical Geography 24*, 515–542.

Gleason, H. A. 1926. The individualistic concept of the plant association. *Bulletin of the Torrey Botanical Club 53*, 7–26.

Gray, A. J., Crawley, M. J., and Edwards, P. J. (eds). 1987. *Colonization, Succession and Stability*. Blackwell, Oxford.

Harper, J. L. 1977. *The Population Biology of Plants*. Academic Press, London.

Horn, H. S. 1976. Succession. In R. M. May (ed.), *Theoretical Ecology: Principles and Applications*. Blackwell, Oxford.

Janzen, D. H. 1986. Keystone plant resources in the tropical forest. In M. E. Soule (ed.), *Conservation Biology: The Science of Scarcity and Diversity*. Sinauer Associates, Sunderland, MA.

Keeley, J. E. and Bond, W. J. 1997. Convergent seed germination in South African fynbos and Californian chaparral. *Plant Ecology 133*, 153–167.

Kent, M., Owen, N. W., Dale, P., Newnham, R. M., and Giles, T. M. 2001. Studies of vegetation burial: a focus for biogeography and geomorphology? *Progress in Physical Geography 25*, 455–482.

Krebs, C. J. 1985. *Ecology, the Experimental Analysis of Distribution and Abundance*, 3rd edn. Harper & Row, New York.

Kutiel, P., Eden, E., and Zevelev, Y. 2000. Effect of experimental trampling and off-road motorcycle traffic on soil and vegetation of stabilised coastal dunes, Israel. *Environmental Conservation 27*, 14–23.

van der Meulen, F. and Salman, A. H. P. M. 1996. Management of Mediterranean coastal dunes. *Ocean & Coastal Management 30*, 177–195.

Moore, P. D. 1986. Site history. In P. D. Moore and S. B. Chapman (eds), *Methods in Plant Ecology*. Blackwell, Oxford

Paine, R. T. 1969. A note on trophic complexity and community stability. *American Naturalist 100*, 65–75.

Rubio-Casal, A. E., Castillo, J. M., Luque, C. J., and Figueroa, M. E. 2001. Nucleation and facilitation in salt pans in Mediterranean salt marshes. *Journal of Vegetation Science 12*, 761–770.

Whitaker, R. H. 1975. *Communities and Ecosystems*. Macmillan, London.

Section 5.3

Braun-Blanquet, J. 1932. *Plant Sociology: The Study of Plant Communities* (trans. G. D. Fuller and H. S. Conrad). McGraw Hill, New York.

Franklin, J. 2002. Enhancing a regional vegetation map with predictive models of dominant plant species in chaparral. *Applied Vegetation Science 5*, 135–146.

Rodwell, J. S. (ed.). 1989. *British Plant Communities*, Vol. 1, *Woodlands and Scrub*. Cambridge University Press, Cambridge.

Rodwell, J. S. (ed.). 1991. *British Plant Communities*, Vol. 2, *Mires & Heaths*. Cambridge University Press, Cambridge.

Rodwell, J. S. (ed.). 1992. *British Plant Communities*, Vol. 3, *Grasslands & Montane Communities*. Cambridge University Press, Cambridge.

6 Systems

Section 6.1

Mehta, M. and Baross, J. 2006. Nitrogen fixation at 92°C by a hydrothermal vent archaeon. *Science 31*, 1783–1786.

Cavicchioli, R., Thomas, T. 2000. Extremophiles. In J. Lederberg (ed.), *Encyclopedia of Microbiology*, 2nd edn. Academic Press, London.

Section 6.2

Cooper, J. P., 1975. *Photosynthesis and Productivity in Different Environments*. Cambridge University Press, Cambridge.

Davis, G. W., Richardson, D. M., Keeley, J. E., and Hobbs, R. J. 1996. Mediterranean-type ecosystems: the influence of biodiversity on their functioning. In H. A. Mooney, J. H. Cushman, E. Medina, O. E. Sala, and E.-D. Schulze (eds), *Environmental Physiology of Plants*. John Wiley & Sons Ltd, New York.

Fitter, A. H. and Hay, R. K. M. 1987. *Physiological Plant Ecology*, 2nd edn. Academic Press, London.

Hay, R. K. M. and Walker, A. J. 1989. *An Introduction to the Physiology of Crop Yield*. Longman, Harlow.

Humphreys, W. F. 1979. Production and respiration in animal populations. *Journal of Animal Ecology 48*, 427–474.

Jones, H. G. 1983. *Plants and Microclimate*. Cambridge University Press, Cambridge.

Orshan, G. 1963. Seasonal dimorphism of desert and mediterranean chamaephytes and its significance as a factor in their water economy. In A. J. Rutter and F. H. Whitehead (eds), *The Water Relations of Plants*. John Wiley & Sons, New York.

Osmond, C. B., Winter, K., and Zeigler, H. 1981. Functional significance of different pathways of CO_2 fixation in photosynthesis. In A. Pirson and M. H. Zimmermann (eds), *Encyclopaedia of Plant Physiology*, new series. Springer Verlag, Berlin.

Salisbury, F. B. and Ross, C. 1992. *Plant Physiology*, 4th edn. Wadsworth, Belmont, CA.

Schultz, E. D. 1970. Der CO_2–Gaswechsel der Buche (*Fagus sylvatica* L.) in Abhängigkeit von den Klimafaktoren in Feiland. *Flora Jena 159*, 177–232.

Schultz, E. D., Fuchs, M., and Fuchs, M. I. 1977. Spatial distribution of photosynthetic capacity and performance in a mountain spruce forest in northern Germany. I. Biomass distribution and daily CO_2 uptake in different crown layers. *Oecologia 29*, 43–61.

Schultz, E. D., Fuchs, M., and Fuchs, M. I., 1977b. Spatial distribution of photosynthetic capacity and performance in a mountain spruce forest in northern Germany. III. The significance of the evergreen habit. *Oecologia 30*, 239–248.

Whittaker, R. H. 1975. *Communities and Ecosystems*, 2nd edn. Macmillan, New York.

Whittaker, R. H. and Likens, G. E. 1973. The primary production of the biosphere. *Human Ecology 1*, 299–369.

Woodward, F. I. 1987. *Climate and Plant Distribution*. Cambridge University Press, Cambridge.

Zscheile, F. P. and Comar, C. L. 1941. Influence of preparative procedure on the purity of chlorophyll components as shown by absorption spectra. *Botanical Gazette 102*, 463–481.

Section 6.3

Begon, M., Harper, J. L., and Townsend, C. R. 1990. *Ecology: Individuals, Populations, and Communities*, 2nd edn. Blackwell, Oxford.

Colinvaux, P. 1980. *Why Big Fierce Animals are Rare*. Penguin Books, Harmondsworth, Middlesex.

Cousins, S. 1987. The decline of the trophic level. *Trends in Ecology and Evolution 2*, 312–316.

Golley, F. B. 1960. Energy dynamics of an old-field community. *Ecological Monographs 30*, 187–200.

Heal, O. W. and Maclean, S. F. 1975. Comparative productivity in ecosystems—secondary productivity. In W. H. Van Dobben and R. H. Lowe-McConnell (eds), *Unifying Concepts in Ecology*. Dr W Junk, The Hague.

Lawton, J. H. 1989. Food webs. In J. M. Cherret (ed.), *Ecological Concepts*. Blackwell, Oxford.

Moore, J. C., de Ruiter, P. C., Hunt, H. W., Coleman, D. C., and Freckman, D. W. 1996. Microcosms and soil ecology: critical linkages between field studies and modelling food webs. *Ecology 77*, 694–705.

Odum, H. T. 1957. Trophic structure and productivity of Silver Springs, Florida. *Ecological Monographs 27*, 55–112.

Odum, P. 1989. *Ecology and Our Endangered Life-support System*. Sinauer Associates, Sunderland, MA.

Phillips, J. 1966. *Ecological Energetics*. Edward Arnold, London.

Swift, M. J., Heal, O. W., and Anderson, J. M. 1979. *Decomposition in Terrestrial Ecosystems*. Blackwell, Oxford.

Varley, G. C. 1970. The concept of energy flow applied to a woodland community. In A. Watson (ed.), *Animal Populations in Relation to their Food Resource*. Blackwell, Oxford.

Section 6.4

Bearhop, S., Thompson, D. R., Waldron, S., Russel, I. C., Alexander, G., and Furness, R. W. 1999. Stable isotopes indicate the extent of freshwater feeding by cormorants *Phalocrocorax carbo* shot at inland fisheries in England. *Journal of Applied Ecology 36*, 75–84.

Briand, F. 1983. Environmental control of food web structure. *Ecology 64*, 253–263.

Briand, F. and Cohen, J. E. 1984. Community food webs have scale-invariant structure. *Nature 307*, 264–266.

Cohen, J. E. and Briand, F. 1984. Trophic links of community food webs. *Proceedings of the National Academy of Sciences USA 81*, 4105–4109.

Degale, B. E., Tollit, D. J., Jarman, S. N., Hindell, M. A., Trites, A. W. and Gales, N. J. 2005. Molecular scatology as a tool to study diet: analysis of prey DNA in scats from captive Steller sea lions. *Molecular Ecology 14*, 1831–1842.

Gende, S. M., Edwards, R. T., Willson, M. F., and Wipfli, M. S. 2002. Pacific salmon in aquatic and terrestrial ecosystems. *Bioscience 52*, 917–928.

Jarman, S. N., Gales, N. J., Tierney, M., Gill, P. C., and Elliott, N. G. 2002. A DNA-based method for identification of krill species and its application to analysing the diet of marine vertebrate predators. *Molecular Ecology 11*, 2679–2690.

Kucklick, J. R., Struntz, W. D. J., Becker, P. R., York, G. W., O'Hara, T. M., and Bohonowych, J. E. 2002.

Persistent organochlorine pollutants in ringed seals and polar bears collected from northern Alaska. *The Science of the Total Environment* 287, 45–59.

Peterson, B. J. and Fry, B. 1987. Stable isotopes in ecosystem studies. *Annual Review of Ecological Systematics* 18, 293–320.

Pimm, S. L. 1982. *Food Webs*. Chapman & Hall, London.

Pimm, S. L. and Kitching, R. L. 1987. The determinants of food chain length. *Oikos* 50, 302–307.

Pimm, S. L., Lawton, J. H., and Cohen, J. E. 1991. Food web patterns and their consequences. *Nature* 350, 669–674.

Polischuk, S. C., Nortstrom, R. J., and Ramsay, M. A. 2002. Body burdens and tissue concentrations of organochlorines in polar bears (*Ursus maritimus*) vary during seasonal fasts. *Environmental Pollution* 118, 29–39.

Price, P. W. 1975. *Insect Ecology*. Wiley, New York.

Sheppard, S. K. and Harwood, J. D. 2005. Advances in molecular ecology: tracking trophic links through predator-prey food webs. *Functional Ecology* 19, 751–762.

Shurin, J. B., Gruner, D. S. and Hillebrand, H. 2006. All wet or dried up? Real differences between aquatic and terrestrial food webs. *Proceedings of the Royal Society B* 273, 1–9.

Tilman, D. 1982. *Resource Competition and Community Structure*. Princeton University Press, Princeton, NJ.

Section 6.5

Ammerman, A. and Cavalli-Sforza, L. L. 1971. Measuring the rate of spread of early farming in Europe. *Man* 6, 674–688.

Beaufoy, G. 1998. *The reform of the CAP Olive-Oil Regime: What are the Implications for the Environment*. EFNCP Occasional Publication No. 14 (En), European Forum for Nature Conservation and Pastoralism, Peterborough.

Beaufoy, G. 2001. *EU Policies for Olive Farming*. WWF Europe/Birdlife, Brussels.

Breznak, J. A. 1975. Symbiotic relationships between termites and their intestinal biota. In D. H. Jennings and D. L. Lee (eds), *Symbiosis*. Symposium 29, Society for Experimental Biology, Cambridge University Press, Cambridge.

Bronowski, J. 1973. *The Ascent of Man*. BBC Publications, London.

Caraveli, H. 2000. A comparative analysis on intensification and extensification in mediterranean agriculture: dilemmas for LFAs policy. *Journal of Rural Studies* 16, 231–242.

Duckham, A. N. 1976. Environmental constraints. In A. N. Duckham, J. G. W. Jones, and E. H. Roberts (eds), *Food Production and Nutrient Cycles*. North Holland Publishing Company, Amsterdam.

Evans, L. T. 1993. *Crop Evolution, Adaptation and Yield*. Cambridge University Press, Cambridge.

Fernandez Ales, R., Angle, M., Ortega, F., and Ales, E. E. 1992. Recent changes in landscape structure and function in a mediterranean region of SW Spain. *Landscape Ecology* 7, 3–18.

Giller, K. E., Beare, M. H., Lavelle, P., Izac, A.-M. N., and Swift, M. J. 1997. Agricultural intensification, soil biodiversity and agroecosystem function. *Applied Soil Ecology* 6, 3–16.

Goudie, A. 1993. *The Human Impact on the Natural Environment*, 4th edn. Blackwell, Oxford.

Hungate, R. E. 1975. The rumen microbiological ecosystem. *Annual Review of Ecology and Systematics* 6, 39–66.

Morgan, R. P. C. 1995. *Soil Erosion and Conservation*, 2nd edn. Longman, Harlow.

Nature Conservancy Council 1984. *Nature Conservation in Great Britain*. Nature Conservancy Council, Peterborough.

Polidori, R., Rocchi, B., and Stefani, G. 1997. Reform of the CMO for olive oil: current situation and future prospects. In M. Tracy (ed.), *CAP Reform: The Southern Products*. Agricultural Policy Studies, Belgium.

Richards, B. N. 1974. *An Introduction to the Soil Ecosystem*. Longman, Harlow.

Rodin, L. E. and Bazilevich, N. I. 1967. *Production and Mineral Cycling in Terrestrial Vegetation*. Oliver & Boyd, Edinburgh.

Simmons, I. G. 1989. *The Changing Face of the Earth: Culture and Environment*, 4th edn. Blackwell, Oxford.

Slesser, M. 1975. Energy requirements of agriculture. In J. Lenihan and W. W. Fletcher (eds), *Food, Agriculture and the Environment*. Blackie, Glasgow and London.

Smith, D. F. and Hill, D. M. 1975. Natural agricultural ecosystems. *Journal of Environmental Quality 4*, 143–145.

Spedding, C. R. W. 1975. *Biology and Agricultural Systems*. Academic Press, London.

Swift, M. J., Vandeermeer, J., Ramakrishnan, P. S., Anderson, J. M., Ong, C. K., and Hawkins, B. A. 1996. Biodiversity and agroecosystem function. In H. A. Mooney, J. H. Cushman, E. Medina, O. E. Sala, and E.-D. Schulze (eds), *Environmental Physiology of Plants*. Wiley, New York.

Tivy, J. 1975. Environmental impact of cultivation. In J. Lenihan and W. W. Fletcher (eds), *Food, Agriculture and the Environment*. Blackie, Glasgow and London.

Tivy, J. 1990. *Agricultural Ecology*. Longman, Harlow.

Tivy, J. 1993. *Biogeography: A Study of Plants in the Ecosphere*, 3rd edn. Longman, Harlow.

Tivy, J. and O'Hare, G. 1981. *Human Impact on the Ecosystem*. Oliver & Boyd, Glasgow.

Usher, M. B. and Thompson, D. B. A. (eds). 1988. *Ecological Change in the Uplands*. Blackwell, Oxford.

Vandermeer, H. H. 2003 *Tropical Agroecosystems*. CRC Press, Boca Raton.

7 Balances

Section 7.1

Barry, R. G. and Chorley, R. J. 1970. *Atmosphere, Weather and Climate*. Holt, Rinehart and Winston, New York.

Boli, B. and Cook, R. B. (eds). 1983. *The Major Biogeochemical Cycles and their Interactions*. John Wiley, Chichester.

Bormann, F. H., Likens, G. E. and Melillo, J. M. 1977. Nitrogen budget for an aggrading northern hardwood forest ecosystem. *Science 196*, 981–983.

Brown, C. M., McDonald-Brown, D. S., and Meers, J. L. 1974. Physiological aspects of inorganic nitrogen metabolism. *Advances in Microbial Physiology 11*, 1–52.

Fochte, D. D. and Verstraete, W. 1977. Biochemical ecology of nitrification and denitrification. *Advances in Microbial Ecology 1*, 135–214.

Freedman, B. 1995. Environmental Ecology: The Ecological Effects of Pollution, Disturbance, and Other Stresses, 2nd edn. Academic Press, San Diego, CA.

Ghassemi, F., Jakeman, A. J., and Nix, H. A. 1995. *Salinisation of Land and Water Resources*. New South Wales Press, Sydney.

Jordan, C. F. and Kline, J. R. 1972. Mineral cycling: some basic concepts and their application in a tropical rain forest. *Annual Review of Ecology and Systematics 3*, 33–49.

Lee, J. 1988. Acid rain. *Biological Sciences Review 1*, 15–18.

McNaughton, S. J. 1988. Mineral nutrition and spatial concentrations of African ungulates. *Nature 334*, 343–345.

Miller, R. M. 1987. Mycorrhizae and succession. In W. R. Jordan, M. E. Gilpin, and J. D. Aber (eds), *Restoration Ecology: A Synthetic Approach to Ecological Research*. Cambridge University Press, Cambridge.

Moss, B. 1988. *Ecology of Fresh Waters: Man and Medium*, 2nd edn. Blackwell, Oxford.

Nature Conservancy Council 1990. *On Course Conservation: Managing Golf's Natural Heritage*. Nature Conservancy Council, Peterborough.

Odum, E. P. 1989. *Ecology and Our Endangered Life-support System*. Sinauer Associates, Sunderland, MA.

Paul, E. A. and Clark, F. E. (1996) *Soil Microbiology and Biochemistry*, 2nd edn. Academic Press, San Diego.

Reinoso, J. C. M. 2001. Vegetation changes and groundwater abstraction in SW Donana. Spain. *Journal of Hydrology 242*, 197–209.

Ricklefs, R. E. 1990. *Ecology*, 3rd edn. Freeman, New York.

Serrano, L. and Serrano, L. 1996. Influence of groundwater exploitation for urban water supply on temporary ponds from the Donana National Park (SW Spain). *Journal of Environmental Management 46*, 229–238.

Silvertown, J., Poulton, P., Johnston, A. E., Edwards, G., Heard, M. and Biss, P. M. 2006. The Park Grass Experiment 1856–2006: its contribution to ecology. *Journal of Ecology 94*, 801–814.

Sprent, J. I. 1983. *The Biology of Nitrogen-fixing Organisms*. McGraw Hill, New York.

Section 7.2

Bettinetti, A., Pypaet, P., and Sweerts, J.-P. 1996. Application of an integrated management approach to the restoration project of the lagoon of Venice. *Journal of Environmental Management 46*, 207–227.

Earle, S. 1992. Assessing the damage one year later. *National Geographical 179*, 122–134.

Edmondson, W. T. 1979. Lake Washington and the predictability of limnological events. *Archiv für Hydrobiologie, Beiheft 13*, 234–241.

European Investment Bank 1990. *The environmental program for the Mediterranean: preserving a shared heritage and common resource*. Report number 8504. International Bank for Reconstruction and Development/World Bank and European Investment Bank, Luxembourg.

Lehman, J. T. 1986. Control of eutrophication in Lake Washington. In *Ecological Knowledge and Environmental Problem-solving, Concepts and Case Studies*. National Academy Press, Washington DC.

Piatt, J. F. and Lensink, C. J. 1989. *Exxon Valdez* oil spill. *Nature 342*, 865–866.

Ritchie, W. and O'Sullivan, M. (eds). 1994. *The Environmental Impact of the Wreck of the Braer*. The Scottish Office, Edinburgh.

Schindler, D. W. 1977. Evolution of phosphorus limitation in lakes. *Science 195*, 260–262.

SEO/BirdLife 2003. *The Disaster of Prestige Oil Tanker and its Impact on Seabirds*. BirdLife International, Madrid.

Van Donk, G. and Gulati, R. D. 1991. Ecological management of aquatic ecosystems: a complementary technique to reduce eutrophication-related perturbations. In O. Ravera (ed.), *Terrestrial and Aquatic Ecosystems: Perturbation and Recovery*. Ellis Horwood, Chichester.

Ward, D. M., Atlas, R. M., Boehm, P. D., and Calder, J. A. 1980. Microbial degradation and chemical evolution from the Amoco spill. *Ambio 9*, 277–283.

Section 7.3

Baker, A., Brooks, R., and Reeves, R. 1988. Growing for gold . . . and for copper . . . and zinc. *New Scientist 117*, 44–48.

Bradshaw, A. D. 1987. Restoration: an acid test for ecology. In W. R. Jordan, M. E. Gilpin, and J. D. Aber (eds), *Restoration Ecology: A Synthetic Approach to Ecological Research*. Cambridge University Press, Cambridge.

Bradshaw, A. D. 1983. The reconstruction of ecosystems. *Journal of Applied Ecology 20*, 1–17.

Bradshaw, A. D. 1984. Ecological principles and land reclamation practice. *Landscape Planning 11*, 35–48.

Bradshaw, A. D. 1989. Management problems arising from successional processes. In G. P. Buckley (ed.), *Biological Habitat Reconstruction*. Belhaven Press, London.

Bradshaw, A. 1993. Understanding the fundamentals of succession. In J. Miles and D. H. Walton (eds), *Primary Succession on Land*. Blackwell, Oxford.

Bradshaw, A. D. and Chadwick, M. J. 1980. *The Restoration of Land: The Ecology and Reclamation of Derelict and Degraded Land*. Blackwell, Oxford.

Bradshaw, A. D., Humphreys, R. N., Johnson, M. S., and Roberts, R. D. 1978. The restoration of vegetation on derelict land produced by industrial activity. In M. W. Holdgate and M. J. Woodward (eds), *The Breakdown and Restoration of Ecosystems*. Plenum, New York.

Buckley, G. P. (ed.). 1989. *Biological Habitat Reconstruction*. Belhaven Press, London.

Bunce, R. G. H. and Jenkins, N. R. 1989. Land potential for habitat reconstruction in Britain. In G. P. Buckley (ed.), *Biological Habitat Reconstruction*. Belhaven Press, London.

Department of the Environment 1994. *The Reclamation of Metalliferous Mining Sites*. Her Majesty's Stationery Office, London.

Down, G. S. and Morton, A. J. 1989. A case study of whole woodland transplanting. In G. P. Buckley (ed.), *Biological Habitat Reconstruction*. Belhaven Press, London.

Grime, J. P., Hodgson, J. G., and Hunt, R. 2006. *Comparative Plant Ecology: A Functional Approach to Common British Species*. Castlepoint Press, Dalbeattie.

Helliwell, D. R. 1989. Soil transfer as a means of moving grassland and marshland vegetation. In

G. P. Buckley (ed.), *Biological Habitat Reconstruction*. Belhaven Press, London.

Jordan, W. R., Gilpin, M. E., and Aber, J. D. (eds). 1987. *Restoration Ecology: A Synthetic Approach to Ecological Research*. Cambridge University Press, Cambridge.

Lee, I. W. Y. 1985. A review of vegetative slope stabilisation. *The Journal of the Hong Kong Institution of Engineers* July, 9–22.

Marrs, R. H. and Bradshaw, A. D. 1993. Primary succession on man-made wastes: the importance of resource acquisition. In J. Miles and D. H. Walton (eds), *Primary Succession on Land*. Blackwell, Oxford.

Pywell, R. F., Bullock, J. M., Roy, D. B., Warman, L., Walker, K. J., and Rothery, P. 2003. Plant traits as predictors of performance in ecological restoration. *Journal of Applied Ecology 40*, 65–77.

8 Scales

Section 8.1

Bolger, D. T., Scott, T. A., and Rotenberry, J. T. 2001. Use of corridor-like landscape structures by bird and small mammal species. *Biological Conservation 102*, 213–224.

Dunning, J. B., Danielson, B. J., and Pulliam, H. R. 1992. Ecological processes that affect populations in complex landscapes. *Oikos 65*, 169–175.

Forman, R. T. T. 1995. *Land Mosaics: The Ecology of Landscapes and Regions*. Cambridge University Press, Cambridge.

Harvey, D. S. and Weatherhead, P. J. 2006. A test of the hierarchical model of habitat selection using eastern massasauga rattlesnakes (*Sistrurus c. catenatus*). *Biological Conservation 130*, 206–216.

O'Neill, R. V., DeAngelis, D. L., Waide, J. B., and Allen, T. F. B. 1986. *A Hierarchical Concept of Ecosystems*. Princeton University Press, Princeton, NJ.

O'Neill, E. G., O'Neill, R. V., and Norby, R. J. 1991. Hierarchy theory as a guide to mycorrhizal research on large-scale problems. *Environmental Pollution 73*, 271–284.

Santos, T., Telleria, J. L., and Carbonell, R. 2002. Bird conservation in fragmented Mediterranean forests of Spain: effects of geographical location, habitat and landscape degradation. *Biological Conservation 105*, 113–125.

Turner, M. G., Romme, W. H., Gardner, R. H., O'Neill, R. V., and Kratz, T. K. 1993. A revised concept of landscape equilibrium: disturbance and stability on scaled landscapes. *Landscape Ecology 8*, 213–227.

Section 8.2

Archibold, O. W. 1995. *Ecology of World Vegetation*. Chapman & Hall, London.

Field, C. B., Behrenfeld, M. J., Randerson, J. T., and Falkowski, P. 1998. Primary production of the biosphere: integrating terrestrial and oceanic components. *Science 281*, 237–239.

Whittaker, R. H. 1975. *Communities and Ecosystems*, 2nd edn. MacMillan, New York.

Section 8.3

Adams, J. M., Faure, H., Faure-Denard, L., McGlade, J. M., and Woodward, F. I. 1991. Increases in terrestrial carbon storage from the last glacial maximum to the present. *Nature 348*, 711–714.

Aspinall, R. and Matthews, K. 1994. Climate change impact on distribution and abundance of wildlife species: an analytical approach using GIS. *Environmental Pollution 83*, 217–223.

Baker, J. T. and Allen, L. H. 1994. Assessment of the impact of rising carbon dioxide and other potential climate changes on vegetation. *Environmental Pollution 83*, 223–235.

Behling, H. 2002. Carbon storage increases by major forest ecosystems in tropical South America since the Last Glacial Maximum and the early Holocene. *Global and Planetary Change 33*, 107–116.

Bekkering, T. D. 1992. Using tropical forests to fix atmospheric carbon: the potential in theory and practice. *Ambio 21*, 414–419.

Bierregaard, R. O., Lovejoy, T. E., Kapos, V., dos Santos, A. A., and Hutchings, R. W. 1992. The biological dynamics of tropical rain forest fragments. *BioScience 42*, 859–866.

Clayton, K. 1995. The threat of global warming. In T. O'Riordan (ed.), *Environmental Science for Environmental Management*. Longman, Harlow.

Davey, P. A., Parson, A. J., Atkinson, L., Wadge, K., and Long, S. P. 1999. Does photosynthetic acclimation to elevated CO_2 increase photosynthetic nitrogen-use efficiency? A study of three native UK grassland species in open-top chambers. *Functional Ecology 13* (Suppl. 1), 21–28.

Dormann, C. F. and Woodin, S. J. 2002. Climate change in the Arctic: using plant functional types in a meta-analysis of field experiments. *Functional Ecology 16*, 4–17.

Gordo, O. and Sanz, J. J. 2005. Phenology and climate change: a long term study of a Mediterranean locality. *Oecologia 146*, 484–495.

Houghton, J. T., Jenkins, G. J., and Ephraums, J. J. (eds). 1990. *Climate Change: The IPCC Scientific Assessment.* Cambridge University Press, Cambridge.

Innes, J. L. 1994. Climatic sensitivity of temperate forests. *Environmental Pollution 83*, 237–243.

Leemans, R. and Zuidema, G. 1995. Evaluating changes in land cover and their importance for global change. *Trends in Ecology and Evolution 10*, 76–81.

Lloyd, D. and Jenkinson, D. S. 1995. The exchange of trace gases between land and atmosphere. *Trends in Ecology and Evolution 10*, 2–4.

Lloyd, J. 1999. The CO_2 dependence of photosynthesis, plant growth responses to elevated CO_2 concentrations and their interactions with soil nutrient status, II. Temperate and boreal forest productivity and the combined effects of increasing CO_2 concentrations and increased nitrogen deposition at a global scale. *Functional Ecology 13*, 439–459.

Lloyd, J. and Farquhar, G. D. 1996. The CO_2 dependence of photosynthesis, plant growth responses to elevated atmospheric CO_2 concentrations and their interaction with soil nutrient status. 1. General principles and forest ecosystems. *Functional Ecology 10*, 4–32.

Mitchell, J. F. B., Johns, T. C., Gregory, J. M., and Tett, S. F. B. 1995. Climate response to increasing levels of greenhouse gases and sulphate aersols. *Nature 376*, 501–504.

Niklaus, P. A., Stocker, R., Korner, C. H., and Leadley, P. W. 2000. CO_2 flux estimates tend to overestimate ecosystem C sequestration at elevated CO_2. *Functional Ecology 14*, 546–559.

Norby, R. J., Gunderson, C. A., Wullschleger, S. D., O'Neill, E. G., and McCracken, M. K. 1992. Productivity and compensatory responses of yellow-poplar trees in elevated CO_2. *Nature 357*, 322–324.

Osterkamp, T. E. 2005. The recent warming of the permafrost in Alaska. *Global Planetary Change 49*, 187–202.

Pastor, J. and Post, W. M. 1988. Response of northern forests to CO_2-induced climate change. *Nature 334*, 55–58.

Phillips, O. L., Martinez, R. V., Arroyo, L., Baker, T. R., Killeen, T., Lewis, S. L., Malhi, Y., Mendoza, A. M., Neill, D., Vargas, P. N., Alexiades, M., Cerón, C., Di Fiore, A., Erwin, T., Jardim, A., Palacios, W., Saldias, M., and Vinceti, B. 2002. Increasing dominance of large lianas in Amazonian forests. *Nature 418*, 770–774.

Nemani, R. R., Keeling, C. D., Hashimoto, H., Jolly, W. M., Piper, S. C., Tucker, C. J., Myneni, R. B., and Running, S. W. 2003. Climate-driven increases in global terrestrial net primary production from 1982 to 1999. *Science 300*, 1560–1562.

Sarmiento, J. L. and Orr, J. C. 1991. Three-dimensional simulations of the impact of Southern Ocean nutrient depletion on atmospheric CO_2 and ocean chemistry. *Limnology and Oceanography 36*, 1928–1950.

Schneider, S. H. and Lane, J. 2005. An overview of 'dangerous' climate change. In *Avoiding Dangerous Climate Change: Proceedings of a conference of the UK Meteorological Office/University of Reading*, pp. 7–23.

Sedjo, R. A. 1992. Temperate forests ecosystems in the global carbon cyle. *Ambio 21*, 274–277.

Stott, P. A., Tett, S. F. B., Jones, G. S., Allen, M. R., Mitchell, J. F. B., and Jenkins, G. J. 2000. External control of 20th century temperature by natural and anthropogenic forcings. *Science 290*, 2133–2137.

Thompson, R. D. 1992. The changing atmosphere and its impact on Planet Earth. In A. W. Mannion and S. R. Bowlby (eds), *Environmental Issues in the 1990s.* John Wiley, Chichester.

Watson, R. T. 2001. Climate Change 2001: Summary Report for Policy Makers. Third Assessment Report of the Intergovernmental Panel on Climate Change, London.

Wigley, T. M. L. and Raper, S. C. B. 1992. Implications for climate and sea level of revised IPCC emission scenarios. *Nature* 357, 293–300.

Wigley, T. M. L., Richels, R., and Edmonds, J. A. 1996. Economic and environmental choices in the stabilization of atmospheric CO_2 concentrations. *Nature* 379, 240–243.

9 Checks

Introduction

Raymo, M. E. and Ruddiman, W. F. 1992. Tectonic forcing of the late Cenozoic climate. *Nature* 359, 117–122.

Section 9.1

Hubbell, S. P. 1997. A unified theory of biogeography and relative species abundance and its application to tropical rain forests and coral reefs. *Coral Reefs* 16 (Suppl.), 9–21.

Jackson, J. B. C. 1991. Adaptation and diversity of reef corals. *BioScience* 41, 475–482.

MacArthur, R. H. and Wilson, E. O. 1967. *The Theory of Island Biogeography*. Princeton University Press, Princeton, NJ.

Section 9.2

Coope, G. R. 1995. Insect faunas in ice age environments: why so little extinction? In J. H. Lawton and R. M. May (eds), *Extinction Rates*. Oxford University Press, Oxford.

Jablonski, D. 1995. Extinctions in the fossil record. In J. H. Lawton and R. M. May (eds), *Extinction Rates*. Oxford University Press, Oxford.

Jackson, J. B. C. 1995. Constancy and change of life in the sea. In J. H. Lawton and R. M. May (eds), *Extinction Rates*. Oxford University Press, Oxford.

Labandeira, C. C. and Sepkoski, J. J. 1993. Insect diversity in the fossil record. *Science* 261, 310–315.

May, R. M. 1992. How many species inhabit the Earth? *Scientific American* 261, 18–24.

Myers, N. 1993. Questions of mass extinction. *Biodiversity and Conservation* 2, 2–17.

Stirling, I., Lunn, N. J., and Iacozza, J. 1999. Long-terms trends in the population ecology of polar bears in western Hudson Bay in relation to climate change. *Arctic* 52, 294–306.

Ward, B. B. 2002. How many species of prokaryotes are there? *Proceedings of the National Academy of Sciences USA* 99, 10234–10236.

Section 9.3

Briand, F. and Cohen, J. C. 1987. Environmental correlates of food chain length. *Science* 238, 956–960.

Chandy, S., Gibson, D. J., and Robertson, P. A. 2006. Additive partitioning of diversity across hierarchical spatial scales in a forested landscape. *Journal of Applied Ecology* 43, 792–801.

Chapin, F. S., Schulze, E.-D., and Mooney, H. A. 1992. Biodiversity and ecosystem processes. *Trends in Ecology and Evolution* 7, 107–108.

Collins, S. L. 1995. The measurement of stability in grasslands. *Trends in Ecology and Evolution* 10, 95–96.

Kremen C., Williams, N. M. and Thorp, R. W. 2002. Crop pollination from native bees at risk from agricultural intensification. *Proceedings of the National Academy of Sciences USA* 99, 16812–16816.

McCann, K., Hastings, A., and Huxel, G. R. 1998. Weak trophic interactions and the balance of nature. *Nature* 395, 794–798.

McNaughton, S. J. 1988. Diversity and stability. *Nature* 333, 204–205.

May, R. M. 1986. The search for patterns in the balance of nature: advances and retreats. *Ecology* 67, 1115–1126.

Moore, J. C. and Hunt, H. W. 1988. Resource compartmentation and the stability of real ecosystems. *Nature* 333, 261–263.

Polis, G. A. 1998. Stability is woven by complex webs. *Nature* 395, 744–745.

Schwartz, M. W., Brigham, C. A., Hoeksema, J. D., Lyons, K. G., Mills, M. H., and van Mantgem, P. J. 2000. Linking biodiversity to ecosystem function: implications for conservation ecology. *Oecologia* 122, 297–305.

Solow, A. R. 1993. Measuring biological diversity. *Environmental Science and Technology* 27, 25–26.

Steele, J. H. 1991. Marine functional diversity. *BioScience* 41, 470–474.

Tilman, D. and Downing, J. A. 1994. Biodiversity and stability in grasslands. *Nature 367*, 3633–3635.

Walker, B. H. 1992. Biodiversity and ecological redundancy. *Conservation Biology 6*, 18–23.

Waltho, N. and Kolasa, J. 1994. Organization of instabilities in multispecies systems, a test of hierarchy theory. *Proceedings of the National Academy of Sciences USA 91*, 1682–1685.

Section 9.4

Angel, M. V. 1994. Spatial distribution of marine organisms: patterns and processes. In P. J. Edwards, R. M. May, and N. R. Webb (eds), *Large-Scale Ecology and Conservation Biology*. Blackwell, Oxford.

Auspurger, C. K. 1983. Offspring recruitment around tropical trees: changes in cohort distance with time. *Oikos 40*, 189–196.

Chown, S. L. and Gaston, K. J. 2000. Areas, cradles and museums: the latitudinal gradient in species richness. *Trends in Ecology and Evolution 15*, 311–315.

Clark, D. A. and Clark, D. B. 1984. Spacing dynamics of a tropical rain forest tree: evaluation of the Janzen-Connell model. *American Naturalist 124*, 769–788.

Clarke, A. 1992. Is there a latitudinal diversity cline in the sea? *Trends in Ecology and Evolution 7*, 286–287.

Cook, S. 1998. A diversity of approaches to the study of species richness. *Trends in Ecology and Evolution 13*, 340–341.

Currie, D. J. 1991. Energy and large-scale patterns of animal- and plant-species richness. *American Naturalist 137*, 27–49.

Rex, M. J., Stuart, C. T., Hessler, R. R., Alen, J. A., Sanders, H. L., and Wilson, G. D. F. 1993. Global-scale latitudinal patterns of species diversity in the deep sea benthos. *Nature 365*, 636–639.

Ricklefs, R. E. 1990. *Ecology*, 3rd edn. Freeman, New York.

Vincent, A. and Clarke, A. 1995. Diversity in the marine environment. *Trends in Ecology and Evolution 10*, 55–56.

Section 9.5

Andren, O. and Balandreau, J. 1999. Biodiversity and soil functioning—from black box to can of worms. *Applied Soil Ecology 13*, 105–108.

Biswas, M. R. 1994. Agriculture and environment: a review, 1972–1992. *Ambio 23*, 192–197.

Deacon, L. J., Pryce-Miller, E. J., Frankland, J. C., Bainbridge, B. W., Moore, P. D., and Robinson, C. H. 2006. Diversity and function of decomposer fungi from a grassland soil. *Soil Biology & Biochemistry 38*, 7–20.

Cragg, R. G. and Bardgett, R. D. 2001. How changes in soil faunal diversity and composition within a trophic group influence decomposition processes. *Soil Biology & Biochemistry 33*, 2073–2081.

Crosson, P. R. and Rosenberg, N. J. 1989. Strategies for agriculture. *Scientific American 261*, 128–135.

Huston, M. 1993. Biological diversity, soils and economics. *Science 262*, 1676–1679.

Kendall, H. W. and Pimentel, D. 1994. Constraints on the expansion of the global food supply. *Ambio 23*, 198–205.

Lawler, S. P. 1993. Species richness, species composition and population dynamics of protists in experimental microcosms. *Journal of Animal Ecology 62*, 711–719.

Li, Q., Allen, H L. and Wollum, A. G. 2004. Microbial biomass and bacterial functional diversity in forest soils: effects of organic matter removal, compaction and vegetation control. *Soil Biology & Biochemistry 36*, 571–579.

Liiri, M., Setätä, H., Haimi, J., Pennanen, T., and Fritze, H. 2002. Soil processes are not influenced by the functional complexity of soil decomposer food webs under disturbance. *Soil Biology & Biochemistry 34*, 1009–1020.

Naeem, S., Thompson, L. J., Lawler, S. P., Lawton, J. H., and Woodfin, R. M. 1994. Declining biodiversity can alter the performance of ecosystems. *Nature 368*, 734–737.

Neilson, R., Robinson, D., Marriot, C. A., Scrimgeour, C. M., Hamilton, D., Stocking, M. 1995. Soil erosion and land degradation. In T. O'Riordan (ed.), *Environmental Science for Environmental Management*, Longman, Harlow.

Tuckwell, H. C. and Koziol, J. A. 1992. World population. *Nature 359*, 200.

Wishart, J., Boag, B., and Handley, L. L. 2002. Above-ground grazing affects floristic composition and modifies soil trophic interactions. *Soil Biology & Biochemistry 34*, 1507–1512.

INDEX

Numbers in italics refers to figures, those in bold to tables